# SPACETIME, GEOMETRY, COSMOLOGY

## William L. Burke

DOVER PUBLICATIONS, INC.
Garden City, New York

*Bibliographical Note*

This Dover edition, first published in 2020, is an unabridged republication of the work originally printed by University Science Books, Mill Valley, California, in 1980.

*Library of Congress Cataloging-in-Publication Data*

Names: Burke, William L., author.
Title: Spacetime, geometry, cosmology / William L. Burke
Description: Garden City, New York: Dover Publications, Inc., 2020. | This Dover edition, first published in 2020, is an unabridged republication of the work originally printed by University Science Books, Mill Valley, California, in 1980. | Includes bibliographical references and index. | Summary: "This novel interpretation of the relationship between space, time, gravitation, and their cosmological implications is based on the author's discovery of a small but significant value in gravitation that was overlooked by both Newton and Einstein. Dubbed the 'Burke Potential,' it resolved an issue with gravitation measurements and rates of entropy related to the discovery of a time-varying part of the local gravitational field. 1980 edition"—Provided by publisher.
Identifiers: LCCN 2020029860 | ISBN 9780486845586 (paperback)
Subjects: LCSH: Space and time. | Geometry. | Cosmology.
Classification: LCC QC173.59.S65 B87 2020 | DDC 530.11—dc23
LC record available at https://lccn.loc.gov/2020029860

Manufactured in the United States by LSC Communications
84558301
www.doverpublications.com
2 4 6 8 10 9 7 5 3 1
2020

To J. L. Synge

# Contents

*Acknowledgments*    ix
*Preface*    xi
*Introduction*    xiii
*Study Guide*    xix

## CHAPTER ONE   SPECIAL RELATIVITY    3

*1. Spacetime Structure*    4
*2. Clocks*    11
*3.    Euclidean Geometry*    15
*4. Inertial Reference Frames*    18
*5.    Physical Clocks*    22
*6. Light Signals and Canonical Reference Frames*    26
*7. Special Relativity*    31
*8.    Simultaneity*    33
*9.    Relative Velocity*    41
*10.    Lorentz Invariance*    44
*11. The Consistency of Special Relativity*    52
*12.    4-Vectors*    55
*13.    Doppler Shift*    61
*14.    4-Momentum*    70

## CHAPTER TWO   GEOMETRY    75

*15. Vectors and Covectors*    77
*16. Tangent Vectors and 1-Forms*    83
*17.    Coordinate Basis Vectors*    91
*18.    Example: Static Equilibrium*    95
*19.    Example: Dispersive Waves*    101

20. *Maps* 110
21. *Tensors* 114
22. *Basis Tensors and the Tensor Product* 122
23. *Minkowski Spacetime* 129
24. *Index Notation* 135
25. *Vector Fields* 141
26. *Manifolds* 149

CHAPTER THREE  GRAVITATION  161

27. *The Motion of Wave Packets* 162
28. *Water-Wave Relativity* 171
29. *The Interaction of Wave Packets* 182
30. *Gravitation* 193
31. *Spacetime Near the Earth* 199
32. *Gravitational Redshift* 205
33. *Huygens' Construction and the Falling Apple* 210
34. *Geodesics* 220
35. *Spacetime Curvature* 228

CHAPTER FOUR  COSMOLOGY  237

36. *Whole Universe Catalog* 237
37. *Robertson-Walker Spacetimes* 241
38. *The Global Structure of the 3-Sphere* 247
39. *The Metric Structure of the 3-Sphere* 255
40. *Light Propagation* 259
41. *Friedmann Universes* 261
42. *The Closed Dust Universe* 264
43. *Observations in a Closed Dust Universe* 269
44. *The Pseudosphere* 276
45. *Dust and Radiation Friedmann Universes* 282
46. *The Hubble Diagram* 289
47. *Newtonian Cosmology* 293
48. *Do-It-Yourself Cosmology* 296
49. *The Big Bang* 302
50. *Observations of the Real Universe* 308
51. *Singularities* 314
   *Bibliography* 323
   *Index* 325

# Acknowledgments

I am fortunate to have learned my relativity from J. L. Synge through my teacher Frank Estabrook. A student's responsibility is to use in his own way what he has been taught. Both are blameless for my zealous devotion to 1-forms.

The idea that I should write this book came from my students and, more directly, from a biography of Mozart: while reading it, I asked myself why I was reading biographies instead of doing something. I started writing notes for a cosmology class that I had just finished, and, to my surprise, in three weeks I had painlessly written four hundred pages. This book is an expanded and much rewritten version of that manuscript. The final rewrite took six months and was painful.

The students in my most recent cosmology class helped greatly in getting the bugs out of the manuscript and suffered with early versions of the problems. Drafts were read by Al Kelley, Peter Renz, and John Faulkner. Brian Hatfield checked most of the problems. Aidan Kelly did an excellent job of editing. Their help was invaluable. Specific sage advice that I wish both to credit and to pass on came from John Letcher ("Keep the pencil moving") and Ralph Baierlein ("You don't finish a book; you just get sick of it"). Carol Fairhurst, Judy Rose, and Dottie Hollinger did a fine job typing the final manuscript. And many thanks, of course, to Kip and Paco and especially Pat.

# Preface

*"In short, I tried to use all my experience in
research and teaching to give an appropriate opportunity
to the reader for intelligent imitation and for
doing things by himself."*

G. POLYA

This is a textbook on the geometric aspects of spacetime and their application to cosmology. It provides an introduction with many examples and applications to the geometric language and ideas that are essential in modern theoretical physics. It is written for the serious student who wants to be able to calculate in the standard cosmological models and to understand those calculations. Problems have been provided at the end of almost every section. In the field of relativity this book lies midway between *Spacetime Physics,* by Taylor and Wheeler, and *Gravitation,* by Misner, Thorne, and Wheeler. In the field of cosmology this book is unusual. It spends little time on the descriptive aspects of the subject, and does not pay much attention to the complex physics of the big bang. Instead, it spends most of its time on the fundamental ideas and the geometric language needed to discuss them. This is a genuine advantage for the student, who can become actively involved, and should acquire tools that can be safely used. The common descriptive treatments can lead too easily to vain and foolish speculation by students who have been carried over their heads by excellent and persuasive prose.

The reader should expect a level of difficulty roughly comparable to that found in a junior-level electromagnetism course, say, that by

Lorrain and Corson. The reader should have had at least a brief previous introduction to linear algebra and ordinary differential equations, such as comes up in a good advanced calculus course. A classical mechanics course would be helpful, but the degree of familiarity with mechanics and electromagnetism acquired in the introductory physics courses should suffice. A brief, three- or four-lecture introduction to special relativity is assumed. You should have heard of the twin paradox, but not necessarily have been happy with its explanation. The difficulty of the material here arises not from complicated analytical manipulations, but from the fact that it demands subtle and fundamental thinking. This demand makes it a healthy complement to the usual courses, which are heavy on technical manipulation. I have tried to make the book suitable both for classroom use and for independent study, perhaps as a supplement to a more traditional cosmology course.

The book starts with special relativity. I do not follow the axiomatic treatment by which special relativity was invented, but a phenomenological treatment that is less mystical and better suited to a critical evaluation. I spend little time on the paradoxes and the arcana, and seek rather to expose what is truly novel as a reasonable possibility rather than as a revolution. The new ideas of special relativity demand a new language. This is developed and then used to discuss Einstein's theory of gravitation, general relativity. Finally, all these tools are applied to study the standard Friedmann models of the universe that are used in most cosmological discussions today.

I was strongly tempted to emphasize the geometric flavor of this book by subtitling it "The 1-Forms Book." The geometric language of the calculus on manifolds is really the "new math" for physicists and should be learned by all. I illustrate it not only with the cosmological applications, but also with more mundane uses, such as dispersive wave propagation. The material here should provide a sound introduction to tensors. One should follow this introduction with a careful, mathematically precise course on the subject. The mathematics used here is the minimum needed to deal honestly with the subject. I hope that some lingering allergy to math will not prevent, say, the eager geophysicist from enjoying the book.

The diligent student will finish this book with a reasonable informal grasp of the calculus on manifolds and of the geometry of curved spacetime, and will know how to calculate in the standard cosmological models. The physical side of cosmology has been neglected: galaxy formation, element synthesis, and so on. Unfortunately, these are far more complicated subjects, and much of what is known about that can be derived only by use of enormous and mysterious computer codes Learning them is in many ways less enjoyable and less challenging than learning the geometric cosmology described here.

# Introduction

This is a book on spacetime. Within small regions, spacetime is approximately described by special relativity, just as a small region on the surface of the Earth can be approximately described by a plane. The phenomenon of gravitation is described in Einstein's general relativity by the behavior of spacetime in the large. Gravitation will be described by the curvature of spacetime, which is closely analogous to the curvature of the surface of the Earth. As we go to larger and larger regions of spacetime, gravitation becomes the dominant physical process. The model of the universe that we describe here is that of a self-gravitating collection of mass-energy embedded in curved spacetime. This book aims to explain precisely what the above description means, and to provide you with the language and tools needed to use such a description to predict and explain the results of astronomical observations.

*Spacetime*

What does it mean to make a model of the universe? We must ask such questions, even though they seem to belong in the realm of philosophy, because of the fundamental nature of the subject itself. A good analogy for this model making is the construction of a map of a city. The city and the map of the city are two distinct things that no one would confuse. This confusion is harder to avoid in cosmology. To be useful, the map must be faithful to some of the city's interesting properties. Furthermore, the map is constructed according to some specific conventions: major roads are in red; north is at the top; and so on. Also, a map does not try to define what a road is. That you learn by demonstration. People point out roads and non-roads to you, and you acquire the idea of a road by induction. Such basic ideas, which are not defined but are learned by demonstration, are called *primitive notions*.

*Models*

> **Primitive notion: A fundamental element in a physical theory that is not defined within the theory but is presumed to be known, either by description or from a more fundamental theory.**

The same things will be true of the models of physical reality that we make. We will construct a map of the world according to some specific but arbitrary conventions. The image is constructed not on paper, but in terms of some mathematical structure, such as a linear vector space. Because we can talk about physical reality only in terms of some model, it is easy for us to forget the distinction between the object itself and our model of the object. As in our map of the city, there are some undefined elements in our model. For spacetime, these primitive notions are *event*, *free particle*, *clock*, and *light signal*. We will describe in words what these correspond to, much as a roadmap might discuss what it does in fact consider to be an acceptable road.

*Coarse-graining*    An important feature of any model is *coarse-graining*. The model ignores a lot of things. The exact width, surface, and color of a road cannot be deduced from the ordinary roadmap. In fact, the roadmap is more useful than a photograph precisely because it omits such irrelevant detail. Similarly, there is a coarse-graining in our primitive notions for spacetime. For example, we will not describe a clock by making a detailed model of its internal construction. Such details will not be mentioned. The model presumes that they are irrelevant, and it is the task of experiment to verify this.

Our model and its rules of manipulation provide us with a language in which to discuss spacetime. Without some language, it cannot be discussed at all. We can consider different models, of course, and these may have different degrees of coarse-graining. A certain amount of the content of a physical theory is contained in the language itself. This is a tricky point, and is both good and bad. On the one hand, assumptions hidden in the very language used to discuss a situation often go unnoticed. This can be dangerous. Such assumptions often go unchecked. On the other hand, a language in which most of the physical content is automatically included is usually an efficient language to use. As we go from a critical discussion of special relativity to explicit computations, we will shift to models that have more of the structure built in.

The process by which we model a physical situation mathematically is called *representation*. The reverse process, by which we take a mathematical structure and find physical situations that are described by it, we will call finding a *realization* for the mathematics. There are usually many different but nonequivalent representations of a given physical situation. In our roadmap analogy, one could draw a map with north on the left side and with east at the bottom, for example. Such a map would be an equally faithful representation, but it would require different rules of manipulation, such as for which way to turn at an intersection. Also, the symbols for buildings would receive an inessential change. These rules of change between different equivalent representations I call *covariance*.

[Caution! Not everyone uses covariance with this precise meaning.]

---

**Covariance: The changes in the rules and mathematical structures as one goes from one equivalent representation to another.**

---

The idea is that the representation is changing, whereas the physical situation is not. This is sketched in Figure 0.1.

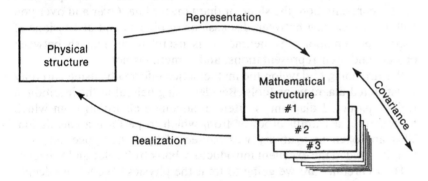

**Figure 0.1.** The idea of covariance as the relations between several different mathematical structures that all model the same physical situation.

What mathematical structure will we use to represent spacetime? The roadmap uses the Euclidean geometry of a piece of paper. We will also use a geometric structure to represent spacetime, but we will need a more flexible structure. We need to describe not only static figures but also motion and rates. The appropriate mathematical structures are called manifolds, and we will discuss them at some length. One should not, of course, confuse the mathematics with the physics. General relativity is not a branch of mathematics. We will provide here only a barely adequate treatment of the mathematics. We will try to find the middle ground between too much fancy mathematics, and so little mathematics that the student is worked into a descriptive blind alley, unable to proceed further in the subject without relearning it.

*Manifolds: A language for spacetime*

Thus, although the appropriate geometric structure is the manifold, we present here only an informal definition of it. The precise definition is too tricky for us to get into, and it contributes little either to intuition or to computation. If you cannot grasp the idea informally, then you probably are not ready for the abstract idea either. Also, even though topological matters will come up, we will not go deeply into them. The informal ideas are intuitive and straightforward: that a coffee cup and a doughnut are alike, and not like an orange or Swiss cheese. The formal development of these ideas involves formidable complications that are of no use to us here.

*The idea behind calculus*          The aim of calculus on manifolds is to extend the familiar ideas of calculus to sets other than Euclidean space. The basic idea of calculus is simple. It is the technology of linear approximations to nonlinear objects: straight lines as linear approximations to curves, planes for surfaces, and so on. This simple idea underlies all the calculus on manifolds that we develop. We will find that it is simpler to concentrate on the linear approximation itself, the tangent line, for example, than on its representation, the slope of that tangent line. Over and over you will find us moving between discussions of objects and discussions of their representations. Sometimes it is useful to distinguish between objects and their representations, and sometimes not.

*Diagrams*          We are going to develop the mathematics informally, using carefully constructed diagrams as tools. Besides being helpful to the intuition, a well-constructed diagram is often an adequate algorithm from which to construct a careful proof or from which to perform a calculation. This approach to relativity was pioneered by J. L. Synge, and was heavily used in the excellent introductory book by Taylor and Wheeler.

*How deep into physics?*          How carefully are we going to treat the physics? We will go deeply into the relativity, but we will not be able to give a broad discussion of the astrophysics. Our discussion of special relativity will be as fundamental and critical as any that I know of. We will not be able to go anywhere nearly as deeply into general relativity, since calculations in general relativity are hard, and require technical skills that need a full year's course to master. The aim here is to give you an understanding of the end result of a general relativity calculation: a spacetime. You will be shown how to describe clocks, free particles, and light signals in any given spacetime. The equations that govern the spacetime itself are beyond us. They will be mentioned, and some intuitive ideas about them discussed, but formidable technical difficulties prevent our dealing with them honestly. We will be able to display these equations only in the context of homogeneous and isotropic cosmological models. We will be able to generate the standard cosmological models, to explore their properties, and to discuss the choices between them. To step outside this limited class of models will be beyond us. However, this limitation is not as serious as it might seem, since we are in fact so ignorant about the universe that we have no justification for choosing any more detailed model.

Behind all our discussions there will be the recurring idea of symmetry. All the mathematics of special relativity is really just the exploration of those mathematical structures that are compatible with Lorentz symmetry. For practice, we will also discuss symmetry in more common physical situations, such as in water waves. Our cosmological models will be constructed to have the maximum amount of symmetry that is compatible with the observed evolutionary nature of the universe. There is a contrary idea, that symmetric situations are special and are therefore very unlikely. This conflict between form and chaos has been with us for centuries. Observations indicate that our universe in the large is very symmetric. Whether this symmetry has its source in known physical laws, or reflects peculiar initial conditions, is not known.

*Symmetry*

This book is novel in several ways. I develop special relativity from the observed behavior of moving clocks rather than from the assumption of Lorentz symmetry. Thus for us a problem such as the twin paradox can be resolved by appealing to experiment rather than to authority. Ths discussion of water-wave relativity is also new. It gives us another realization of the idea of a relativity symmetry.

The book contains a large number of problems, and they form an important part of the book. They span a wide range in difficulty, and some require much initiative. I have provided an estimate of their difficulty according to a scale devised by D. Knuth:

*Problems*

(00) Extremely easy exercise, work in your head;

(10) simple problem, about one minute;

(20) average problem, ten to twenty minutes;

(30) moderate difficulty and complexity, say, two hours;

(40) term-paper topic.

"*A writer who wishes to be read by posterity*
*must not be averse to putting hints*
*which might give rise to whole books,*
*or ideas for learned discussions,*
*in some corner of a chapter*
*so that one should think he can afford to*
*throw them away by the thousand.*"

G. C. LICHTENBERG

# Study Guide

*"It will seem difficult at first,
  but everything is difficult at first."*

MUSASHI

*"How can I tell what I think
  till I see what I say?"*

E. M. FORSTER

Many readers will use this book for self-study. Let me discuss here the strategy and tactics of learning this material that a teacher would normally provide.

*Physics*

This is a physics book, and the reader is expected to be aiming for the skills of a physicist. These I see to be threefold. First, you must be able to consider a situation and form an intuitive picture of it. You should be able to "wiggle it in your mind." Second, you should be able to make a mathematical model of the situation, translating the relevant physical features into mathematical structures and operations. Third, you must have the technical facility to manipulate the mathematical model. All three skills are important. Only the third is easy to teach. Do not forget the others.

*A new language*

It is important for you to realize that you are going to learn a new language here. This new language contains not only new names for familiar things, but also new ideas that cannot be easily expressed without it. You will have some difficulty at first, because in this new language we recycle some common words to give them new, precise

xix

meanings. For example, we will use the word "clock" for a precise notion that is a bit different from common usage, but so close that it would be crazy to invent a new word for it and call it a "ckloc," say. Also, some things that are simple ideas in common usage will not be simple in the new language. Ideas involving the speed of light, the rate of a moving clock, and so on require careful discussion. It is inevitable that a language contains in its very structure significant information about the things that it describes. A language is a little bit of a theory. Stay alert for this. An old language ties you to an old theory.

One useful cognitive skill here is being able to treat hypothetical situations properly. If I ask you, "Suppose there are four six-legged dogs in the front yard. How many legs do they have altogether?" can you play the game? Or do you grind to a halt, saying, "That cannot be"? The example is trivial, but the problem is not. Things only mean what we agree they mean, even if we give them familiar names. It is easy to agree with this statement in principle. Be humble about the practice, however. Recall the difficulties that even very intelligent people had with $\sqrt{-1}$. You will have similar problems.

*Representations*     A tricky area that requires a balanced approach involves the representations of objects. In order to deal efficiently with physical objects, one usually finds a mathematical representation of them. These mathematical structures may themselves have representations, and now the degree to which one distinguishes between an object and its representation becomes important.

*Example* | A square matrix is an array of numbers which can be used to represent a linear transformation. One can think of the determinant of the array either as a particular set of operations on the array of numbers, or as a property of the linear transformation, specifically, the ratio by which the transformation changes volumes. The theorem about the determinant of a product being the product of the determinants is a very messy calculation in the first view, and an obvious result in the second.

A problem similar to this example comes up with tensors. Physicists used to think of tensors as arrays of numbers. Now it is realized that it is easier to think of the linear operators represented by those numbers, and that is the view that I follow here.

One way to free yourself from the peculiarities of a particular representation is to have several representations. We will do this quite often, usually giving a graphical representation as well as an algebraic one. We will find structures in our spacetime diagrams that faithfully represent tensors, and with no risk of confusion I will call these graphical representations tensors as well.

Although no specific mathematical knowledge beyond calculus and linear algebra is required here, the material to be developed does demand quite a bit of abstract thinking. Students at the level of this books are just acquiring such skills. The material here should provide useful practice. The goal of abstraction is to find common structure in different concrete situations, and then to discuss this structure independently of the specific situations. Done properly, this increases one's knowledge by pooling information from diverse situations. But be careful. Excessive abstraction is a very common error. Keep the references to concrete instances in mind, and do not give the abstract structures an existence of their own. Treating ideas as things (reification), done to excess in politics and religion, has no doubt murdered more people than any other cognitive error.

*Background*

*Abstraction*

A useful learning tactic, when one is faced with new abstract material, is to proceed slowly from the specific to the general, from the concrete to the abstract. The tremendous success of Euclidean geometry as a deductive system has led people to feel that a subject should be developed in the deductive mode, from general rules to specific examples. They are mistaken, especially when it comes to learning. Our brains seem to be very good at spotting patterns and generalizing from specific instances. Logic, on the other hand, seems to be a very unnatural activity. This gradual approach that I recommend, from specifics and special cases to generality, is well shown in the way I develop the idea of tangent vectors. I first define the tangents to straight lines, then to curves in vector spaces, and then finally deal with tangents to curves in manifolds. You should try to use this same style on smaller problems as well. As you try to follow an argument or to grasp a new definition, work out examples and draw some diagrams. You cannot expect to learn this material by reading this book. A more active involvement is needed. Scribble notes in the margins, work out intermediate steps in calculations, draw more diagrams, work problems, work problems, and even invent some more problems. A tactic that is often useful is to calculate things twice. An initial graphical calculation is especially useful. The errors in graphical calculations are usually more intelligent than the errors in algebraic calculations.

*Be concrete*

Graphical illustrations are used prominently in this book. Besides the graphical representations that are complements to numerical representations, many of our arguments use diagrams as essential parts. A good diagram is often an adequate algorithm for a proof, and a detailed proof would often be clumsy and boring. Most of the diagrams are two-dimensional, using one space and one time dimension. This two-dimensionality reflects the structure of paper and the architecture of the visual system. Usually the essential points can be shown in diagrams with only two dimensions, and the extensions to more dimensions is a useful but routine exercise.

*Visualize*

[See the second epigraph.]

*Complications*      Tensor analysis has a reputation as a complicated and difficult subject, partly because few people's intuition can cope with the representation of tensors as arrays of numbers. Instead I will ask you to start with geometric representations of tensors. For example, a metric tensor will be represented in a spacetime diagram by a hyperbola. Now, a hyperbola in general position is analytically complicated, but it is not at all a difficult concept. So too with tensors, I hope. My most optimistic attempt to skirt the complexities of tensors is the use of the geodesic square to discuss curvature.

There are some people who do not take diagrams seriously. I assure you that the diagrams here were very carefully drawn. They are meant to be carefully examined. If you are someone who does not usually look carefully at diagrams, you should be careful, or you will miss an important part of this book.

*Do problems*      As you work through this material, you should observe your own thinking and problem solving. Find your weak points and work on them. Be careful about abstraction. Too little is better than too much. Too much abstraction usually shows up in the use of jargon and the uncritical manipulation of words instead of concepts. Such bad habits often come from trying to learn too much too fast.

*Other reading*      What other reading do I suggest? For background in astronomy, you should read one or another of the popular astronomy books. *Astronomy and Cosmology* by Fred Hoyle is a very good one. The special-relativity background is well covered in Taylor and Wheeler. If you want to read more about thinking and problem solving, I recommend the books by Weinberg and Polya. To continue on from here, I would recommend Rindler and Ohanian for more conventional treatments at about this level. Further progress can be made by looking in the encyclopaedic Misner, Thorne, and Wheeler for material and further references.

# Spacetime, Geometry, Cosmology

# CHAPTER ONE

# Special Relativity

Information comes to us from the universe in two main ways. The sky is full of light raining down on us. The ground is full of rocks, their natural radioactivity providing us with clocks that were set running by special events in the distant past. To understand this information, we need a theory for clocks and light signals. The physical theory called special relativity is such a theory, useful where effects of gravity are negligible.

Special relativity will be discussed in three different ways in this book. In this chapter we will develop it from some fundamental postulates combined with observations on moving clocks. In this development, Lorentz symmetry is not assumed but discovered. The style is similar to plane geometry *à la* Euclid. In the next chapter we will adapt a special language to special relativity, and discuss it more in the style of analytic geometry. Finally, in Chapter III we will go back and tinker with the foundations, redefining and extending the idea of a free particle in order to smoothly introduce general relativity.

The description of special relativity given here is a mock-historical presentation. It is not how special relativity was discovered, nor is it how I feel that special relativity should have been discovered. It is, I hope, a clear and efficient way to learn special relativity and to appreciate its subtle logical structure.

## 1. Spacetime Structure

*"Nothing puzzles me more than time and space;*
*and yet nothing troubles me less,*
*as I never think about them."*

CHARLES LAMB

This book is about spacetime and its structure. Spacetime is the arena in which things move, the collection of all possible positions at all possible times. Spacetime has structure that is shown in the regularities in the behavior of objects moving through it. For us, the structure of spacetime will mean the most fundamental regularities in the behavior of free particles, clocks, and light signals. Within this fundamental structure, we can explore the detailed behavior of clocks, and describe the experimental evidence for the theory of clocks that, so far, has adequately described all such experiments; this theory is special relativity. The spacetime structure discussed here provides the language in which the theory of special relativity can be discussed.

All the structure discussed in this first chapter of the book is that in small regions of spacetime within which gravitation may be ignored. Chapter II develops the mathematical language that, in Chapter III, lets us generalize this structure to accommodate gravitation, thus allowing a consistent discussion of the self-gravitating universe, which we pursue in Chapter IV.

*Event*     The most basic idea of spacetime is that of an *event*. By an event we mean some particular place at some particular time. The event is to spacetime as the point is to Euclidean geometry. We call it an event rather than a point to emphasize that it involves both space and time. Real events have a finite but uninteresting extent, of course. The coarse-graining of our model ignores this internal structure.

[See the remarks on coarse-graining in the Introduction.]

*Spacetime Diagram*     We will represent physical events and their relations by making a representation of them, a four-dimensional map, called a *spacetime diagram*. In practice, some of the dimensions may be uninteresting or ignorable; so most of our spacetime diagrams will be two-dimensional, space plus time, and will represent events along a single line in space. For situations with more dimensions, we will either attempt perspective drawings, or present sections through the situation, or project the situation (forgetting the extra dimensions). Objects that have a continuous existence are represented by a continuous line of events called a *world line*.

***Example*** | Consider the problem of representing the traffic flow on a street with traffic lights. Here we have events, such as a particular traffic light turning red just as a certain car drives up, and so on. Figure 1.1 is a spacetime diagram for such a situation. Some events, such as traffic lights changing or cars starting, are indicated. World lines for two cars and two traffic lights are shown. The coordinates used here are the familiar time and distance coordinates.

[Here we have a projection. The cars do not collide.]

The basic idea of a spacetime diagram is quite straightforward, as I hope the example shows. It is a graphical version of familiar time-tables, such as an airline timetable. Remarkably, such a simple notion can be developed into a powerful tool that will be indispensable for our work here. As we have defined it, an event need not have an actual happening at it. This is just like Euclidean geometry; there is nothing going on at all at most of the points in the plane.

To make clear and unequivocal what we mean by a spacetime diagram as a representation of physical events, we should give an operational definition, that is, a sequence of operations and physical measurements for constructing it. However, we will postpone doing so until a later section, because the operations will not make sense until we have discussed clocks and light signals. We also need to be precise about the spacetime diagram itself. It is a mathematical model, and it is based on the mathematical idea of a linear vector space, which idea will be used repeatedly in this book. It is a generalization of the familiar 3-vectors of ordinary mechanics and electromagnetism.

Any set of objects will be called a linear vector space if it obeys certain axioms. There must exist a rule for adding two elements of the vector space to produce another vector. There must also be a rule that allows one to scale the vectors, i.e., to multiply them by real numbers. The axioms demand a certain reasonableness for these rules.

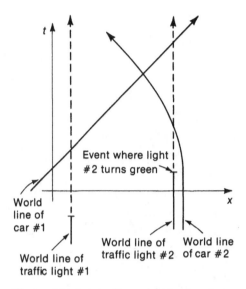

**Figure 1.1.** A two-dimensional spacetime diagram representing traffic flow. The spatial dimension across the street has been left out.

**Linear Vector Space Axioms.** If $a$, $b$, and $c$ are vectors from a given linear vector space, and if $k$ and $m$ are real numbers, then the rules for addition and scaling must satisfy

*Linear vector space*

$$a + (b + c) = (a + b) + c, \qquad (1.1)$$

$$a + b = b + a. \qquad (1.2)$$

There must exist a zero vector, called 0, such that

$$a + 0 = a \qquad (1.3)$$

for all vectors $a$. For each vector $a$ there must exist a vector $(-a)$ such that

$$a + (-a) = 0. \qquad (1.4)$$

The scaling must satisfy

$$(km)a = k(ma), \qquad (1.5)$$

$$(k + m)a = ka + ma, \qquad (1.6)$$

$$k(a + b) = ka + kb. \qquad (1.7)$$

Finally, the real number 1 must not change the vector

$$1a = a. \qquad (1.8)$$

---

[$\mathbb{R}$ is the universally used symbol for the set of all real numbers. $\mathbb{R}^2$ is used for the Cartesian plane, the set of all number pairs.]

**Example 1**

Look at the set of number pairs $(x, y)$, where $x$ and $y$ are any real numbers. The rules for addition,

$$(x, y) + (z, w) \equiv (x + y, z + w), \qquad (1.9)$$

and scaling,

$$k(x, y) \equiv (kx, ky), \qquad (1.10)$$

obey the axioms above, and define a linear vector space structure for the set of number pairs, $\mathbb{R}^2$. This is the standard linear vector space structure for this set. It is easily generalized to any number of dimensions.

**Example 2**

The following set of rules,

$$[x, y] + [z, w] \equiv \qquad (1.11)$$

$$\left[ \frac{x(z^2 + w^2) + z(x^2 + y^2)}{(x + z)^2 + (y + w)^2}, \frac{y(z^2 + w^2) + w(x^2 + y^2)}{(x + z)^2 + (y + w)^2} \right],$$

and

$$k[x, y] \equiv \left[ \frac{x}{k}, \frac{y}{k} \right], \qquad (1.12)$$

also turns the space $\mathbb{R}^2$ into a linear vector space. We use square brackets to indicate that these rules are different from the standard ones. That the axioms are followed is by no means obvious. In fact, $\mathbb{R}^2$ does not contain a zero vector, and we must add to $\mathbb{R}^2$ the points at infinity. The zero vector is to be any number pair $[x, y]$ such that $x^2 + y^2 = \infty$. All such pairs must be considered a single vector. Also, the point $(0, 0)$ must be taken out of $\mathbb{R}^2$ or it will violate the axioms. Although this second example seems bizarre, it is also a natural linear vector space structure for this set. We will find better representations for it in Section 15.

[The idea of a linear vector space is fundamental. To become familiar with it, you should check that the above spaces really do satisfy the axioms. Invent some more examples. See Loomis and Sternberg if you wish more mathematical details.]

These vectors are like the familiar 3-vectors, except that they all have their tails at the origin. They are sometimes called bound vectors to distinguish them from free vectors, which can start from any point.

Our spacetime diagrams are to be drawn in a four-dimensional linear vector space, usually the set of number 4-tuples with the standard structure. For this linear vector space to be useful, there must be a rule that relates the mathematical structures defined in it to the physical world. (Such a relation stands between the mathematical and physical worlds and is quite complicated. We leave its details to Section 4, and here merely assume that some such maps exist.) Such relations between sets are called *maps*, and we will use the word "map" in the technical mathematical sense. The mathematical notation for maps is also very useful, and deserves to be more often used.

*Maps*

---

**Map: An operation which assigns to each element of one set a unique element of some other set.** In mathematical shorthand, we write the sentence, "$f$ maps elements of the set $A$ onto the set $B$," as

$$f: A \rightarrow B. \qquad (1.13)$$

In addition, we will use the stopped arrow, $\mapsto$, either to specify the notation used for the element in the second set or to actually define the map. Thus we write, "the element of $B$ that results from the map $f$ acting on the element $a$ of $A$ will be written $f(a)$," as

$$f: A \rightarrow B; \ a \mapsto f(a). \qquad (1.14)$$

---

*Example*

The map $g$ taking real numbers and doubling them is written

$$g: \mathbb{R} \rightarrow \mathbb{R}; \ x \mapsto 2x. \qquad (1.15)$$

A map is a generalization of the idea of a function. A function is a map from real numbers to real numbers.

The next important idea after that of an event is the idea of a *free particle*. The physical idea is of a body whose size is so small that it is unimportant, and which is moving freely, subject to no forces whatsoever. The world lines of such particles form a special class among all possible world lines in spacetime. Recall the law given by Galileo that free particles move with constant velocity. In a spacetime diagram, constant velocity is equivalent to a straight world line. We will put this law into our model by using the mathematical structure of a vector space to model spacetime. We introduce the following postulate.

*Free particles*

---

**Free-Particle Postulate: One can find a map of events into a spacetime diagram such that the world lines of free particles are straight lines.**

---

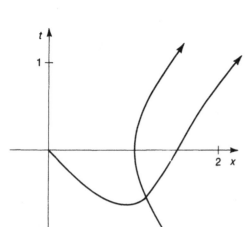

**Figure 1.2.**

A straight line is a mathematical notion that is well-defined in a linear vector space. It is any set of vectors $v$ that can be generated by letting the number $s$ range over all values in the expression

$$v = a + sb, \qquad (1.16)$$

where $a$ and $b$ are given fixed vectors.

[Primitive notions were discussed in the Introduction.]

The idea of a free particle is a primitive notion in our system. It cannot be defined within our system. The idea can only be demonstrated to you. This is why physics courses need laboratories.

It might seem that the free-particle postulate is only a statement about the map, and does not say anything about the actual world. This is wrong. Look at Figure 1.2. It shows two curved lines in a vector space. Suppose these are the images of two world lines under some map. Can you conclude that these particles are not free particles? Not at all. The lines might be curved just because of the map used. On the other hand, consider the map which results in Figure 1.3. Two different straight lines can intersect only once. No map will straighten out both of these lines at once. Since not all sets of world lines can be mapped into straight lines simultaneously, to assert that the world lines of free particles can be is not an empty assertion.

**Figure 1.3.**

A map which represents free particles as straight lines is far from unique. There are an enormous number of maps which are all satisfactory. We will see how to deal with this situation in Section 4, on inertial reference frames.

The next important idea to add to spacetime is that of a clock. A clock is a device for measuring world-line segments. It might involve a repetitive phenomenon, such as the ticking of a clock, or the regular indication of a standard interval, as by the lifetime of an unstable particle. The rate of the clock should be determined by some intrinsic property. Most of our clocks have their rates determined by atomic or nuclear processes. When we discuss water-wave relativity, we will see another physical realization of this idea of a clock with an intrinsic rate. To be called a clock, we demand that any interval-measuring algorithm satisfy the following two postulates.

**Clock Universality Postulate: The relative rate of any two different clocks carried along the same world line does not depend on where in spacetime or along what world line it is measured.**

This is another coarse-graining postulate. It asserts that we do not need separate theories for differently constructed clocks, say, atomic clocks versus nuclear clocks. It is easy to see the physical content of this postulate, and every improvement in technology allows new experiments to check its validity. One future experiment being discussed would send a collection of different clocks very close to the Sun to see if the presence of a massive body affects their relative rates.

The second postulate both describes the behavior of clocks and restricts further the map representing physical events in a spacetime diagram.

**Clock Uniformity Postulate: There exists a representation of events in a vector space with free particles moving along straight lines and with uniform clock rates.**

By uniform clock rates, we mean that the intervals measured by clocks agree with the linear structure of the vector space. A world-line segment twice as long should measure twice the time interval. The time intervals measured on opposite sides of a parallelogram should be equal. See Figures 1.4 and 1.5.

Free particles involve only the straight-line structure of a vector space. Clocks bring in also the idea of parallelism. Note that we do not need the idea of perpendicularity, nor does a vector space have a

*Clocks*

**Figure 1.4.** If the vector $AC$ is $k$ times the vector $AB$, then a clock carried along the segment of world line $AC$ measures $k$ times the interval that it would measure along $AB$. Here $k$ is about 4.

[I have intentionally left the axes off these figures to emphasize the independence of the ideas from any particular coordinate system.]

natural definition of perpendicularity. There are many clock structures compatible with the above postulates. It is the task of experiment to discover what particular structure describes real clocks. A spacetime-diagram representation in which free particles have straight world lines and in which clock rates are uniform will be called here an *inertial reference frame.*

*Inertial reference frame*

### PROBLEMS

[The numbers in parentheses indicate the approximate difficulty of the problem; see the scale on page xvii.]

1.1. (15) List some of the idealizations that went into Figure 1.1.

1.2. (15) If the traffic lights in Figure 1.1 are a block apart, did I leave a reasonable delay for the driver of car 2 to react to the light changing?

1.3. (24) Discuss, using a spacetime diagram, why one cannot synchronize the traffic lights for traffic going in both directions along a street if the lights are not uniformly spaced.

1.4. (12) Verify that the axioms for a vector space are satisfied by the standard vector space structure given in Example (1) on page 6.

1.5. (18) Show that the coordinate transformation

$$(t,\, x) \mapsto (u,\, v) = \left( \sqrt{x^2 - t^2},\, \tanh^{-1}\!\left(\frac{t}{x}\right) \right)$$

straightens the lines in Figure 1.2.

1.6. (35) Trace some more free-particle lines in the curvilinear representation of Figure 1.2. In particular, what happens in the neighborhood of the origin?

1.7. (15) Show that no transformation can straighten out the family of circles shown in Figure 1.6, given by

$$(x - a)^2 + (y - b)^2 = 1.$$

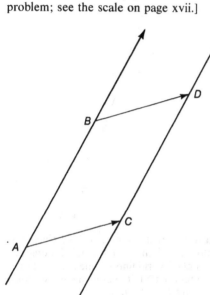

**Figure 1.5.** If the vectors *AC* and *BD* are equal, then a clock carried along the segment of world line *AB* measures the same time interval as one carried along the segment *CD*. This is one aspect of uniformity.

Different circles correspond to different values for *a* and *b*.

1.8. (30) Find a coordinate transformation that maps the family of parabolas shown in Figure 1.7, given by

$$x = a + bt + t^2,$$

into straight lines.

[The (15) rating for this problem assumes that the results of Problem 1.8 are available to you, and does not include the difficulty of Problem 1.8.]

1.9. (15) Consider electrons moving in a region of uniform electric field. Can they be considered free particles? What about a mixture of particles with different ratios of charge to mass?

1.10. (12) Discuss the spacetime diagrams shown in Figures 19.6, 19.7, and 29.15. Describe the situation at different times.

1.11. (34) Invent a three-dimensional analog of the bizarre vector space on $\mathbb{R}^2$ given in Example (2) on page 6.

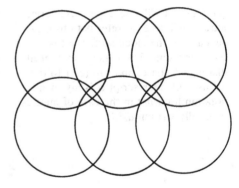

## 2. Clocks

*"If everybody minded their own business,"*
  *the Duchess said in a hoarse growl,*
*"the world would go round a great deal*
  *faster than it does."*

LEWIS CARROLL

Figure 1.6.

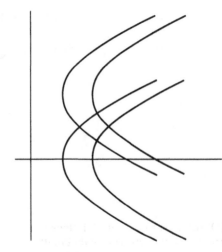

Figure 1.7.

*Representation of clock structure*

The fundamental spacetime structure discussed in Section 1 is compatible with many different clock behaviors. We will use the structure of an inertial reference frame as the language in which to discuss the behavior of a clock. For this we will need an efficient representation to describe the behavior of any clock that is compatible with our postulates. Such a representation is easy to find. We will first apply this representation to the familiar clock behavior of Newtonian mechanics, so that we can practice our thinking on familiar ground before going on to the new and subtle clock behavior of special relativity. Since the representation of spacetime in an inertial reference frame is not unique, we must discuss the different representations of clocks in the different inertial reference frames. This introduces the important idea of covariance, which is fundamental to the entire development of tensor analysis.

How, then, will we represent any particular clock behavior? Assume only that the behavior is compatible with our postulates, and that we have picked somehow some particular inertial reference frame in which to work. Although the uniformity postulate constrains the behavior of clock rates along any single world line and on all those parallel to it, it says nothing about different parallel families. These correspond to families of clocks moving with different velocities. Our different clock theories are characterized by the different behaviors of moving clocks.

How do we represent this additional information? We can do so in a straightforward manner. Consider different clocks carried through the origin (the zero vector) with all possible velocities. Mark the distance along each of their world lines which is a unit time interval away from

[Notation: I will use upper-case script letters such as $\mathscr{G}$ to refer both to these sets and to the clock structure that they represent. The logician may need to distinguish these, and can refer to the set as "$\mathscr{G}$." This script $\mathscr{G}$ is my own notation for what in the case of metrics is usually written $ds^2$.]

the origin. The set $\mathscr{G}$ of all such marks for all possible clocks will be our representation for the clock behavior.

Once we know the set $\mathscr{G}$ of all events that are a unit time interval from the origin, we have a complete description of the behavior of clocks. We can use the clock-uniformity postulate to translate the standard interval anywhere we wish, and thus we can measure any time interval that can be measured. This is sketched in Figure 2.1. Let me emphasize that knowing $\mathscr{G}$ in one inertial reference frame is sufficient to completely describe the behavior of the clocks. It is not necessary to describe each clock in an inertial reference frame in which it is "at rest." We have not yet even defined the concept "at rest." Note that our representation is set up in such a way that it is easy to study experimentally. The experimental physicist can provide us with measurements of the set $\mathscr{G}$, complete with error bars.

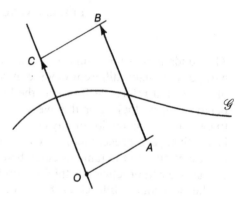

[Observe that in Figure 2.1 clocks moving to the left behave differently from clocks moving to the right.]

**Figure 2.1.** The use of uniformity to measure the time interval $AB$. Here it is about 1.5 time units.

*Absolute-time clocks*    Let us practice with the above idea, by using it to represent the familiar clock behavior of Newtonian mechanics. We will call such clocks absolute-time clocks. They form a consistent mathematical model, with a useful physical realization in the behavior of real clocks at low speeds. In addition, they provide us with the conventional background against which we can view the novelty of special-relativity clocks. The usual space and time coordinates of Newtonian mechanics provide us with an inertial reference frame here. Clock readings are given directly by the time coordinate. Let us use $\mathscr{A}$ to denote the set of events which represents the behavior of absolute-time clocks: $\mathscr{A}$ is a horizontal line at a unit distance above the origin (see Figure 2.2). In more dimensions it will be a plane or hyperplane. This is a consistent theory of clocks, one that is even "pretty" in some sense. We will see that is has its own symmetry, called Galilean symmetry, analogous to the Lorentz symmetry of special relativity. Special relativity was actually invented because the Galilean symmetry of absolute time was not compatible with the Lorentz symmetry of Maxwell's electrodynamic theory.

We now must face the problem that our postulates do not uniquely determine the inertial reference frames. We will go into this problem deeply in Section 4, on inertial reference frames; but it is subtle, and it is well for us to start thinking about it right now. A convenient way to discuss the differences between two different inertial reference frames, that is, two different maps, say, $\psi$ and $\psi'$, into our vector space $V$, is to discuss the transformation of $V$ which takes the spacetime diagram that results from one map and transforms it into the spacetime diagram that results from the other map.

---

**Transformation: A map from a set into itself.**

---

*Many representations possible*

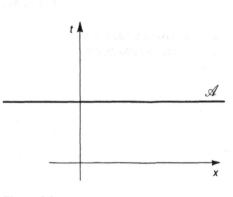

**Figure 2.2.**

In this way we stay inside our mathematical model. The maps $\psi$ and $\psi'$ are complicated, because they deal at one end with the physical world and at the other end with the mathematical world. If the two maps $\psi$ and $\psi'$ are both to give us inertial reference frames, then in each spacetime diagram the free-particle world lines must be straight lines. The transformation between these spacetime diagrams must take straight lines into straight lines. Also, the clock rates must be uniform in both; so (and this is not so hard to see) we must take parallel lines into parallel lines. Maps which do this are called linear transformations. Because they are important and will come up often, let me give explicitly the linear transformations in two dimensions, sketched in Figure 2.3. They are:

*expansions,* $\quad\quad (x, y) \mapsto (kx, ky);$ $\quad\quad\quad\quad$ (2.1)

*rotations,* $\quad (x, y) \mapsto (\cos\theta\, x + \sin\theta\, y, \cos\theta\, y - \sin\theta\, x);$ (2.2)

*shears,* $\quad\quad\quad (x, y) \mapsto \left(\alpha x, \dfrac{y}{\alpha}\right),$ $\quad\quad\quad\quad$ (2.3)

$$(x, y) \mapsto \frac{1}{\sqrt{1-v^2}}\,(x - vt, t - vx).$$ (2.4)

All linear transformations are combinations of the above transformations. In addition, we can also merely translate our spacetime diagrams,

*Linear transformations*

[The stopped arrow, $\mapsto$, notation was explained on page 7.]

[When $k = 2$, Equation (2.1) becomes $(1, 0) \mapsto (2, 0)$; that is, the transformation takes the point that is a unit distance out from the origin along the $x$-axis and moves it out to a distance of two units.]

[These two shears differ only by a 45° rotation.]

moving the origin to a new location. Linear transformations can also be represented by square matrices.

In each different representation, the set $\mathscr{G}$ representing clock behavior will be different, even though it represents the same clocks. Figure 2.3 also shows how the structure of absolute-time clocks $\mathscr{A}$ changes under the above transformations. This change in the representation of the clock-rate structure is called *covariance*. Throughout this book we will deal with representations for which linear transformations, or a subset of them, change one acceptable representation into another acceptable one. The idea of covariance involves these changes.

*Covariance*

[It may take a while for this idea of covariance to become clear to you.]

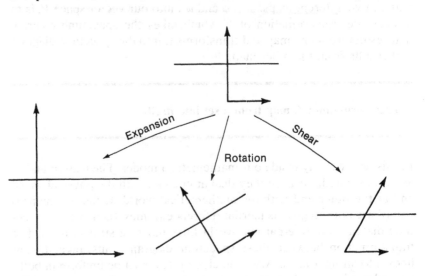

**Figure 2.3.** Examples of linear transformations in two dimensions: the expansion $(x, y) \mapsto (2x, 2y)$; the rotation $(x, y) \mapsto \frac{1}{2}(\sqrt{3}x - y, x + \sqrt{3}y)$; and the shear $(x, y) \mapsto (x + \sqrt{3}y, y)$.

PROBLEMS

2.1. (10) Find an explicit linear transformation that takes Euclidean perpendicular lines to ones that are not perpendicular.

2.2. (9) What time interval would be read by a clock carried at uniform speed from event $O$ to event $B$ in Figure 2.1?

2.3. (20) Show that for an absolute-time clock, the time interval read by a clock carried between two events is independent of the path followed.

## 3. Euclidean Geometry

The covariance of a representation under linear transformations is an important and subtle idea. We study it here to contrast the usual view of Euclidean geometry with another that is covariant under linear transformations. Each view is a valid representation. One's choice between them should be based not on prejudice, but on the practical details of the computations being attempted. A similar choice will confront us when we return to special relativity. The measurement of spatial distances is quite similar to the measurement of time intervals. Both the similarities and the differences are important. We will see that the clock representation we have introduced also describes Euclidean geometry.

We will discuss first the conventional view of Euclidean geometry. Nothing essential is lost by restricting this discussion to two dimensions. We can turn a vector space into Euclidean geometry most easily by defining a dot product, as follows.

*Euclidean geometry as usual*

**Euclidean Dot Product:**

$$a \cdot b \equiv a^x b^x + a^y b^y, \qquad (3.1)$$

where $a$ is a 2-vector with components $a^x$ and $a^y$ and similarly for $b$.

[The reasons for using superscripts for the components will become clear in the next chapter, on geometry. Do not confuse them with exponents.]

We will use parallel displacement to calculate the dot products of vectors which do not start at the origin (remember uniformity). Algebraically, this is reflected in the distributive rule,

$$(a + b) \cdot c = a \cdot c + b \cdot c, \qquad (3.2)$$

[Here $a$, $b$, and $c$ are all vectors.]

which follows directly from the definition 3.1. The familiar ideas of length and angle can be represented by using this dot product. The length of a vector $a$, call it $|a|$, is found by

[Dot power!]

$$|a|^2 = a \cdot a, \qquad (3.3)$$

and the angle $\theta$ between two vectors $a$ and $b$, is found by

$$\cos \theta = \frac{a \cdot b}{|a| \, |b|}. \qquad (3.4)$$

*Orthonormal coordinates*

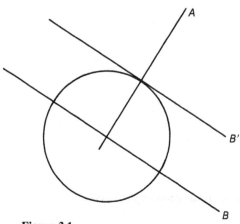

**Figure 3.1.**

The dot-product rule given by Equation 3.1 represents Euclidean geometry only if we are using properly chosen coordinates. The axes must be at right angles, and the scales on the axes must be of unit length. Such coordinates are called *orthonormal*. They are not unique. If one rotates coordinates, then the same rule again describes the geometry. The simplicity of this rule makes it efficient usually to use only orthonormal coordinates. We will find a similar situation in special relativity, where the clock structure $\mathcal{G}$ can be represented by a different dot-product rule. Special-relativity calculations can be done most efficiently in representations where the dot product is simple. Unfortunately, general relativity cannot be discussed in such specialized coordinates. In fact, the presence of gravity is shown by our inability to find such representations. Fortunately, our representations that are covariant under linear transformations are not as restricted and can be suitably generalized. This more general representation is "cleaner," in that properties of a special representation are not confused with properties of $\mathcal{G}$ itself.

*General coordinates*

How do we represent Euclidean geometry in a manner that is covariant under linear transformations? Proceed as we did for clocks. Take the set of points, call it $\mathcal{E}$, that are at a unit distance from the origin. In orthonormal coordinates, $\mathcal{E}$ is a circle. After we perform any amount of expansion, shear, and rotation, we end up with an ellipse of arbitrary size, eccentricity, and orientation. A rule for measuring line segments is called a metric, and I will call sets like $\mathcal{E}$ *metric figures*. In general, the metric figure for Euclidean geometry is an ellipse. How do we use $\mathcal{E}$ to measure lengths and angles? Watch carefully how we do this. The method is subtle, and one that we will use many times. We know what length and angle mean in orthonormal coordinates. If we can describe them in terms of operations that behave properly under linear transformations, such as parallelism, then the same operations can be used in any linear coordinates, and they will produce the same result.

*Metric figures*

*Angles*

Consider perpendicularity. In Figure 3.1, which is in orthonormal coordinates, the lines $A$ and $B$ are perpendicular; $B'$ is parallel to $B$, and is tangent to the unit circle at the point where $A$ intersects it. Since parallelism and tangency are preserved by linear transformations, we can use this construction to prove perpendicularity in any linear representation, not just in orthogonal ones. A typical case is shown in Figure 3.2. If this construction seems peculiar to you, try to stretch the figure in your mind until it becomes circular. Better yet, hold the paper at an angle until the ellipse projects into a circle.

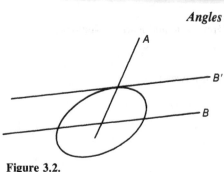

**Figure 3.2.**

*Lengths*

To find the length of a line segment, proceed as in Figure 3.3 Find a line that is parallel to the given segment and that passes through the origin. Then use parallel lines to transfer a unit length out to the segment, and use this to measure its length (uniformity). Again, doing so involves only operations that are preserved by linear transformations; so it can be done in general. Because the dot operation is linear, having a rule for the lengths of vectors also tells us about the dot product of two different vectors:

$$2(a \cdot b) = (a + b) \cdot (a + b) - a \cdot a - b \cdot b. \qquad (3.5)$$

We have succeeded in our aim. We have extended the usual representation of Euclidean geometry, which is covariant only under rotations, to a representation that is covariant under all linear transformations. We have extended the idea of the dot product also. Equation 3.1 is to be used only in orthonormal coordinates; otherwise use the generalization given above, using the metric figure to compute the lengths of vectors, and Equation 3.5 to compute the dot product. You should practice with this representation. It is very similar to the ones that we will use for special relativity and general relativity. Also, this practice should make the idea of covariance and linear transformations familiar to you. You need to develop an eye for what is "covariant under linear transformations."

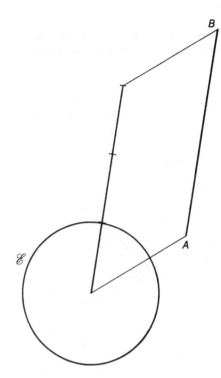

**Figure 3.3.** The line segment $AB$ is transferred to the metric figure $\mathscr{E}$ for measurement. Here $AB$ is about three units long.

*Example*

Figure 3.4 is drawn in a general linear frame. If the parallelogram $A$ is really a Euclidean square, let us show graphically that angle $B$ is a right angle.

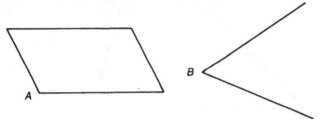

**Figure 3.4.**

One way to proceed would be to find the ellipse that represents Euclidean geometry in this frame. An easier approach is to change the representation so that the geometry is represented by a circle. We will do that by making a series of linear transformations to both figures. It is important to make the *same* transformation of each figure. This is the "co" in covariant. The aim is to turn the parallelogram into a square. In such a frame the geometry is represented in the usual manner, and we will be able to see directly if angle $B$ is a right angle.

Let us first shear both figures, turning Figure 3.4*A* into the rectangle shown in Figure 3.5. Then let us

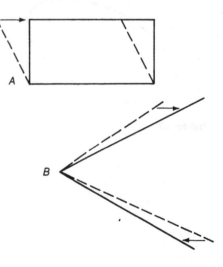

**Figure 3.5.** A shear of the form $\binom{x}{y} \mapsto \left(\begin{smallmatrix} 1 & q \\ 0 & 1 \end{smallmatrix}\right)\binom{x}{y}$.

[Strictly speaking, these shears also involve some expansion and rotation.]

shrink the *x*-axis to turn the rectangle into a square. This is also a shear. The result is shown in Figure 3.6. The angle *B* is indeed a right angle.

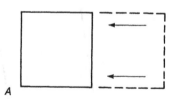

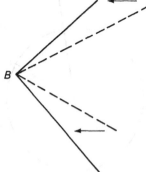

**Figure 3.6.** A shear of the form $\left(\begin{smallmatrix} x \\ y \end{smallmatrix}\right) \mapsto \left(\begin{smallmatrix} b & 0 \\ 0 & 1 \end{smallmatrix}\right)\left(\begin{smallmatrix} x \\ y \end{smallmatrix}\right)$.

## PROBLEMS

**3.1.** (17) In a general linear frame in which the unit circle is as shown in Figure 3.7, draw squares in several different orientations.

**3.2.** (20) Convert the data in Figure 3.4 into numbers, and repeat that example numerically.

**3.3.** (22) If the angles in Figure 3.8 are both right angles, draw a square that has one side parallel to line *L*. (This is not a ruler and compass construction.)

**3.4.** (15) Sketch the construction shown in Figure 3.3 in a general linear frame.

**3.5.** (25) Can one always find a frame such that any two angles like those in Problem 3.3 will appear to be right angles?

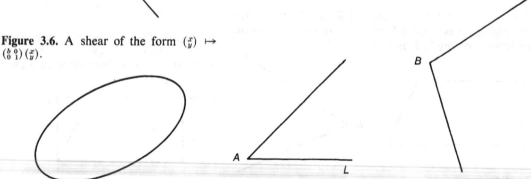

**Figure 3.7.**                         **Figure 3.8.**

## 4. Inertial Reference Frames

We are now ready to take a careful look at the construction of an inertial reference frame. An inertial reference frame is a map $\psi$ from the events of the physical world into a spacetime diagram, one with special properties relating to free particles and clocks. At the most basic level, we can look at maps which do no more than assign four numbers to each event in some smooth and systematic fashion. One such realization can be found in the study of earthquakes. Let the events be earthquakes. To locate an event in time and space, we can record the arrival times of particular elastic waves at four different locations on the Earth. Such a rule associates to each event four numbers, the arrival times. It is, then, an example of one of our maps $\psi$. We know that spacetime is four-dimensional, because it takes four numbers in general to always assign different numbers to different events, and more than four are never needed. The stations can be located anywhere, except that they should not have any special relationship, such as all lying on the same great circle.

*Earthquakes*

One would normally transform such raw coordinates into a more useful form before using them. Let us imagine that some computer sits between the four raw coordinate numbers and the final coordinates. This computer can make whatever transformations we wish on the raw coordinates. We shall choose this transformation to simplify the resulting spacetime representation. Our example gets a bit strained here, since there is no obvious seismic analog for either free particles or intrinsic clocks. A fanciful sketch of an apparatus for mapping spacetime diagrams appears in Figure 4.1.

Let us first ask of our transformation that it turn the world lines of free particles into straight lines. If no transformation of the raw coordinates can do this, then we must conclude that there were some unnoticed forces acting on the supposedly free particles. After this requirement has been satisfied, there are still many transformations which preserve the straightness of the free-particle world lines. Until we add further requirements, we are free to use any of these that we like, generating different but equally satisfactory representations. Besides the linear transformations discussed in Section 3, there are other transformations that preserve straight lines but which do not preserve the uniformity of intervals. These are the *projective* transformations familiar in perspective drawing and photography. Think about a photograph of railroad tracks vanishing into the distance: in the photograph, straight lines appear straight, but uniform distances, like the railroad ties, appear smaller the further away they are. An example of a projective transformation in two dimensions is

*Inertial reference frames*

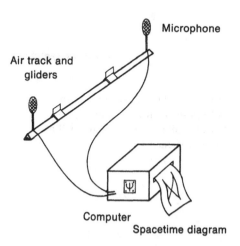

**Figure 4.1.** A fanciful illustration of a two-dimensional spacetime diagram apparatus. Events would be glider collisions. The computer could correct for the travel time of sound.

$$(x, t) \mapsto \left( \frac{x}{t-1}, \frac{t+1}{t-1} \right). \tag{4.1}$$

We next use the freedom of these projective transformations to make the clock intervals uniform. In general, the coordinates chosen to satisfy only the free-particle postulate will not satisfy the clock-uniformity postulate. We must search through projective transformations until we find one that does represent the clock intervals uniformly. Having done this, we still have the freedom of linear transformations. Recall our definition:

---

**Inertial Reference Frame: A spacetime-diagram representation fixed up to linear transformations by the requirements that free particles move in straight lines and that clock intervals be uniform.**

---

One can work satisfactorily in representations that are specified only up to linear transformations. That was the point of the Euclidean-geometry example. On the other hand, we can specialize the representation further if we want to use specific information about the behavior of clocks. Let us pursue this specialization for absolute-time clocks. Doing so will bring up some important points about the symmetry of these clocks.

*Canonical frames*

We can decide to place the set $\mathscr{A}$ horizontal and at a unit distance above the origin. We can certainly do so by using linear transformations. The set $\mathscr{A}$ must be some straight line. Merely rotate, and then expand or contract as needed. We will call an inertial reference frame that has been further specialized according to some rule or convention a *canonical* reference frame (*canon* means rule). The above canonical reference frame is still not uniquely determined, nor can we add additional rules to make the frame unique by using only the clock structure $\mathscr{A}$.

[This definition is not in general use.]

*Galilean symmetry*

The remaining freedom in the canonical representation of our absolute-time clocks comes from the fact that the set $\mathscr{A}$, like the clock structure it represents, has a symmetry. The transformation

[Although this transformation involves both pure shear and rotation, it is loosely called a shear, and I will also call it a shear.]

$$(x,\ t) \mapsto (x + vt,\ t) \tag{4.2}$$

for any fixed value of $v$ takes the set $\mathscr{A}$ into itself. It takes, therefore, one canonical representation into another one. These particular transformations are called Galilean transformations.

---

**Symmetry: An object has a symmetry if there is a transformation that leaves the object unchanged.**

---

If the set $\mathscr{A}$ had not been symmetric, but instead had bumps or kinks, then we could have made further rules about putting the bump here and the kink there until the representation was uniquely specified. We describe this symmetry of absolute-time clocks by saying that they are *invariant* under Galilean transformations. We are going to use the words invariance and covariance in a precise manner to mean different things.

*Invariance*

[N.B. again: Some authors are not so careful and use these words interchangeably.]

---

**Invariance: The symmetry of some object under transformations.**

**Covariance: The behavior of the representation of any object, not necessarily symmetric, under changes in the representation of spacetime.**

---

For an example of the action of Galilean transformation, consider Figure 4.2. The world-line segment *I* can be carried into the segment *I'* by some particular Galilean transformation. Therefore, one absolute-time clock carried along *I* and another carried along *I'* must both record the same time interval. Why? The clock reading is a physical quantity which can be calculated in any representation. We can compute the interval for the segment *I* in any representation, in particular in the one in which the segment coincides with the segment *I'*. The clock rule is the same in both representations. Therefore, the segments must represent the same time interval.

*An invariance argument*

In addition to the symmetries found among the linear transformations, there is also the symmetry of translations. We are free to pick any event for our origin. If spacetime is homogeneous, then we will see the same behavior around every event. These inhomogeneous transformations are not included among the linear transformations, nor are they represented by matrices. We will often use this translational symmetry to simplify our calculations.

*Translations*

## PROBLEMS

4.1. (10) Criticize my suggestion about Figure 3.2 that you should look at the paper at an angle to make the ellipse a circle.

4.2. (12) Show that absolute-time clocks are invariant under the space expansions

$$(x, t) \rightarrow (kx, t),$$

where $k$ is a fixed constant.

4.3. (18) Discuss the projective transformation given as an example on page 19.

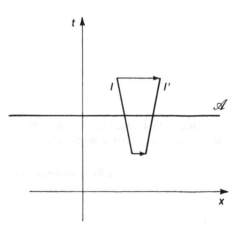

**Figure 4.2.** The interval $I$ is carried into the interval $I'$ under a Galilean transformation, here $(x, t) \mapsto (x + 2t/5, t)$.

4.4. (20) If the clock structure is given in two dimensions by a set $\mathcal{G}$ such that

$$x + y = 1,$$

find the linear transformation that puts this into canonical form for an absolute-time clock.

4.5. (35) Explain how intelligent scientists can disagree about whether the velocity of light should be measured or given a defined value.

4.6. (27) A nice example of hidden covariance is the vector integral

$$\oint r \times dr.$$

Show that the value of this integral does not depend on the location of the origin. Work in Euclidean 3-space so that the cross product is defined.

## 5. Physical Clocks

*"An ellipse is fine for as far as it goes,*
  *But modesty, away!*
  *If I'm going to see Beauty without her clothes*
  *Give me hyperbolas any old day."*

**ROGER ZELAZNY**

[For simplicity I continue to discuss the case of only one space dimension.]

*Clock experiments*

We have seen how one can in principle set up an inertial reference frame. In practice, the usual time and distance coordinates measured by clocks and measuring rods produce an approximately satisfactory inertial reference frame. Within this framework we can now study the behavior of actual clocks. Will they be described by the mathematical model of absolute-time clocks? No, only for low velocities will the model of absolute-time clocks agree with the behavior of real clocks.

The experimental procedure is to carry a clock from the event $(0, 0)$ along a straight world line until it records a unit time interval (a "tick"), and to note the event at which this happens. Repeat with clocks moving at different velocities. They have world lines of different slopes in the spacetime diagram. You will have to use the parallelogram construction to transfer all the time intervals to the origin, since all the clocks cannot pass through the same event. This is sketched in Figure 5.1. The set of events found in this way provides a complete description of the behavior of physical clocks.

Of course, actual experiments provide us with only a finite number of events, and our clocks can cover only a limited range of velocities. It is an inductive generalization to describe the set by a mathematical equation. When the experiment is actually done, the clock events corresponding to one-second time intervals lie (to within experimental error) on the hyperbola given by

$$t^2 - \left(\frac{x}{c}\right)^2 = 1. \tag{5.1}$$

The constant $c$ appearing in this expression reflects the choice of units for length and time measurements. In SI units (the international metric system) it has the approximate numerical value of $3 \times 10^8$ meters/second. The events given by the above equation form the set $\mathscr{G}$ that describes physical clocks. This is sketched in Figure 5.2. On the scale of this drawing this set is not distinguishable from that for absolute-time clocks. This is why absolute-time clocks provide a useful approximation to the behavior of real clocks. Only if you look over much larger distances, of the size of $10^8$ meters for clock intervals of seconds, do you see any difference.

*Results for real clocks*

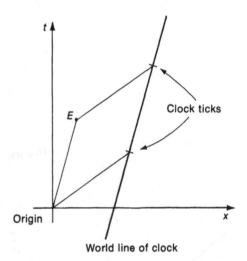

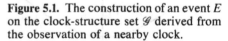

**Figure 5.1.** The construction of an event $E$ on the clock-structure set $\mathscr{G}$ derived from the observation of a nearby clock.

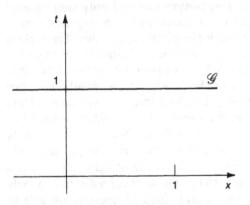

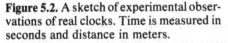

**Figure 5.2.** A sketch of experimental observations of real clocks. Time is measured in seconds and distance in meters.

To see what is really happening, we need to compress our drawings in the $x$ direction so that these large distances will appear in our diagram. To do this we adopt a new unit of length: we call it the light-second, for reasons that will become obvious later. It is a distance of

*Light seconds*

$2.998 \times 10^8$ meters. Nearly all lengths in this book will be measured in terms of this unit. Figure 5.3 replots $\mathscr{G}$ using these new units. The difference between real clocks and absolute-time clocks is now readily apparent. We leave it as a problem for you to show that the hyperbola in Figure 5.3 implies that the time interval read by a clock carried at constant velocities from the event $(x_0, t_0)$ to the event $(x_1, t_1)$ measures a time interval $\tau$ given by

[Here $x_0$ and $x_1$ are measured in light-seconds.]

$$\tau^2 = (t_0 - t_1)^2 - (x_0 - x_1)^2. \qquad (5.2)$$

*Acceleration*

What about the transport of clocks along curved world lines? Many clocks, and especially high-precision clocks, do not keep time if subjected to more than small accelerations. Even poor clocks like people (three score and ten) fail when dropped from buildings. Some clocks, such as unstable elementary particles, seem to be quite insensitive to acceleration. Experiments involving acceleration-sensitive clocks must be arranged so that measurements are made only on the straight sections of their world lines. For acceleration-insensitive clocks, we can calculate time intervals for curved paths by taking the limit of a series of short straight-line segments, just as one computes the lengths of curves using calculus.

*History*

We have followed here a mock historical development. In actual historical fact, the hyperbolic rule was guessed by looking at the symmetry of Maxwell's theory of electrodynamics and only later verified by experiment. The approach to understanding the theory used here is straightforward, and does not require inspiration or genius. The earliest actual experiments involved measurements of unstable elementary particles, $\mu$ mesons, that are created in the upper atmosphere by cosmic rays. These mesons were known to be moving at very nearly the speed of light, and their mean lifetimes, the clock interval, was known from other experiments. A spacetime diagram for the experiment is sketched in Figure 5.4. We have tilted it on its side so that up is up. Naively (knowing about $\mu$-mesons, but not about special relativity), one would expect the meson, traveling at nearly the speed of light, and having a lifetime of only $2 \times 10^{-6}$ seconds, to be able to cover a distance of only $2 \times 10^{-6}$ light-seconds before it decays. Instead, mesons are able to penetrate the entire thickness of the atmosphere, a distance of around $40 \times 10^{-6}$ seconds. (Note the short time interval taken by the experiment. For such short times, gravity can be neglected, and a reference frame set up on the Earth acts like an inertial reference frame. We will discuss this more deeply when we describe gravitation.)

Further experiments were, of course, necessary. Two points do not sufffice to determine an entire curve. Extensive experimental con-

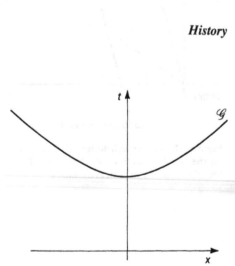

**Figure 5.3.** Figure 5.2 replotted using the light-second as the unit of distance.

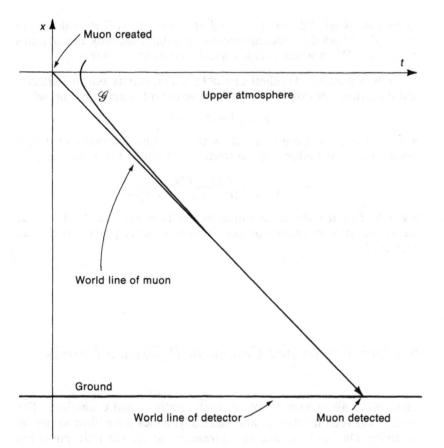

**Figure 5.4.** Sketch of the muon-decay observations. The curve $\mathscr{G}$ is given by $t^2 - x^2 = \tau^2$, the distance $x$ is measured in light-seconds, and $\tau$ is the muon lifetime, $2 \times 10^{-6}$ seconds, not to scale.

firmation of the clock-rate hyperbola comes from the experiments done with high-energy particle accelerators. The peculiar behavior of clocks must be taken into account in their very design.

PROBLEMS

5.1. (10) If Figure 5.4 were drawn to scale, how far above the ground would that particular muon have been created?

5.2. (16) A clock is carried at constant velocity from the event $(0, 0)$ to the event $(1, 2)$, and then to the event $(0, 4)$. Describe in words what happened. What did the clock read at event $(0, 4)$ if it read zero at event $(0, 0)$? All distances are measured in light-seconds and all times in seconds. What would a clock which stayed at $x = 0$ read?

5.3. (28) Consider hypothetical clocks which satisfy our universality and uniformity postulates, and whose structure is given by the set $\mathscr{G}$,

$$t^4 = x^4 + Ax^2 + 1,$$

$A$ being some given constant. Show that the time interval $\tau$ read by a clock carried at uniform speed from the origin to the event $(x, t)$ is

$$\tau^2 = \frac{2t^4(1 - v^4)}{Ax^2 + [4t^4 + (A^2 - 4)x^4]^{1/2}}.$$

5.4. (16) In our units acceleration has units of seconds$^{-1}$. Show that the acceleration of gravity on the Earth is about $(1 \text{ year})^{-1}$. Is that an accident?

## 6. Light Signals and Canonical Reference Frames

Light plays an important role in both relativity and cosmology. The conflict between mechanics and electrodynamics gave birth to special relativity. Our main source of information about the universe is the electromagnetic radiation collected by our telescopes. Out of all this complexity, we abstract the idea of a light signal, meaning "a short burst of light going in a particular direction," such as one might make with a laser and a fast shutter. We are going to ignore the spatial extent of the light pulse, and look at the light signal as only a special kind of particle. We decline to call it a photon, because that term connotes a specific quantum-mechanical idea. In our mathematical model, light signals will be represented by world lines. The fundamental rules for light signals are expressed in the following two postulates.

---

**Light-Signal Inertial Postulate:** Light-signal world lines are straight lines in an inertial reference frame.

**Light-Signal Unique Velocity Postulate:** The world lines of light signals going in the same spatial direction are parallel in an inertial reference frame.

---

The idea is that the propagation of light is independent of the source of the light and its motion. Here spatial direction is being used in the familiar sense. Along a line there are two spatial directions: left and right. In a plane there is a one-parameter family of directions: north, east, etc. In our 1+1-dimensional spacetime diagrams, the light-signal world lines form two families of parallel lines. We shall conventionally draw light-signal world lines as dashed lines. Just as our fundamental postulates on clocks needed to be supplemented with a detailed clock structure $\mathscr{G}$ that resulted from actual experimental measurements, so too our light-signal postulates describe only their fundamental properties; we need experiments to tell us where the actual light-signal world lines are.

[Notation: To avoid confusion, we refer to a spacetime diagram with one space and one time dimension as 1+1-dimensional.]

Figure 6.1 shows a possible situation in 1+1 dimensions that I just made up. Figure 6.2 shows a possible situation in 2+1 dimensions. In Figure 6.2 we have sketched all possible light signals coming from a given event $E$. They form a cone-shaped surface called the *light cone*.

*Light cones*

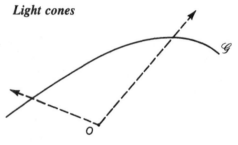

**Figure 6.1.**

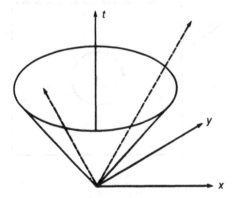

**Figure 6.2.** Light signals in two space dimensions and one time dimension.

How can we think about the situation in 3+1 dimensions? One trick is to section the diagram, much as an architect draws successive plans for each floor in a building. For practice, let us section the 2+1 diagram. Figure 6.3 shows a pair of sections taken at different times, like frames of a movie. Each frame shows us a ring of light signals, which are expanding in time. What will successive frames of the 3+1 slices look like? Expanding spheres of light signals, certainly.

These fundamental postulates for light signals are continually being rechecked. For example, one experiment now being considered would compare "old light" from distant galaxies with "new light" made locally. To be fair, the comparison must be made above the atmosphere, since passage through all that material might slowly convert "old light" to "new." Now, no one *expects* to find any difference between "old

*Check your postulates*

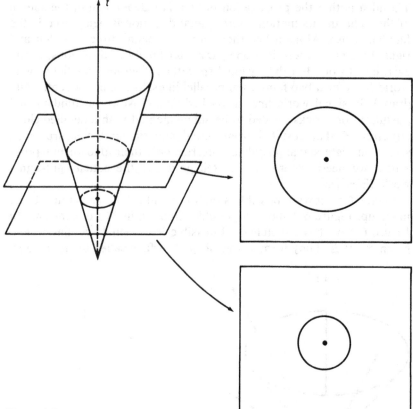

**Figure 6.3.**

light" and "new light," but, also, no one expects our simple ideas to be true forever, either.

*Spacetime diagram summary* Our representation of spacetime structure can be summarized in a spacetime diagram showing the origin, the set $\mathscr{G}$, and the allowed directions for light signals. My nightmare theory of Figure 6.1 was an example of this. Figure 6.4 is another, and should be familiar. It is our good friend the absolute-time clock in a general inertial reference frame.

With the additional spacetime structure of light signals, we can now go further with our canonical reference frames. The clearest discussion uses our absolute-time clocks. They are a very useful instructional tool, even if they are not nature's clocks. One way to specialize an inertial reference frame is to orient it so that some particular world line is in a special location. Often the world line of the observer setting up the reference frame is aligned vertically in the spacetime diagram. We will

[The earlier discussion of canonical reference frames was in Section 4.]

describe such observer-specialized frames as having the observer "at rest." There are now two different ways to further specialize the reference frame. Either one can simplify the description of clocks, or one can simplify the description of light signals. In general, one cannot do both.

A time-specialized reference frame will be one arranged so that the set $\mathscr{A}$ is a horizontal line, perpendicular to the observer's world line. A light-signal-specialized frame will be one arranged so that light signals go along the lines whose slopes are $+1$ and $-1$ in the diagram. In Figure 6.5 we sketch how the situation shown in Figure 6.4 is transformed into a time-specialized frame. W is the world line of the special observer. We first make a rotation, and then a Galilean transformation. In Figure 6.6 we do the same for the generation of a light-signal-specialized frame.

*Time-special frames and light-signal-special frames*

[See page 20 for the Galilean transformation.]

Which of these is the right frame? That, I hope you realize, is the wrong question to ask. Both of them are valid and equivalent representations of the *same* physical situation. You might prefer one or the other because of a particular application that you have in mind, but any problem can be done in any frame, including the original unspecialized one. Taking a representation too seriously is a common source of error. One often does so without noticing it when one's language forces one into using a particular frame. A person looking only at a time-specialized frame would say that the speed of light is different for right-going and left-going light signals. Someone else who looks only at light-signal-specialized frames will think this first person is crazy, because to him the speed of light is constant by definition. Think carefully about how these people are both right and both wrong at the same time. To someone steeped in the physics and philosophy of Newtonian mechanics, time is a more basic idea than light, and to choose any but a time-specialized frame would seem bizarre. **The revolutionary idea of special relativity is that light is more basic than time.**

*Any frame*

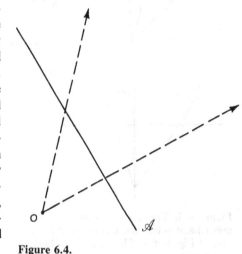

Figure 6.4.

*Special-relativity clocks*

Special relativity resolves this conflict between clocks and light signals in a beautiful and pleasing manner. In special relativity, there is no conflict. The same systems that simplify clock behavior also simplify light signals. The light-signal world lines turn out to be the asymptotes of the clock hyperbola $\mathscr{G}$. See Figure 6.7. Had we not discussed absolute-time clocks, you might not have appreciated this unification.

---

**Canonical Reference Frame Based on World Line $W$: An inertial reference frame in which light signals have a slope of unity and the world line $W$ is vertical.**

[No precise definition such as this is in common use. This one is my own.]

---

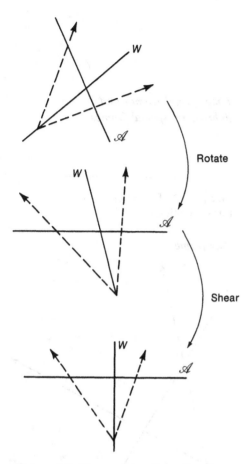

In the hypothetical situation of absolute-time clocks, one can specialize the reference frames even further. There is just one canonical reference frame in which the clock structure $\mathscr{A}$ is perpendicular to that frame's special world line. We will call this the *preferred reference frame*. In it the physics is particularly simple.

Our current theory, in which particles and light signals are the primitive elements, is not the most fundamental theory at our disposal today. A better description of light can be found by using Maxwell's electrodynamics. A better description of particles can be found by using quantum mechanics. But it would be a poor idea for us to try to start out with those theories. Their very language rests on the ideas of inertial frames and Lorentz symmetry of clocks and light signals. In Chapter III we will go part way toward these better theories by taking the idea of a wave packet as a primitive notion. This is called "semi-classical theory."

**Figure 6.5.** The construction of a time-specialized reference frame for the situation of Figure 6.4. The shear is a Galilean transformation.

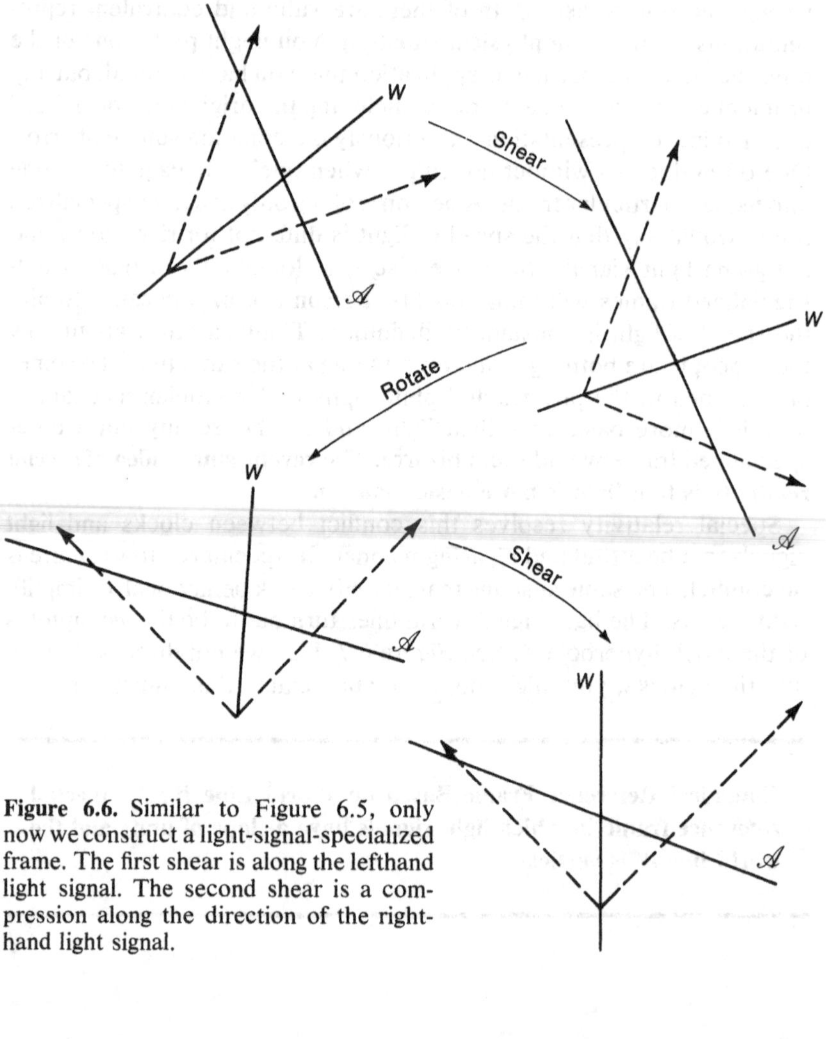

**Figure 6.6.** Similar to Figure 6.5, only now we construct a light-signal-specialized frame. The first shear is along the lefthand light signal. The second shear is a compression along the direction of the right-hand light signal.

PROBLEMS

6.1. (25) Find the matrices which represent the transformations shown in Figure 6.5. Angles are 30°, 45°, and 60°.

6.2. (25) Do the same as in 6.1 for Figure 6.6.

6.3. (20) For the light signals and the world line $W$ shown in Figure 6.8, find the transformation to a canonical reference frame, both graphically and analytically.

6.4. (30) Find a light-signal canonical reference frame for hypothetical clocks described by a $\mathscr{G}$ given by

$$t^4 - x^4 = 1,$$

with light signals given by

$$t^2 = x^2,$$

relative to a world line $W$ that is 30° off the vertical.

6.5. (35) In the hypothetical situation where clocks are absolute-time clocks, invent an operational procedure involving only light signals and clocks that one can use to measure one's velocity relative to the preferred reference frame described on page 30.

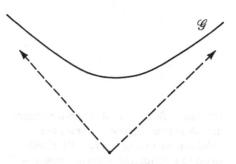

**Figure 6.7.** Real clocks as described by special relativity. Axes have been left off to emphasize coordinate independence.

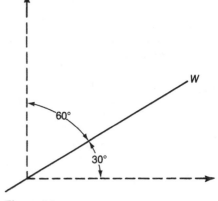

**Figure 6.8.**

# 7. Special Relativity

*"I do not wish to break the fold in your head differently,*
*Sir, but I can only tell you it is not true."*

G. C. LICHTENBERG

We have finished the basic structure of special relativity. The details that we will now develop are all consequences of this basic structure. There will be no new physical ideas until we discuss momentum and energy in the last section. Here is a quick review of this basic structure.

The theory of special relativity models the part of physical reality described by the primitive ideas of events, free particles, clocks, and light signals. Our fundamental postulates assert that one can represent these things by mathematical objects in a vector space in such a manner that the world lines of free particles and light signals are straight lines,

[There is a fundamental shift in this list of primitive notions. It is not *time*, but *clock* that is basic in special relativity.]

clock rates are uniform, and light signals have a unique velocity independent of their source. Such a representation we call an inertial reference frame. A more specialized representation can be chosen if we make the speed of light equal to 1, and single out some particular world line to be at rest. This special representation we call a canonical reference frame for that world line. In one such canonical reference frame, the behavior of clocks is found experimentally to be described by a set $\mathscr{G}$, those events a unit time interval from the origin, given by

$$t^2 - x^2 = 1 \qquad (7.1)$$

or, in 3+1 dimensions, by

$$t^2 - x^2 - y^2 - z^2 = 1. \qquad (7.2)$$

This is all sketched in Figure 7.1.

[Warning: Both inertial reference frames and canonical reference frames are valid representations. The only reason to go to a canonical reference frame is convenience.]

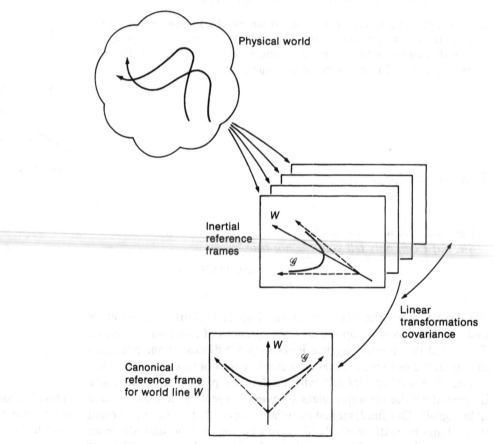

**Figure 7.1.**

The sections to follow will use this basic structure to discuss simultaneity, clock synchronization, time dilation, relative velocity, Lorentz symmetry, and 4-vectors. All these ideas are direct consequences of this basic structure. In particular, we will show that if the clock structure $\mathscr{G}$ is really a hyperbola in any one canonical reference frame, then it is the same hyperbola in all canonical reference frames. Most presentations of special relativity start from this equivalence as their basic assumption. We have instead set up a framework in which this equivalence is discovered. Such a framework is needed to discuss the experiments which verify the equivalence. In fact, such a discussion would be quite technical, and we do not pursue it, but to properly understand special relativity, one should understand how in principle one would do it.

### PROBLEM

7.1. (30) One of the general systems laws of Gerald Weinberg is the Count-to-Three Principle: "If you cannot think of three ways of abusing a tool, you do not understand how to use it." Apply this to special relativity.

## 8. Simultaneity

As attention shifts from time to clocks, the concept of simultaneity plays a less important role in special relativity than it does in Newtonian physics. This shift is surprising, and leads to lengthy discussions of simultaneity in most textbooks. I had considered the radical step of leaving it out entirely, just to emphasize that one can do without it, but its discussion will bring in a number of useful ideas, and provide more examples of the kind of geometric thinking I find so useful.

We will discuss simultaneity not as another primitive notion, but as a concept defined in terms of the theory already given. There is no physical process that intrinsically defines simultaneity—no natural simultaneity meter—except one built out of clocks and light signals. For us simultaneity will be only a defined notion. A striking feature of special relativity is that the most reasonable definition of simultaneity is not consistent. This failure of intuition leads to what is usually called the twin paradox.

*Simultaneity: a defined notion*

Our intuitive notion of time corresponds to an absolute-time clock. This is entirely reasonable, since we saw that such clocks provide

*Absolute-time simultaneity*

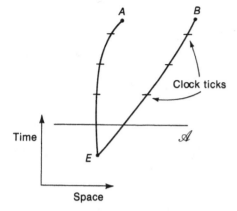

**Figure 8.1.** Finding two simultaneous events, *A* and *B*, by transporting absolute-time clocks from some other event *E*.

### Special-relativity simultaneity

[The path dependence of clocks is discussed in Problem 5.2.]

good approximations to the actual behavior of clocks at low velocities. How would one use an absolute-time clock to define simultaneity? For such clocks the time interval between two events is independent of the path followed by a clock carried between them. Thus it is possible to define simultaneity by transporting clocks around, as sketched in Figure 8.1.

---

**Clock-Transport Simultaneity: If for two events *A* and *B*, and a third earlier event *E*, a clock carried from *E* to *A* records the same time interval as one carried from *E* to *B*, the events *A* and *B* are called clock-transport simultaneous.**

---

For real clocks described by the special-relativity hyperbola, clock transport is path-dependent; so we are forced to abandon the clock-transport definition of simultaneity. To what shall we turn? When trying to define a physical quantity, one should look to the intended use. Unfortunately, the main use for simultaneity is in setting up a reference frame, and we will find other ways of doing that. The applications give us no clue.

When we discuss Lorentz invariance, we will prove that we cannot define simultaneity if we use only the primitive notions of our theory. Something else must be brought in. One way to proceed is to define simultaneity relative to a specific canonical reference frame. In this specific frame, events lying along a horizontal line are defined to be simultaneous events, as shown in Figure 8.2. This, in fact, is the best that can be done. We can simplify things somewhat by giving an operational definition equivalent to the above. The canonical reference frame depends on the choice of some particular world line to be "at rest." We can find an operational definition of simultaneity involving clocks, light signals, and this one special world line.

Again we use the standard argument. We know what we want to say in a canonical reference frame. If we can state this in terms that make no reference to any special frame, then the definition can be used in any frame. Such a definition is called *covariant*, meaning that it shares the same independence of representation as the rest of our theory. This is the same line of argument that we used in the discussion of Euclidean geometry to define perpendicularity in the general linear frame. To turn the definition sketched in Figure 8.2 into a covariant definition, we need to say that these events *A* and *B* lie on a line perpendicular to the special world line *W*, but we need to say so only in terms of straight

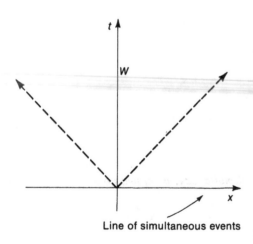

**Figure 8.2.** The line of simultaneous events in a light-signal canonical reference frame specified by a world line *W*.

lines, light signals, and clock intervals. This is straightforward. Look at Figure 8.3. Here we have drawn all the light signals that one can draw. Perpendicularity is assured if the intervals $\tau_1$ and $\tau_2$ along the world line $W$ are equal. This gives us our definition.

---

**Light-Signal Simultaneity: Two events $A$ and $B$ are simultaneous relative to a world line $W$ if light signals from $A$ and $B$ into the past and the future cut-off equal time intervals on $W$.**

---

This definition can be applied in any reference frame, as shown in Figure 8.4. Clearly the dependence on the world line $W$ is needed, since the relation defined by using the moving world line $W'$ in Figure 8.5 does not agree with the one defined by using the world line $W$, which appears stationary in that frame.

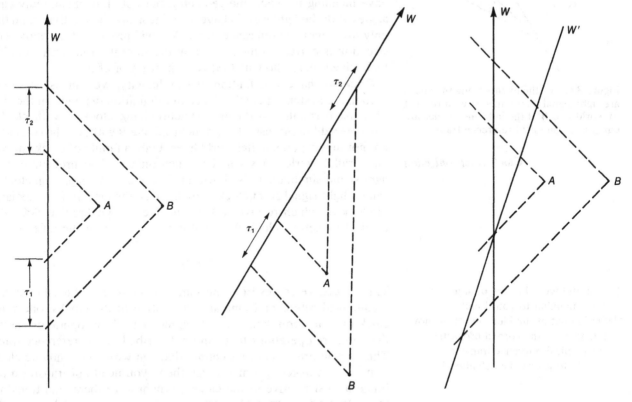

**Figure 8.3.** Covariant operations defining light-signal simultaneity.

**Figure 8.4.** Figure 8.3 redrawn for the general linear frame.

**Figure 8.5.**

*Simultaneity depends only on light signals*

Notice that the only comparison of intervals was along a single world line. This means that the definition uses only the uniformity of clocks, and is independent of any particular clock theory $\mathscr{G}$. This is why I called it light-signal simultaneity rather than special-relativity simultaneity. This definition could be used with absolute-time clocks. It wasn't, because the dependence on the world line $W$ is there an avoidable complication. It is only because clock-transport simultaneity is inconsistent that we are driven to accept this weaker definition. Any further mention of simultaneity in this book refers to light-signal simultaneity.

*Graphics*

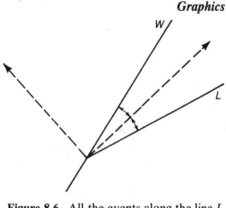

There is a simple relationship between the world line $W$ and the line of events $L$ that are light-signal simultaneous with respect to $W$, and it holds in any canonical reference frame: the world line $W$ and the line $L$ make equal Euclidean angles with the light-signal world lines, as shown in Figure 8.6. Step back for a moment and look at what we just did. We have a model for the geometry of spacetime constructed out of Euclidean geometry. Some of the concepts of Euclidean geometry have meaning in spacetime geometry, but not all of them. Only equal angles with the light signals have meaning in special relativity, and then only in a canonical reference frame. We will prove that the above construction is correct in the section on the Lorentz transformation. The tools given there allow us to give a short proof of it.

**Figure 8.6.** All the events along the line $L$ are light-signal simultaneous with respect to world line $W$ if the indicated angles are equal in a canonical reference frame.

Once we have a definition of simultaneity, we can use it to synchronize a system of clocks. Since our definition depends on the slope of some particular world line, synchronizing clock $A$ with clock $B$ will give a different result from synchronizing $B$ with $A$ if the two clocks are not moving on parallel world lines. Only a family of clocks moving on parallel world lines can be synchronized. The procedure using light-signal simultaneity is shown in Figure 8.7. At time $t_1$ on clock $A$, send a light signal from clock $A$ to clock $B$, then return it immediately to clock $A$, which receives it back at time $t_2$. Synchronize clock $B$ to clock $A$ by setting clock $B$ so that it reads time $t_0$ at the reflection:

*Clock synchronization*

$$t_0 = \tfrac{1}{2}(t_1 + t_2). \tag{8.1}$$

[Be careful with this line of argument if it is unfamiliar to you. I pick a definition out of the blue, and then show you that it has the correct properties. The canonical reference frame is used here only to justify the definition.]

This procedure comes from the same line of argument used before. In a canonical reference frame at rest relative to the clocks, one wants clock time and coordinate time to agree. The above operations assure that, and are operations that can be described in *any* reference frame. Thus if you are given a spacetime diagram with some moving clocks in it and are asked to synchronize them, you need not transform to a frame at rest relative to the clocks, synchronize them, and transform back. Instead, you simply perform the above covariant operations.

You should begin to appreciate how advantageous it is to have a co-variant formalism that makes it unnecessary to repeatedly change reference frames.

The idea of clock synchronization is crucial to a discussion of the rates of moving clocks. Much of the confusion about the issue (and the issue is only important because there is so much confusion about it) comes from the assumption that one can compare the rate of a moving clock and the rate of a stationary clock. Not so. We take the comparison of clock readings at the same event (nearly) to be possible. A rate measurement requires *two* readings, however. What one can do is to compare the rate of a moving clock with a *family* of synchronized stationary clocks, and for this we need at least two stationary clocks. Only if one doesn't realize this does the situation appear to be symmetrical and hence paradoxical. Look at the situation in Figure 8.8. There $C_1$ and $C_2$ are stationary clocks, and $C$ is the moving clock. The calculation is most easily done in this canonical reference frame, where $C_1$ and $C_2$ are at rest. Because all our operations are covariant, any frame can be used, and Problem 8.3 explores this.

Synchronize the stationary clocks by using the above procedure. For simplicity, set the origin of the coordinates at the event $E$ where clock $C$ passes clock $C_1$. We will set our clocks so that both $C$ and $C_1$ read zero time at the event $E$. $C_2$ will read zero at event $G$, synchronous with $E$. We can compare the relative rates if we can find the readings of the two clocks at the event $F$. We first find the coordinates of the event $F$. The world line of the moving clock will be given by

$$x = vt. \tag{8.2}$$

The world line of the clock $C_2$ will be given by

$$x = L, \tag{8.3}$$

where $L$ is some constant that measures the distance between the two stationary clocks. The event $F$ lies on both of the world lines and must satisfy both of the above equations. It has coordinates $(L, L/v)$.

We can now calculate the time intervals between events $E$ and $F$ and between events $G$ and $F$. The time interval $\tau_2$ read by clock $C_2$ is just the coordinate interval, because it is stationary:

$$\tau_2 = \frac{L}{v}. \tag{8.4}$$

*Rates of moving clocks*

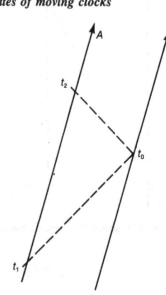

Figure 8.7.

[Remember that we are measuring distances in light-seconds; so velocities are measured relative to the speed of light and are dimensionless.]

[The way to calculate time intervals between two known events is given by Equation 5.2.]

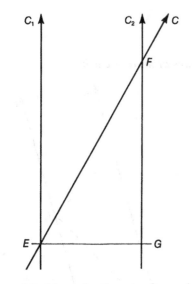

**Figure 8.8.** Measuring the rate of a moving clock.

*Spatial distance*

The time interval $\tau$ recorded by the moving clock is

$$\tau = \sqrt{\frac{L^2}{v^2} - L^2} = \frac{L}{v}\sqrt{1 - v^2}, \qquad (8.5)$$

and the ratio of these is the relative clock rate,

$$\frac{\tau}{\tau_2} = \sqrt{1 - v^2}, \qquad (8.6)$$

which is independent of $L$. We say that the moving clock runs slowly compared with the stationary clock. It is not a simple "look over your shoulder" observation, however, as I hope you can see.

We can use our definition of simultaneity to give an operational definition of the distance between two events. Look at Figure 8.9. We take the distance between the events to be that measured in a canonical reference frame in which the two events are simultaneous. Again we can translate this into a covariant operational definition, one which looks like the radar ranging that is actually used to measure many distances. Take any world line such that the events $E$ and $F$ are simultaneous with respect to. it. Use the equal-angles construction given

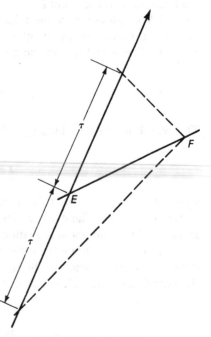

**Figure 8.9.** Measuring the interval between two events $E$ and $F$ to be $\tau$ light-seconds.

earlier. Define the distance between $E$ and $F$ to be half the time interval between round-trip light signals. The events that are a unit spatial distance away from the origin according to the above definition are given by

$$x^2 - t^2 = 1. \tag{8.7}$$

You may be surprised that we do not include spatial distance as one of our primitive notions, corresponding to the physical idea of a rigid rod. One can do that. However, a clock is a simpler idea. In a space-time diagram a clock is represented by a single world line with tick marks on it. A rigid rod is represented by an entire 2-surface. Further-more, we see no reason why clocks cannot be made more and more accurate, but the very nature of materials prevents one from making a rod that is more and more rigid.

*Length not fundamental for us*

If we adopt this clock-based definition of length, then we must use a different logical interpretation of experiments said to measure the velocity of light, such as the Michelson-Morley experiment, which compared light propagating in two perpendicular directions. The clock-based interpretation sees them as discovering a property of elastic bodies: that under some conditions their light-signal lengths are con-stant. Since we have set up reference frames using only clocks and free particles, the rigid rod plays no part in any of our discussions in this book.

[This shift in language should be compared with the shift between time-specialized reference frames and light-specialized frames discussed on page 29.]

## PROBLEMS

8.1. (08) In Figure 8.4, show that all events along the line $AB$ are simultaneous.

8.2. (10) One twin is an office worker; one is a bus driver. When they retire, all other things being equal, which one is older and by how much? Give a numerical estimate.

8.3. (10) Redraw Figure 8.8 so that it shows a canonical reference frame in which clock $C$ is at rest.

8.4. (12) Show that the construction sketched in Figure 8.6 does indeed agree with our previous definition of light-signal simultaneity.

8.5. (13) Find a counterexample showing that the equal-angles con-struction for simultaneity does not work in a general inertial reference frame.

8.6. (12) Verify that the construction sketched in Figure 8.7 assigns equal clock readings to simultaneous events.

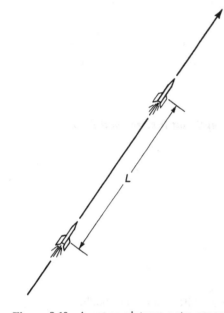

**Figure 8.10.** A space picture, not a space-time diagram.

8.7. (24) Two identical rocket ships are at rest, separated by a distance $L$, and pointed in the same direction along the line joining them. See Figure 8.10. They each accelerate along this line with identical thrusts, starting their engines simultaneously. Draw a spacetime diagram for this. When they are coasting afterward, how far apart are they? Would a cable of length $L$ be slack?

8.8. (23) Show that the relation "$A$ is simultaneous with $B$" for events $A$ and $B$, which we will write $A =_s B$, is an equivalence relation. That is, show that it satisfies the three conditions:

  (i) reflexive,                        $A =_s A$;

  (ii) symmetric,           $A =_s B$ implies $B =_s A$;

  (iii) transitive,      $A =_s B$ and $B =_s C$ implies $A =_s C$.

For (iii) do not assume that the events $A$, $B$, and $C$ lie along a line. Use the time-reflection symmetry $t \to -t$ to simplify the argument.

8.9. (30) The path dependence of special-relativity clocks could be eliminated if we agree to transport clocks only at constant velocity, that is, as free particles. Use Figure 8.11 to show that clock-transport simultaneity would still be inconsistent.

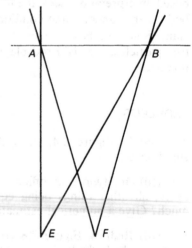

**Figure 8.11.**

8.10. (28) Find the time dilation for hypothetical clocks whose structure is given by the set $\mathscr{G}$

$$t^4 - x^4 = 1.$$

8.11. (21) Look at Figure 8.12, which shows the construction of simultaneous events in 2+1 dimensions. Find the equation for the interaction of these two cones and show that it is a plane curve.

**Figure 8.12.**

8.12. (35) Use the above construction to show how a moving disc appears contracted along its direction of motion.

## 9. Relative Velocity

We will follow the style of the previous section in this discussion of relative velocity. We know what we mean by velocity of a world line relative to an observer at rest in a canonical reference frame. We will translate this idea into a covariant operational definition which can then be used in any inertial reference frame. Note that we can define the relative velocity between two world lines before we discuss the Lorentz transformation. Most books confuse the idea of relative velocity by discussing it as a specific application of the Lorentz transformation. It does not depend on the specific clock structure $\mathscr{G}$.

In any canonical reference frame, we use the familiar definition of velocity as distance per unit time. We just compare the slope of a world line to the slope of a light-signal world line. In Figure 9.1 the world line $W$ has a velocity $v$ with respect to that frame given by

$$v = a/b. \tag{9.1}$$

*Familiar definition*

[Remember that lengths are measured in light seconds and velocities in terms of light velocity.]

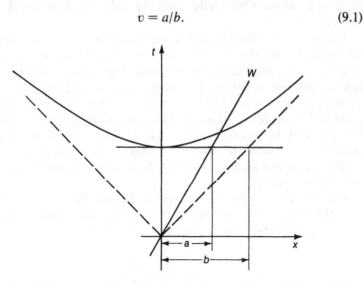

**Figure 9.1** Velocity measured in a canonical reference frame. If the world line $W$ makes an angle of 30° with the vertical, then the velocity is $1/\sqrt{3}$.

How do we convert this into a covariant operational definition? First, replace the idea of a canonical reference frame with its special world line, and speak now of the relative velocity between two world lines. We must convert our slope measurements into time-interval measurements. Since light signals are at 45° in a canonical reference frame, the lengths $a$ and $b$ can both be transferred to the world line at rest as shown in Figure 9.2. We can measure the two time intervals $\tau_1$ and $\tau_2$

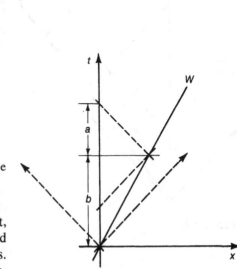

**Figure 9.2.** Using light signals, the lengths $a$ and $b$ of the preceeding diagram can be measured as time intervals.

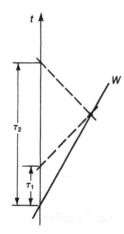

**Figure 9.3.**

shown in Figure 9.3; they are related to $a$ and $b$ by

$$\tau_1 = b - a, \tag{9.2}$$

$$\tau_2 = b + a. \tag{9.3}$$

From these equations, we find that

$$v = \frac{a}{b} = \frac{\tau_2 - \tau_1}{\tau_2 + \tau_1}. \tag{9.4}$$

Thus we have expressed the relative velocity in terms of time intervals defined by light signals. It is a covariant definition and can be used in any frame. Again we see that we need to compare time intervals only along a single world line; hence relative velocity does not depend on the clock structure $\mathscr{G}$ but only on the spacetime structure of an inertial reference frame. One could properly call this light-signal relative velocity.

*Addition of velocities*

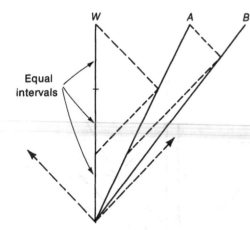

**Figure 9.4.**

The manner in which these relative velocities compound is surprising. In Figure 9.4 we sketch a situation in which world line $A$ has a speed of $\frac{1}{2}$ relative to the world line $W$, which is at rest in that frame. World line $B$ has been constructed to have a velocity of $\frac{1}{2}$ relative to $A$. The velocity of $B$ relative to $W$ is not $\frac{1}{2} + \frac{1}{2} = 1$, but appears from the diagram to be roughly $\frac{4}{5}$. This peculiar compounding of relative velocity is a property of this light-signal definition of relative velocity, not a property of special relativity. Of course, if physical clocks were absolute-time clocks, then we would have adopted a different definition of relative velocity.

The above geometric operations can be translated into an algebraic rule. If you see two world lines that correspond to motion with velocities $v_1$ and $v_2$, then their relative velocity $v$ is given by

$$v = \frac{v_2 - v_1}{1 - v_1 v_2}. \tag{9.5}$$

This expression is easily derived by using 4-vectors and the four-dimensional dot product, as we will show when we discuss them in section 13.

You should think of velocity as being just the slope of the world line in a spacetime diagram. Since in Euclidean geometry slopes are just tangents of angles, it should not surprise you that this addition law (really it is a subtraction law) looks something like the addition law for tangents of angles:

$$\tan (A - B) = \frac{\tan A - \tan B}{1 + \tan A \tan B}. \qquad (9.6)$$

It is not the same expression, of course, because the geometry of special relativity is not Euclidean geometry.

The similarity has led people to seek out a quantity like the measure of angle in Euclidean geometry. After all, you usually do not add slopes, which is a messy job, but angles, which is just simple addition. Can we find a function of velocity, call it $\psi$, such that velocity addition becomes simple addition? Yes, we can, and the function $\psi$ is called *rapidity*. The analogy with Euclidean geometry,

$$\mu = \tan \theta, \qquad (9.7)$$

suggests

$$v = \tanh \psi, \qquad (9.8)$$

and this works. It is a simple exercise in hyperbolic functions (notice the name) to show that the above definition leads to the law

$$\psi = \psi_2 - \psi_1. \qquad (9.9)$$

One useful tool in descriptive Euclidean geometry is a protractor, a circle calibrated in angle units. There is a similar tool useful in the graphical geometry of spacetime. Take the one-second hyperbola, and mark on it different slopes in rapidity units. You should be able to invent many uses for such a protractor. It is sketched in Figure 9.5. A certain amount of playing around with it will help you to get a feel for the relations in spacetime geometry. Note that a velocity scale can be added along the line

$$t = 1. \qquad (9.10)$$

The scaling of $\psi$ is similar to that for radians; so for small rapidity we have

$$\psi \sim v. \qquad (9.11)$$

*Rapidity*

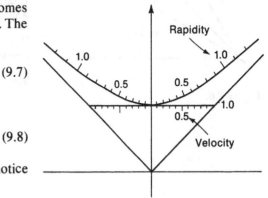

*Special relativity protractor*

**Figure 9.5.** Special-relativity protractor.

[We will meet rapidity again when we study the pseudosphere in Section 44. The space of all possible velocities in all directions is a pseudosphere, and rapidity is a useful coordinate for it.]

PROBLEMS

9.1. (10) Suppose that in some general inertial reference frame you are given the hyperbola $\mathscr{G}$ and a point on it corresponding to some arbi-

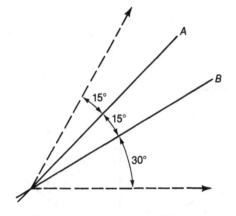

**Figure 9.6.**

trary rapidity. Find a construction using Euclidean geometry for the point on it corresponding to twice that rapidity. Explain.

9.2. (14) Verify that relative velocity is symmetric.

9.3. (18) Is relative velocity invariant under projective transformations?

9.4. (19) For the spacetime diagram shown in Figure 9.6, what is the relative velocity between $A$ and $B$? Do this both graphically and analytically.

9.5. (11) For the spacetime diagram shown in Figure 9.7, draw world lines that have a relative velocity of $\pm\frac{1}{2}$ with respect to the world line $W$.

9.6. (27) Write a set of instructions for the special-relativity protractor.

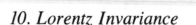

**Figure 9.7.**

# 10. Lorentz Invariance

*"We understand change
only by observing what remains invariant,
and permanence only by what is transformed."*

GERALD M. WEINBERG

We are now ready to discover an extremely important property of our special-relativity clocks. Since we view the clock-structure hyperbola $\mathscr{G}$ as the result of experimental measurements, we can only claim that it represents clocks in the particular canonical reference frame in which the clock experiments were described. We are now going to show that if $\mathscr{G}$ is a hyperbola in one frame, it is the same hyperbola in all other canonical reference frames.

The line of argument used here is a general method for showing that something has a symmetry. We find a transformation which leaves the thing unchanged. A sphere has symmetry because it is unchanged by rotations. A sphere with a dot marked on it has less symmetry. Fewer transformations leave it unchanged. This argument was used in the discussion of the rotational symmetry of Euclidean geometry, and will be used here to find the Lorentz symmetry of special-relativity clocks. It will be used later to discuss the symmetries of water waves and of the pseudospheres.

We start with the idea of a canonical reference frame, which depends on the properties of light signals, on the uniformity of clocks, and on the choice of a special world line. It does not depend on the details of clock behavior. Just as in the last sections, we can find a covariant operational definition of a canonical reference frame. This definition will allow us to transform any one canonical reference frame into any other. We show that, starting with a canonical reference frame in which $\mathscr{G}$ is the hyperbola

$$t^2 - x^2 = 1, \tag{10.1}$$

in any other canonical reference frame $\mathscr{G}$ is also the same hyperbola. This is a special property of the hyperbola clock structure, and the problems give examples which lack this symmetry. This symmetry of $\mathscr{G}$ is called Lorentz invariance, and the transformations are called Lorentz transformations.

The inertial reference frames and the special ones called canonical reference frames play a central role in special relativity. So far we have only an intrinsic definition of inertial reference frames. We have defined them in terms of their properties, which is a more abstract approach, and may be harder to follow, than a definition in terms of explicit operations. We are now in a position to follow the style of the preceding sections and to give a covariant operational definition of the canonical reference frame associated with a world line $W$. For simplicity we continue to work in only 1+1 dimensions. The style of the argument follows the familiar pattern. We know what we want to do in some particular canonical reference frame. We rephrase this in terms of covariant operations, and so get a definition that can be used in any inertial reference frame.

We seek covariant operations that measure the coordinates of any event $E$, which may be anywhere in spacetime. All we have available are light signals, the special world line $W$ associated with the particular

[I will not distinguish between $\mathscr{G}$, the clock structure, and "$\mathscr{G}$," the set in a spacetime diagram representing the clock structure.]

*Proving symmetry*

*Operational definition of a canonical reference frame*

[Remember that covariant operations are physical operations that make sense independent of any particular representation. Examples are sending out a light signal or measuring a time interval.]

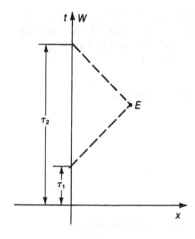

**Figure 10.1.** The time intervals $\tau_1$ and $\tau_2$ are the light-signal coordinates of event $E$ relative to the world line $W$.

*Light signal coordinates*

[You should try to step back and look over what you are doing, so that questions like this *do* naturally occur to you.]

[Relative velocity was defined and discussed in Section 9.]

canonical reference frame of interest, and the ability to measure time intervals. Figure 10.1 puts these operations together. The light signals drawn there are the only ones that can be associated with the event $E$, and $\tau_1$ and $\tau_2$ are the only time intervals singled out. In terms of the coordinates in our canonical reference frame, we have

$$\tau_1 = t - x, \tag{10.2}$$

$$\tau_2 = t + x. \tag{10.3}$$

Thus we can define our coordinates in terms of these time intervals by

$$x = \tfrac{1}{2}(\tau_2 - \tau_1), \tag{10.4}$$

$$t = \tfrac{1}{2}(\tau_2 + \tau_1), \tag{10.5}$$

and this is the sought-for covariant operational definition of the coordinates $x$ and $t$ of the canonical reference frame associated with the world line $W$ .

In fact, the numbers $\tau_1$ and $\tau_2$ themselves make perfectly good coordinates. We will call them light-signal coordinates. They form a system of coordinates rotated by 45° from the usual $(x, t)$ coordinates.

A natural question to ask at this point is how the coordinates of two different canonical reference frames are related. This can be answered easily, now that we have an explicit definition of our coordinates.

We are going to transform from the special canonical reference frame, where $\mathscr{G}$ is given by

$$t^2 - x^2 = 1, \tag{10.6}$$

to another canonical reference frame, one based on a world line that represents motion with a speed $v$ relative to the special world line $W$ in the first frame. This is shown in Figure 10.2. Let $\tau_1$ and $\tau_2$ be the light-signal coordinates of the frame associated with $W$, and $\tau_1'$ and $\tau_2'$ be coordinates associated with $W'$. Let us proceed as if $\tau_1$ and $\tau_2$ are known, and find expressions for $\tau_1'$ and $\tau_2'$ in terms of $\tau_1, \tau_2$, and $v$, the relative velocity of $W'$ with respect to $W$. We need now to find the coordinates of the event $A$, which defines the coordinates in the frame associated with the world line $W'$. We can find either its $(\tau_1, \tau_2)$ coordinates or its $(x, t)$ coordinates. The light-signal coordinates are simpler but less familiar. We calculate here first in $(x, t)$ coordinates, and then again in $(\tau_1, \tau_2)$ coordinates.

The event $A$ lies on the world line $W'$; so its coordinates, call them $\bar{x}$ and $\bar{t}$, must satisfy

$$\bar{x} = v\bar{t}. \tag{10.7}$$

It also lies on the light-signal world line; so we have

$$\bar{t} = \bar{x} + \tau_1. \tag{10.8}$$

The event $A$ thus has coordinates

$$\bar{x} = \frac{v\tau_1}{1-v}, \tag{10.9}$$

$$\bar{t} = \frac{\tau_1}{1-v}. \tag{10.10}$$

The time interval $\tau_1'$ is just the time interval between the event $A$ and the origin. We have

$$(\tau_1')^2 = \bar{t}^2 - \bar{x}^2, \tag{10.11}$$

$$(\tau_1')^2 = \frac{(\tau_1)^2}{(1-v)^2}(1-v^2); \tag{10.12}$$

that is,

$$\tau_1' = \sqrt{\frac{1+v}{1-v}}\,\tau_1. \tag{10.13}$$

A similar argument gives

$$\tau_2' = \sqrt{\frac{1-v}{1+v}}\,\tau_2. \tag{10.14}$$

It is now just a tedious bit of algebra to convert these expressions to ones involving $x$'s and $t$'s, using our definitions

$$\tau_1 = t - x, \tag{10.15}$$

$$\tau_2 = t + x, \tag{10.16}$$

$$\tau_1' = t' - x', \tag{10.17}$$

$$\tau_2' = t' + x', \tag{10.18}$$

[The way to calculate the time interval between two known events is given by Equation 5.2.]

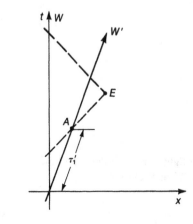

**Figure 10.2.** The light-signal coordinate $\tau_1'$ of $E$ relative to the world line $W'$.

and then solving for $t'$ and $x'$. We find

*Lorentz transformation*

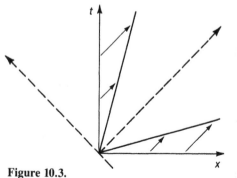

**Figure 10.3.**

$$t' = \frac{1}{\sqrt{1 - v^2}}\,(t - vx), \qquad (10.19)$$

$$x' = \frac{1}{\sqrt{1 - v^2}}\,(x - vt). \qquad (10.20)$$

This transformation between canonical reference frames is called the Lorentz transformation.

Our aim in this book is to give a covariant treatment of as much of the subject as possible. Since covariance allows us to work in any reference frame, we will not have to make Lorentz transformations to special frames, as one often has to in the usual treatments. I could easily have avoided the Lorentz transformations altogether.

*Simultaneity*

As an example of the usefulness of Lorentz transformations, we now provide the proof promised earlier for the equal-angle construction for the line of events $L$ simultaneous with respect to a given world line $W$. Start in the canonical reference frame based on the world line $W$. There the line $L$ is perpendicular to $W$, and these lines do indeed make equal angles with the light-signal world lines. Now look at the Lorentz transformation in terms of light-signal coordinates, Equations 10.13 and 10.14. We must shrink one light-signal coordinate, and expand the other one by the same factor. But it should be obvious that stretching one light-signal coordinate does not change the quality of the angles, since they are symmetrically disposed around it, as sketched in Figure 10.3. Therefore, the angles between $W$ and $L$ and the light signals remain equal under Lorentz transformations. Thus in every canonical reference frame accessible by a Lorentz transformation, we have the equal-angle construction. Now, no world line for any particle has ever been observed outside the light cone. All world lines inside the light cone can be Lorentz-transformed into one another. This should be clear from Figure 10.4.

[The light cone was described on page 27 in Section 6.]

Note the simplicity of the Lorentz transformation in light-signal coordinates. One familiar with linear algebra would say that light-signal coordinates are useful because they diagonalize the Lorentz transformation. To really learn special relativity, you need to draw spacetime diagrams for the same situation in different canonical reference frames. The Lorentz transformations involved are easily carried out by using light-signal coordinates. For practice, let us repeat our earlier calculation of the Lorentz transformation, but using light-signal coordinates this time. We sought the coordinates of the event $A$, shown in Figure 10.2. We now calculate its light-signal coordinates, call them $\tilde{\tau}_1$ and $\tilde{\tau}_2$. The condition 10.7 becomes (using Equations 10.4 and 10.5)

$$(1 - v)\tilde{\tau}_2 = (1 + v)\tilde{\tau}_1. \tag{10.21}$$

That events $A$ and $E$ lie on the same light signal means that

$$\tilde{\tau}_1 = \tau_1, \tag{10.22}$$

where $\tau_1$ is the light-signal coordinate of $E$. Together these relate the coordinates of $A$ and $E$:

$$\tilde{\tau}_2 = \frac{1 + v}{1 - v}\tau_1. \tag{10.23}$$

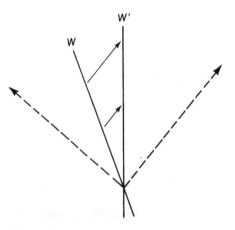

To proceed we need to convert the time-interval rule into light-signal coordinates. It collapses neatly into the expression

$$\tau^2 = \tau_1\tau_2, \tag{10.24}$$

Figure 10.4.

where $\tau$ is the time interval between the origin and the event $(\tau_1, \tau_2)$. From this we find that the time interval between the origin and event $A$, the new coordinate $\tau_1'$, is

$$\tau_1' = \sqrt{\frac{1 + v}{1 - v}}\tau_1, \tag{10.25}$$

in agreement with our previous calculation.

Note the simplicity of the expression for $\mathscr{G}$ in light-signal coordinates:

$$\tau_1\tau_2 = 1. \tag{10.26}$$

Using the Lorentz transformation, Equations 10.13 and 10.14, we clearly have

$$\tau_1'\tau_2' = 1 \tag{10.27}$$

as the equation for the same set of events in the new frame. Clearly this is the same hyperbola. Thus the moving canonical reference frame has the same rule for the description of clocks. The frames are equivalent. This is the Lorentz invariance that we have promised to demonstrate. As an exercise you should show that the light cone is given by

$$\tau_1\tau_2 = 0 \tag{10.28}$$

and that it too is invariant. You already know this, because the constancy of the speed of light is built into the definition of a canonical reference frame.

I promised earlier to show that no definition of simultaneity was possible if one used only clocks and light signals. Both clocks and light signals are Lorentz-invariant. What can be done in one canonical reference frame could have been done in any other, but a line of simultaneous events will not be Lorentz-invariant. A covariant definition of simultaneity using only clocks and light signals cannot be consistent. One must somehow single out some particular canonical reference frame. We did that by making simultaneity depend also on the choice of some special world line.

*The grand generalization*

After finding the Lorentz invariance of clocks as an experimental result, we naturally want to generalize it. The most sweeping generalization, and one that still seems to be true, is that all physical processes are Lorentz-invariant. No physical system can measure an absolute velocity. We will talk about an observer's velocity relative to some physical system, of course, and doing so makes perfectly good sense. The universe is filled with microwave radiation, and our velocity with respect to that radiation can be measured and is an important quantity.

For pedagogical reasons, we will often consider theories that violate Lorentz invariance. I remind the reader that no such modifications of special relativity are necessary. All experimental observations are adequately explained by special relativity. Still, it is useful to consider these alternatives in order to fully understand what special relativity really means. They should be used as test cases to debug your arguments.

## PROBLEMS

10.1. (12) Consider the simple Lorentz transformation in which $\tau_1$ is doubled and $\tau_2$ is halved. What rapidity does this correspond to?

10.2. (13) Provide the calculation of $\tau_2{'}$ that was skipped in the text.

[You should be able to invent such routine exercises as these two for yourself.]

10.3. (13) Provide the missing steps leading from Equations 10.13 and 10.14 to 10.19 and 10.20.

10.4. (23) Draw three spacetime diagrams showing an observer measuring the time dilation for a moving clock, one in a frame at rest with the observer, one at rest with the clock, and one in which the observer and clock have equal and opposite velocities. Show in each diagram why the moving clock appears to run slower.

10.5. (20) Find a Lorentz transformation that turns the parallelogram in Figure 10.5 into a Euclidean square in the spacetime diagram. This is only an exercise. Such a square has no physical significance. What rapidity does this transformation correspond to?

**10.6.** (26) Read about the length contraction in any standard special-relativity book. Write an explanation in our language. Use the three spacetime diagrams described in Problem 10.4.

**10.7.** (20) Write the set

$$t^4 - x^4 = 1$$

in light-signal coordinates.

**10.8.** (15) Show that the hypothetical clocks described by the $\mathcal{G}$

$$t^4 - x^4 = 1$$

do not have any symmetry like Lorentz symmetry. Show that there is no transformation which both takes one observer into another moving at a different velocity and preserves the set $\mathcal{G}$.

**10.9.** (36) Discuss canonical coordinates and the Lorentz transformation for the hypothetical clocks described in Problem 10.7. The Lorentz transformation is useful even if the clocks are not Lorentz-invariant!

**10.10.** (28) A rod at rest and of length $L$ lies at an angle $\theta$ with respect to the $x$-axis. Describe its world lines and its appearance in a frame moving along the $x$-axis.

**10.11.** (33) Find and discuss transformations which preserve the light-signal structure but not the $\mathcal{G}$ of special relativity.

**10.12.** (22) Show that the four-dimensional transformation

$$t' = \frac{1}{\sqrt{1-v^2}}\,(t - vx),$$

$$x' = \frac{1}{\sqrt{1-v^2}}\,(x - vt),$$

$$y' = y,$$

$$z' = z,$$

preserves the four-dimensional clock structure given by

$$t^2 - x^2 - y^2 - z^2 = 1.$$

**10.13.** (24) Are there any symmetries for the clock structure $\mathcal{G}$ shown in Figure 10.6?

**10.14.** (30) Are there any symmetries for $\mathcal{G}$ if it is a parabola through the origin?

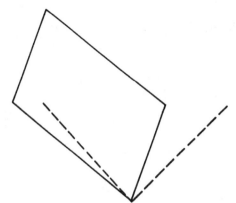

**Figure 10.5.**

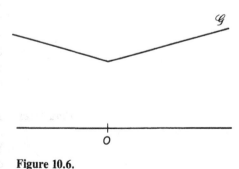

**Figure 10.6.**

10.15. (30) Are there any symmetries for $\mathscr{G}$ if it is the curve

$$t^n = x^n,$$

$n$ being an integer?

10.16. (30) Are there any symmetries for $\mathscr{G}$ if it is the curve

$$t^n = x^m,$$

$n$ and $m$ being integers?

## 11. The Consistency of Special Relativity

[This epigraph is from "Direction of the Road," which is a fine story.]

*"If the human creatures will not understand Relativity, very well . . ."*

URSULA LEGUIN

Of all the branches of physics, only thermodynamics attracts more cranks than special relativity. Is it really scientific to just dismiss them? Should we not examine each case on its scientific merit, lest our conservatism lead us to miss out on the next scientific revolution? No, not at all.

The reason we can give such a flat answer is that most critics of special relativity are concerned with its consistency, not with the experiments that verify it. All such claims of inconsistency can be ignored, because the logical consistency of special relativity can be demonstrated. Although a physical theory is a correspondence between things in the physical world and structures in mathematics, the question of consistency is a question about the mathematical model; and, like many mathematical questions, it can be answered decisively.

*Paradoxes*    Our model of special relativity is constructed out of concepts from Euclidean geometry but with different rules of manipulation. If these rules are consistent, then our model will be just as consistent as Euclidean geometry. Any logical paradox in special relativity can be converted into a logical paradox involving Euclidean geometry.

*Non-Euclidean geometry*    This is the same line of reasoning used by Henri Poincaré to settle the question of the consistency of non-Euclidean geometry. Non-Euclidean geometry is a generalization of Euclidean geometry in which the parallel postulate is not true. The parallel postulate states that, for any given line and point not on that line, there is a unique parallel line that passes through the given point. There were many attempts to prove this postulate from the others, that is, to prove that its denial would be inconsistent. Poincaré settled the question by constructing a model for non-Euclidean geometry out of Euclidean geometry. Actually, he made two, but one of them is of special interest for us. It will turn out to be a model for an open, expanding universe.

The model uses the upper half of the *x,y*-plane (points for which *y* > 0), and is called the Poincaré half-plane. Points of the non-Euclidean geometry are just the points of the plane. Straight lines of the non-Euclidean geometry are not the straight lines of the plane. To distinguish them I will call them "straight lines." A "straight line" is any Euclidean circle whose center lies on the x-axis. Such "straight lines" obey all the axioms and postulates of Euclidean geometry except for the parallel postulate. Many different "straight lines" can be drawn through an exterior point but do not intersect a given "straight line," and so can all be called parallels (see Figure 11.1). Were the parallel postulate necessary, then the above model would have an inconsistency, which we could translate into an inconsistency about circles in Euclidean geometry. Thus Poincaré showed that non-Euclidean geometry was as consistent as Euclidean geometry, which is as far as anyone has been able to go. The consistency of Euclidean geometry has never been established or seriously questioned. Our model of special relativity is just like the above model of non-Euclidean geometry. It differs from Euclidean geometry not in having a new definition of a straight line, but in using a different curve for measurements, the hyperbola instead of a circle.

*Poincaré half-plane*

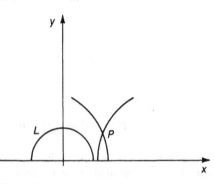

**Figure 11.1.**

The typical special-relativity paradox takes a verbal description of a situation and discusses it in two different inertial frames. Even though we should realize that we need not discuss a situation in more than one frame, such a discussion is often educational and certainly cannot hurt. We have shown that all the elements of special relativity — events, free particles, clocks, and light signals — obey the same rules in any canonical reference frame. A clock interval will be computed by different numerical operations in different frames, but it will have the same value. To find different results in different frames, which is the usual form of paradox, there must be a mistake. The mistake must lie in a mistranslation of the situation. The operation by which a situation is translated from words into mathematics is difficult and lacks formal rules. It is the most likely place to look for an error.

Another fruitful source of relativity paradoxes is the use of a concept which has a familiar meaning, but which has not been carefully defined in the context of special relativity. Common offenders here are rigid rods and potential energy.

*Relativity paradoxes*

Armed with the courage of mathematics, you should fearlessly take on any relativity paradox, sure at least that the theory is consistent. Now, none of this should be taken to mean that I think special relativity to be true beyond doubt or beyond the need for further experimental study. Far from it. I have taken some pains to set up special relativity in such a way that the means of experimental verification are clear, and to give you a formalism adequate to describe the case where special relativity needs some small correction. The space-time structure described here can describe the fall of special relativity without a revolutionary shift of ideas.

*Water-wave model*

Further confidence in the consistency of special relativity comes from the discussion of water waves in Section 28. The familiar physical system of waves on deep water provides another realization of the logical structure of relativity. The primitive notions of clocks, light signals, and free particles all appear in that realization and obey the same postulates as in special relativity. The clock structure is different, but again it has a relativity symmetry such that all moving observers are equivalent. Many special-relativity paradoxes can be directly translated into this model, although not those involving rigid rods.

*Faster-than-light motion*

[These faster-than-light ideas differ from the tachyons that are sometimes discussed. See Section 29 for some remarks on tachyons.]

The water-wave model of relativity also points out the flaw in the usual discussion of faster-than-light (FTL) motion. True, FTL motion is not compatible with special relativity—true, but irrelevant. The trouble with a mathematical argument is that small changes in the assumptions do not lead necessarily to only small changes in the conclusions. The water-wave model can be improved by including in it the small effects of surface tension. These are enough to allow the equivalent of FTL motion. Similarly, the clock structure $\mathscr{G}$ of special relativity could be slightly changed so that it would still agree with our present experiments, yet allow FTL motion.

*"Proofs" of special relativity*

There are other writers, more dangerous even than the relativity cranks, who "derive" special relativity and the Lorentz transformation from seemingly innocuous assumptions. They seriously confuse the logical structure of the theory. If you are tempted by their arguments, then you should read the history of non-Euclidean geometry. Seemingly harmless assumptions, such as that there can be triangles of arbitrary area, are in fact inconsistent with the axioms of the non-Euclidean geometries. The idea is not to reduce a theory to the smallest number of very clever assumptions, but to reduce it to testable assumptions.

*Twin paradox*

[The answer to Problem 5.2 is considered intolerable by some, and is called the twin paradox or the clock paradox. It is discussed well in Taylor and Wheeler, badly in a great many places. Both kinds of discussion are instructive.]

The difference between my approach here and that often found is well shown by the twin paradox. I have tried to show here how this behavior can be traced back to initially surprising experiments. There is no advantage in an axiomatic resolution of the twin paradox. The axioms must be only clever and obscure ways of restating the effect. We take it here to be the foundation of special relativity.

PROBLEM

11.1. (28) A mouse is sitting in a windowsill whose open dimension is $L$. A cruel young schoolchild flies by in a spaceship and pushes a rod of length $L$ at the mouse. The mouse sees a shorter rod, because of the Lorentz contraction, and despairs. The schoolchild sees the opening in the window contracted, and does not expect to kill the mouse. One expects the death of the mouse to be an invariant idea. What is wrong? (*Scientific American*, April 1975, p. 126.)

## 12. 4-Vectors

We need better mathematical tools to deal with quantitative problems, especially if they involve more than one space dimension. We need a vector analysis for spacetime. Just as there is a vector algebra adapted to Euclidean geometry, we can adapt a vector algebra to the special features of special-relativity geometry. This vector algebra will be used in this section to give an easy calculation of the velocity-addition law, in Section 13 to discuss the Doppler shift and the aberration of light, and in Section 14 to discuss momentum and energy.

*Vector algebra*

We use vectors in two different ways. On the one hand, we represent the events of spacetime by displacement vectors that begin from some arbitrarily chosen origin. On the other hand, we use free vectors to represent displacements from one event to another. The linear structure of a vector space allows one to slide a vector around to any event by using the parallelogram construction. The free vectors in spacetime are what we will call 4-vectors.

*Vector-space model for spacetime*

*Free vectors in spacetime*
[To distinguish these 4-vectors from 3-vectors, we do not put them in boldface type. In this section $a$, $b$, $\sigma$, the various types of $\lambda$, and $\hat{x}$, $\hat{y}$, and $\hat{t}$ are are all 4-vectors.]

To represent vectors quantitatively, we pick a set of basis vectors and represent any vector as a linear combination of these basis vectors. In special relativity, the most useful set of basis vectors is one adapted to the canonical reference frames. Each canonical reference frame has a different set of basis vectors. We will call these special basis vectors $\hat{x}$, $\hat{y}$, $\hat{z}$, and $\hat{t}$. The 4-vector $\hat{x}$ represents a displacement from the origin to the event $(t, x, y, z) = (0, 1, 0, 0)$. It also represents a displacement from any event $(t, x, y, z)$ to the event $(t, x + 1, y, z)$. Similarly with the other three. A basis adapted to a canonical reference frame in this manner will be called an *orthonormal basis*.

*Basis vectors*

Any 4-vector can be expanded in this basis, and we write the coefficients in the following way:

**Components**

[Here $a^x$ is a symbol for the $x$-component of the vector $a$, not for the $x$th power of some number. In the equation, $a$, $\hat{x}$ $\hat{y}$, $\hat{z}$, and $\hat{t}$ are the 4-vectors.]

$$a = a^x\hat{x} + a^y\hat{y} + a^z\hat{z} + a^t\hat{t}. \tag{12.1}$$

To multiply a vector by a scale factor $k$, multiply all its components by $k$. To add two 4-vectors, add their components.

*Example*  |  The vector $b$ representing a displacement from the origin to the event $(1, 1, 0, 0)$ is written

$$b = \hat{x} + \hat{t} \tag{12.2}$$

and has components

$$b^x = 1, \qquad b^z = 0,$$
$$b^y = 0, \qquad b^t = 1. \tag{12.3}$$

**Tangent vectors**

We can specify the direction of a world line by giving a 4-vector tangent to it. The tangent vector to a straight world line is any vector that connects different events of that world line.

*Example*  |  The world line given by

$$x = y = z = 0 \tag{12.4}$$

has a tangent 4-vector $\hat{t}$ and the 4-vector $b$ in the example above is a tangent to the light-signal world line

$$x = t,$$
$$y = z = 0. \tag{12.5}$$

**Covariance under Lorentz transformations**

Different canonical reference frames provide different representations of events. This means that the same 4-vectors will also have different representations in different canonical reference frames. The different frames use different orthonormal basis vectors. Since our 4-vectors are defined in terms of events, their law of transformation must be the same as the transformation of events, the Lorentz transformation. A vector represented in one canonical reference frame by

$$a = a^x\hat{x} + a^y\hat{y} + a^z\hat{z} + a^t\hat{t} \tag{12.6}$$

will be represented in a frame moving at a velocity $v$ along the $x$-axis by

[See Problem 10.11.]

$$a = \frac{1}{\sqrt{1-v^2}}\left[(a^x - va^t)\hat{x}' + (a^t - va^x)\hat{t}'\right] + a^y\hat{y}' + a^z\hat{z}'. \tag{12.7}$$

Here the $\hat{x}'$ are the basis vectors in the moving frame.

**Dot product**

The special properties of Euclidean geometry were completely summarized by a dot-product operation. We can use a similar trick to represent the geometry of special relativity. We will define a 4-vector dot product, again written $a \cdot b$, according to the following definition.

---

**4-Vector Dot Product:**

$$a \cdot b \equiv a^x b^x + a^y b^y + a^z b^z - a^t b^t, \qquad (12.8)$$

where $a^x$, etc., are the components in an orthonormal basis.

---

One can see why it was defined this way. The time-inverval rule of special relativity can be simply written.

*Time intervals*

---

**Special-Relativity Time-Interval Rule:** The time interval $\tau$ read by a clock carried along the world-line segment given by the vector $a$ is given by

$$\tau^2 = -a \cdot a. \qquad (12.9)$$

---

This dot product, like the Euclidean dot product, is covariant. Its value depends only on the vectors themselves, and may be calculated in any canonical reference frame. This covariance should be clear, since time intervals have this same covariance. One can also verify directly that the value of the dot product is not changed by a Lorentz transformation. If we had followed the conventional route and postulated the Lorentz transformation, then we would have followed the logic in reverse order and would have chosen the definition of the dot product so that it would be invariant under Lorentz transformations. Either approach is valid.

*Lorentz transformations*

This dot product allows us to concisely state the laws of special relativity. For example, a light-signal world line is characterized by a tangent 4-vector $\sigma$ satisfying

*Light signals*

$$\sigma \cdot \sigma = 0. \qquad (12.10)$$

As we have defined it, a tangent vector to a world line can have any length. We can rescale the tangent vector to a clock world line to satisfy some useful conditions; such a special tangent we will call 4-velocity.

*4-velocity*

[We will use λ for the 4-velocities of clock world lines and σ for the tangents to light-signal world lines whenever we can]

**4-Velocity:** A tangent 4-vector λ normalized to satisfy

$$\lambda \cdot \lambda = -1 \qquad (12.11)$$

and

$$\lambda^t > 0. \qquad (12.12)$$

It follows directly from the special-relativity time-interval rule that if λ is a 4-velocity, then a segment of the world line given by $\tau\lambda$ corresponds to a time interval $\tau$.

[By the square of $v$ I mean $v \cdot v$.]

Vectors whose squares are positive are tangents to lines that represent motion at speeds greater than the speed of light. They are called spacelike vectors. Vectors whose squares are negative are called timelike. Vectors whose squares are zero are called null vectors. They are the tangents to light-signal world lines. They are not the zero vector.

*Examples*

What is the 4-velocity of a world line representing motion with velocity $v$? Let the world line be given by

$$x = vt,$$
$$y = z = 0. \qquad (12.13)$$

A tangent vector $a$ is:

$$a = v\hat{x} + \hat{t}, \qquad (12.14)$$

which is not properly normalized to be a 4-velocity, since

$$a \cdot a = v^2 - 1. \qquad (12.15)$$

The vector λ,

$$\lambda = \frac{1}{\sqrt{1-v^2}}(v\hat{x} + \hat{t}), \qquad (12.16)$$

is properly normalized and parallel to $a$; hence it is the 4-velocity of the curve. A world line at rest has a 4-velocity $\hat{t}$.

*Relative velocity*

The dot product of any two 4-velocities is an invariant quantity and must be related to their relative velocities. The explicit relationship is

$$\lambda_1 \cdot \lambda_2 = -\frac{1}{\sqrt{1-v^2}}. \qquad (12.17)$$

Where did this come from? Since the dot product can be computed in any frame, let us compute it in a simple frame, one in which one 4-velocity represents a world line at rest. Note that this is not a zero 4-vector! Explicit forms of the two 4-velocities are given in the example above. We find the dot product by using the special-relativity dot-product rule. Because the dot product is covariant, a computation done in any other canonical reference frame will give the same answer.

[See Problem 12.8 for an exercise in this kind of reasoning in the context of Euclidean geometry.]

We can use this result to quickly find the velocity-addition law. Suppose I see two world lines moving with velocities $v_1$ and $v_2$. Their 4-velocities in a canonical reference frame in which I am at rest will be

*Velocity-addition law*

$$\lambda_1 = \frac{1}{\sqrt{1 - (v_1)^2}} (v_1 \hat{x} + \hat{t}), \tag{12.18}$$

$$\lambda_2 = \frac{1}{\sqrt{1 - (v_2)^2}} (v_2 \hat{x} + \hat{t}), \tag{12.19}$$

from Equation 12.16. The relative velocity $v$ which an observer on world line 1 sees for world line 2 will be given by Equation 12.17:

$$-\frac{1}{\sqrt{1 - v^2}} = \lambda_1 \cdot \lambda_2, \tag{12.20}$$

$$-\frac{1}{\sqrt{1 - v^2}} = \frac{v_1 v_2 - 1}{\sqrt{1 - (v_1)^2} \sqrt{1 - (v_2)^2}}, \tag{12.21}$$

and routine algebra (along with choosing the correct sign for the square root) gives us

$$v = \frac{v_2 - v_1}{1 - v_1 v_2}, \tag{12.22}$$

the result given earlier in Section 9.

We can extend the idea of a tangent vector to curved world lines. We use a limit process for this, finding tangents to segments so short that they are nearly straight. It is easiest to work with a curve given in *parametric form*; that is, we have four functions giving the coordinates of the events of the curve:

*Tangents to curved world lines*

$$x = X(u), \qquad z = Z(u),$$

$$\tag{12.23}$$

$$y = Y(u), \qquad t = T(u).$$

In the notation of maps (see Section 1, page 7), a parametrized curve is specified by a map $P$ such that

$$P: \mathbb{R} \to V; \, u \mapsto P(u). \tag{12.24}$$

The tangent vector $v$ is defined by

$$v \equiv \lim_{\epsilon \to 0} \frac{P(u+\epsilon) - P(u)}{\epsilon}. \tag{12.25}$$

Each $P(u)$ is an event, and the difference between two events is a displacement vector; thus the righthand side is indeed a vector. This limit process is sketched in Figure 12.1. In terms of our basis, we have

[Quite a few steps are left out at this point, since they are straightforward, but not particularly instructive. You can safely take the result on faith. A better derivation will come up in Section 26.]

$$v = \frac{dX}{du}\hat{x} + \frac{dY}{du}\hat{y} + \frac{dZ}{du}\hat{z} + \frac{dT}{du}\hat{t}. \tag{12.26}$$

Different parametrizations can describe the same curve. The tangent vector will change its length as the parametrization is changed. If we have

$$v \cdot v = -1, \tag{12.27}$$

then the tangent is a 4-velocity. The special parameter for which this

***Proper time***    happens is called *proper time*, because clock intervals are then the same as parameter intervals.

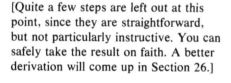

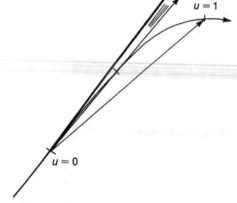

**Figure 12.1.** Limit process for constructing the tangent vector at the point $u = 0$.

### PROBLEMS

12.1. (14) Verify the covariance of the special-relativity dot product under Lorentz transformations.

12.2. (20) Verify the transformation law for the components of a 4-vector under a Lorentz transformation by computing the tangent to a world line before and after a Lorentz transformation.

12.3. (22) Show how spacetime 2-vectors can be written in a null basis corresponding to the $(\tau_1, \tau_2)$ light-signal coordinates.

12.4. (34) Give a definition of 4-acceleration analogous to that for 4-velocity, so that it is the rate of change of 4-velocity with respect to proper time. What are the 4-accelerations of the curves:

$$u \mapsto (\cos \Omega u, 0, 0, u),$$
$$u \mapsto (0, \sin \Omega u, 0, u),$$
$$u \mapsto (\cos \Omega u, \sin \Omega u, 0, u),$$

at $u = ?$ [Caution! This $u$ is not necessarily proper time.]

12.5. (18) Show that the dot product of the 4-acceleration and the 4-velocity of a curve is identically zero.

12.6. (30) What curve is followed by a particle moving along a line with a 4-acceleration that is constant in magnitude? If you accelerate at one $g$ for your lifetime, how old is the universe when you die? (Hint: A sinh and cosh substitution simplifies part of this problem.)

12.7. (16) Show that a rocket moving with constant proper acceleration can outrun a light signal indefinitely. How much of a head start does it need?

12.8. (31) In Euclidean geometry let us define the *dihedral product* of three vectors to be

$$Q(a, b, c) \equiv \frac{(b \cdot b)(a \cdot c) - (b \cdot a)(b \cdot c)}{[(b \cdot b)(a \cdot a) - (b \cdot a)^2]^{1/2}[(b \cdot b)(c \cdot c) - (b \cdot c)^2]^{1/2}}.$$

Show that it has the following properties:
(i) invariant under changes in the lengths of $a$, $b$, and $c$;
(ii) invariant under changes of $a$ that keep $a$ in the $a,b$ plane. That is

$$Q(a + kb, b, c) = Q(a, b, c).$$

Thus the dihedral product depends only on the two intersecting planes defined by these vectors. The only geometric quantity associated with two such planes is their dihedral angle $\theta$. Show by working in a special coordinate system that we have

$$\cos \theta = Q(a, b, c).$$

# 13. Doppler Shift

We now have enough tools to give a simple and complete discussion of the Doppler shift. Besides being a good example of a 4-vector calculation, this will be good practice. We will have to do a similar calculation later in curved spacetime. The systematic redshift in the light from distant galaxies is the evidence for a dynamic, evolutionary universe — the most important idea in modern cosmological thinking. We will also briefly discuss aberration, the displacement in the apparent positions of objects caused by the motion of the observer. This is given as a good exercise, even though it has little application in cosmology.

The Doppler shift is the change in frequency of a periodic signal caused by the relative motion between the source and the receiver. We

*Doppler shift*

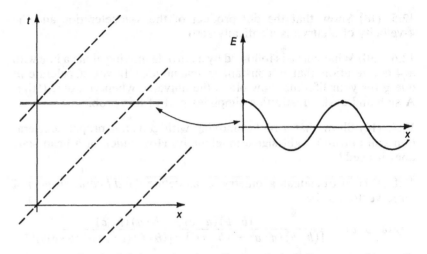

**Figure 13.1.** On the left is a spacetime diagram for a light wave going to the right. The dashed world lines follow the wave crests. On the right is a section along the doubled line showing the wave amplitude $E$ as a function of $x$.

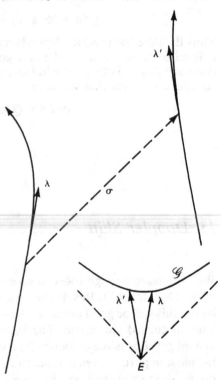

**Figure 13.2.** The geometry of a Doppler-shift measurement. The metric is given at an event $E$, and shows that $\lambda$ and $\lambda'$ are properly normalized to be 4-velocities.

will discuss only the Doppler shift of light. We replace the periodic signal by a sequence of light signals. In fact, only two will be needed. Think of them as successive wave crests in the periodic signal. This idealization is sketched in Figure 13.1.

The situation involves only three different 4-vectors. The motions of the source and the receiver must be specified by their 4-velocities. Their relative positions in spacetime will be described by a null vector $\sigma$ that connects the events of emission and reception of the light. The basic geometry is given in Figure 13.2. We will assume that the period of the signal is so short that the source and receiver do not move very much between one period and the next. Let $\tau$ be the time interval at the source for one period, and $\tau'$ be the time interval measured by an observer at the receiver. We want to compute the ratio of $\tau'$ to $\tau$ as they both become very short. The geometry of two successive light signals is sketched in Figure 13.3. We have a quadrilateral in spacetime. Two opposite sides are made up of light-signal world lines. The other two sides are segments of the world lines of the source and the receiver. It will not be a planar diagram unless the motion is only along the line connecting the source and the receiver.

### Spacetime geometry

[Remember that a null vector is not the zero vector, but a 4-vector satisfying $\sigma \cdot \sigma = 0$.]

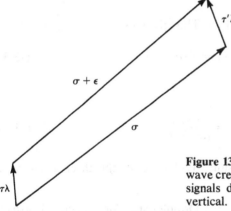

**Figure 13.3.** Geometry of two successive wave crests. Not a plane figure; so the light signals do not make 45° angles with the vertical.

The geometric information is contained in the demand that the quadrilateral be a closed figure:

$$\tau\lambda + \sigma + \varepsilon = \tau'\lambda' + \sigma. \tag{13.1}$$

Here $\sigma$ is the world line of the first light signal. The world line of the second light signal differs from $\sigma$ by only a small amount and we write it $\sigma + \varepsilon$; $\varepsilon$ is a small vector, going to zero as we take $\tau'$ and $\tau$ to zero.

[Remember that if $\lambda$ is a 4-velocity, then a segment of the world line corresponding to a time interval $\tau$ is given by the vector $\tau\lambda$. See page 58.]

We have two 4-velocities,

$$\lambda \cdot \lambda = -1, \tag{13.2}$$

$$\lambda' \cdot \lambda' = -1, \tag{13.3}$$

and two null vectors,

$$\sigma \cdot \sigma = 0, \tag{13.4}$$

$$(\sigma + \varepsilon) \cdot (\sigma + \varepsilon) = 0. \tag{13.5}$$

The special-relativity dot product obeys the usual distributive rule of multiplication; so the last equation can be written

$$(2\sigma \cdot \varepsilon) + (\varepsilon \cdot \varepsilon) = 0. \tag{13.6}$$

As we take the limit where $\varepsilon$ goes to zero, the square of $\varepsilon$ becomes smaller than the other terms, giving us

$$\sigma \cdot \varepsilon \rightarrow 0. \tag{13.7}$$

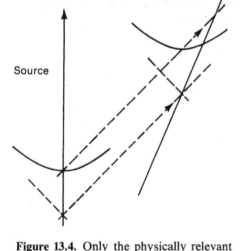

Now dot $\varepsilon$ through Equation 13.1. This gives us

$$\tau(\sigma \cdot \lambda) = \tau'(\sigma \cdot \lambda'); \tag{13.8}$$

**Figure 13.4.** Only the physically relevant world lines and metric figures are drawn here. The axes were deliberately not drawn, since they are not needed. To find the time-like direction, just look at the metric figures.

so our ratio is

$$\frac{\tau'}{\tau} = \frac{(\sigma \cdot \lambda)}{(\sigma \cdot \lambda')}. \tag{13.9}$$

*Result*

As we expected, the expression involves only the three vectors $\lambda$, $\lambda'$, and $\sigma$.

[In Section 23 I will give another derivation of this result.]

*Motion along a line*

*Example 1*

Let us work out the Doppler shift for the special case of motion along a line. This case is even simple enough to do graphically. Let us work in a canonical reference frame in which the source is at rest. A spacetime diagram for this is drawn in Figure 13.4. The hyperbola

$$t^2 - x^2 = \tau^2, \tag{13.10}$$

where $\tau$ is the source period, has been drawn at each end. Figure 13.5 is a blowup of the region around the observer. The times $\tau$ and $\tau'$ are indicated. The velocity is about 0.43 and the Doppler shift for the situation is

$$\frac{\lambda'}{\lambda} = 1.6.$$

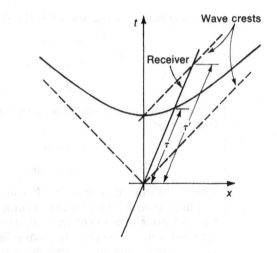

**Figure 13.5.** A blowup of the region around the receiver in Figure 13.4.

We now calculate this analytically. The source 4-velocity is just $\hat{t}$. The 4-velocity of the moving observer is

$$\lambda' = \frac{1}{\sqrt{1-v^2}}(v\hat{x} + \hat{t}).$$   (13.11)   [See page 58 for this.]

Note that the Doppler-shift formula depends only on the direction of $\sigma$ and not on its length. A vector with the correct direction is

$$\sigma = \hat{t} + \hat{x};$$   (13.12)

so we have

$$\frac{\tau'}{\tau} = \frac{\sigma \cdot \hat{t}}{\sigma \cdot \lambda'} = \frac{\sqrt{1-v^2}}{1-v} = \sqrt{\frac{1+v}{1-v}}.$$   (13.13)

*Example 2*  Another problem worth doing involves the Doppler shift between a source on the rim of a flywheel and a receiver on the axis. Let the angular velocity relative to a canonical reference frame be $\Omega$, and let the radius of the rotating wheel be $a$. Such experiments have been done, and they provide experimental confirmation of special relativity.   *Rotating disc*

Again we need to find the three vectors $\lambda$, $\lambda'$, and $\sigma$. The geometry is sketched in Figure 13.6. The world line of the receiver is just the line

$$x = y = z = 0.$$   (13.14)

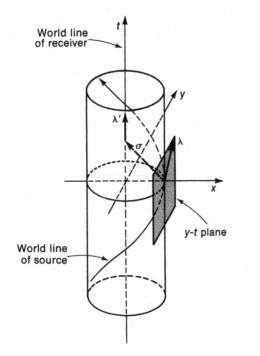

**Figure 13.6.** This is a three-dimensional spacetime diagram.

[See Equation 12.26.]

The world line of the source is the spacetime helix

$$x = a \cos \Omega t,$$
$$y = a \sin \Omega t, \qquad (13.15)$$
$$z = 0.$$

We consider a light signal sent out at $t = 0$. Its world line will be

$$x = a - t,$$
$$y = z = 0. \qquad (13.16)$$

This is the correct world line, because it satisfies three conditions: (i) it has a tangent vector that is null; (ii) it passes through the event $(a, 0, 0, 0)$ where the signal is sent out; (iii) it intersects the world line of the receiver. We can find tangent vectors to these world lines by converting them to the obvious parametric forms. The source world line can be written

$$x = a \cos \Omega u,$$
$$y = a \sin \Omega u,$$
$$z = 0, \qquad (13.17)$$
$$t = u,$$

and so on.

We have the tangent vectors

$$\lambda' \propto \hat{t}, \qquad (13.18)$$
$$\lambda \propto \hat{t} - a\Omega \sin \Omega t\, \hat{x} + a\Omega \cos \Omega t\, \hat{y}, \qquad (13.19)$$
$$\sigma = \hat{t} - \hat{x}. \qquad (13.20)$$

We use proportional signs rather than equals signs because we have not yet checked the normalizations. The normalized 4-velocities are

$$\lambda' = \hat{t}, \qquad (13.21)$$
$$\lambda = \frac{1}{\sqrt{1 - a^2\Omega^2}} (\hat{t} + a\Omega\hat{y}). \qquad (13.22)$$

There is no need to normalize $\sigma$. The dot products are easy to work out, and the Doppler shift is

$$\frac{\tau'}{\tau} = \frac{\sigma \cdot \lambda}{\sigma \cdot \lambda'} = \frac{1}{\sqrt{1 - a^2\Omega^2}}. \qquad (13.23)$$

With our 4-vector formalism, the only difficulty is in working out the geometry of the situation.

The general strategy used in our previous sections was to find covariant descriptions which could be used in any reference frame. Such a strategy eliminates the need for many of the Lorentz transformations usually found in special-relativity books. The Doppler-shift expression is a beautiful example of this strategy. While we are on the subject of light, we will give another example. What angle will an observer see between two light signals? Observers moving with different velocities will see different angles between the same two light signals, an effect called *aberration*. This problem must have an answer in terms of the geometry of the situation, and the geometric information is given by three vectors: the 4-velocity of the observer and the tangents to the light-signal world lines. It is a good exercise in the use of 4-vectors to find the correct expression.

*Aberration*

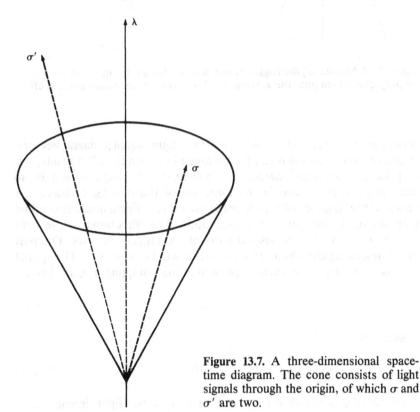

**Figure 13.7.** A three-dimensional space-time diagram. The cone consists of light signals through the origin, of which $\sigma$ and $\sigma'$ are two.

We proceed in the familiar fashion. We first figure out what we want to say in a special reference frame. Then we translate that into 4-vectors, which express the result in a frame-independent fashion. In Figure 13.7 we diagram the situation in the canonical reference frame in which the

*Spacetime geometry*

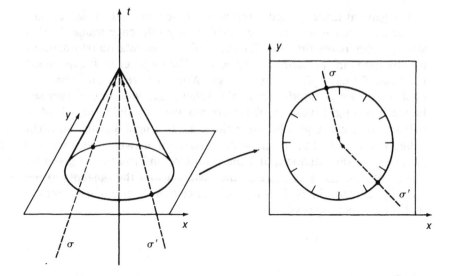

**Figure 13.8.** Measuring the angles between light-signal tangent vectors $\sigma$ and $\sigma'$ by projecting them onto the $x, y$-plane. This is shown separately in the section.

observer is at rest. Here we have two light-signal tangent vectors, $\sigma$ and $\sigma'$. What do we mean by the angle between them? A crude but, in principle, accurate method for measuring the angle would be to sight along a protractor in the direction of the two light sources. A space-space diagram of this, along with a three-dimensional spacetime diagram of it, are given in Figure 13.8. The directions of the light signals are given by the spatial parts of the tangent vectors. The part of $\sigma$ lying along the observer's 4-velocity will be $-(\sigma \cdot \lambda)\lambda$. The spatial part is found by subtracting from $\sigma$ its time component. Call this $\sigma_\perp$:

$$\sigma_\perp = \sigma + (\sigma \cdot \lambda)\lambda. \tag{13.24}$$

[Note the minus sign. It comes from the normalization $\lambda \cdot \lambda = -1$.]

It satisfies

$$\sigma_\perp \cdot \lambda = 0. \tag{13.25}$$

The angle $\theta$ that the observer sees between the two light signals will be the angle between $\sigma_\perp$ and $\sigma_\perp'$. For vectors with no time components, our spacetime dot product is the same as the Euclidean dot product; so we can use the Euclidean expression, Equation 3.4,

$$\cos \theta = \frac{\sigma_\perp \cdot \sigma_\perp'}{\sqrt{(\sigma_\perp \cdot \sigma_\perp)(\sigma_\perp' \cdot \sigma_\perp')}}. \tag{13.26}$$

This simplifies to the expression

$$(\cos \theta - 1) = \frac{\sigma \cdot \sigma'}{(\sigma \cdot \lambda)(\sigma' \cdot \lambda)}. \tag{13.27}$$

This is the covariant expression that we sought. Various applications of it will be developed in the problems.

## PROBLEMS

13.1. (20) The luminosity of a light source depends both on the rate at which photons are received and on the energy of each photon. The energy of a photon is proportional to its frequency. How does the luminosity of a moving source relate to its luminosity at rest?

13.2. (21) Use the graphical construction of the example on page 64 to plot Doppler shift against velocity for motion along a line.

13.3. (19) Use a graphical argument to derive the low-velocity limit of the Doppler shift for motion along a line.

13.4. (10) Show that absolute-time clocks have the same low-velocity limit for their Doppler shift for motion along a line.

13.5. (14) Show that for motion along a line with constant velocity no assumption of small $\tau$ need be made in the argument on page 63.

13.6. (14) What angle does an observer moving in the direction

$$\hat{x} + \hat{y} + \hat{z}$$

with speed $v$ see between two photons coming to him from the $\hat{x}$ and the $\hat{y}$ directions?

13.7. (24) Calculate the angle $\theta'$ that a moving observer sees between a star directly in his line of motion and one that a stationary observer sees at an angle $\theta$ from that line (see Figure 13.9). Use this calculation to find the transformation of the celestial sphere

$$(\theta, \phi) \mapsto (\theta', \phi'),$$

$$\tan \frac{\theta'}{2} = \sqrt{\frac{1 - v}{1 + v}} \tan \frac{\theta}{2},$$

$$\phi = \phi',$$

between two different reference frames.

13.8. (25) What is the relation between Equation 13.6 and the dihedral product defined in Problem 12.8?

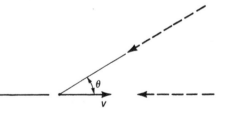

**Figure 13.9.** This is a space diagram, not a spacetime diagram.

13.9. (33) Show that the map defined in Problem 13.7,

$$(\theta, \phi) \mapsto (\theta', \phi),$$

is conformal, that is, that it preserves angles and takes infinitesimal circles into other infinitesimal circles.

13.10. (35) Show that the same map takes finite circles into finite circles.

13.11. (25) Derive the same map using the Lorentz transformation.

13.12. (39) Fed up with it all, you decide to flee the galaxy. You depart perpendicular to the plane of the galaxy at one $g$ acceleration. Describe the changing appearance of the galaxy as a function of time. (See Fred and Geoffrey Hoyle, *Into Deepest Space*.)

## 14. 4-Momentum

*The grand generalization*

The theory of special relativity as we have so far described it is a theory for clocks, light signals, and free particles. The structures describing these objects were all found to have Lorentz symmetry. Because of this symmetry, all states of motion are equivalent in the description of these objects. No preferred state that could be called absolute rest exists. We can now essay a grand generalization. Perhaps all of physics has this symmetry. This is a bold step, and we must admire the vision and courage of Einstein, who made it on far less evidence than we have presented here. So far, science has turned up nothing to contradict this sweeping generalization. We build it into the basic framework of every physical theory. General relativity does slightly modify it. Lorentz symmetry becomes only a local symmetry. Over long times and distances the postulates of special relativity will be modified. Special relativity will still provide a valid local description, a result that is often called the Principle of Equivalence.

*Particle collisions*

The dynamics of particle collisions can easily be made compatible with Lorentz symmetry. In Newtonian mechanics, collisions are governed by the laws of momentum conservation and energy conservation. What will be the spacetime view of these? The generalization is easy. Describe a particle by a 4-vector determined by the direction of its world line, whose spatial component is related to its ordinary momentum and whose time component is related to its energy. This

4-vector will be called the 4-momentum. This 4-momentum is to be a physical quantity with Lorentz symmetry. It must, therefore, have a covariant description. As before, we find covariant descriptions easily by using 4-vector equations. If the 4-momentum $p$ is to depend only on the 4-velocity $\lambda$, then the only relation that we can write is

*4-momentum*

$$p = m\lambda, \tag{14.1}$$

with a constant of proportionality that cannot depend on $\lambda$. Let us look at the implications of this relation for particle velocities that are small compared with the speed of light. The 4-velocity

$$\lambda = \frac{1}{\sqrt{1 - v^2}}(v + \hat{t}) \tag{14.2}$$

can be written to a good approximation

$$\lambda \simeq v + \left(1 + \frac{v^2}{2}\right)\hat{t}, \tag{14.3}$$

and so

$$p = mv + (m + \tfrac{1}{2}mv^2)\hat{t}. \tag{14.4}$$

As defined, if $m$ is the mass of the particle, then the spatial part of $p$ is ordinary 3-momentum in this approximation, and the time part is kinetic energy plus rest mass. In low-velocity collisions, total 3-momentum is conserved, total energy is conserved, and also total mass. Thus for collisions of low velocities, the total vector 4-momentum will be conserved.

If we assert that total 4-momentum is conserved for all collisions, then we will have a generalization of collision dynamics compatible with Lorentz symmetry. This is a weaker law than the law of Newtonian collisions. Mass and energy are not separately conserved. Everyone must be aware of the spectacular verifications of this.

*Rest mass*

[We call this constant $m$. The next few lines show that $m$ is the mass of the particle, usually called the *rest mass* of the particle.]

[Here we write the three spatial components of a 4-vector as an ordinary 3-vector. Thus $v$ is a 4-vector which has no $\hat{t}$ component, with an $\hat{x}$ component of $vx$, and so on. We abbreviate $v^2 \equiv v \cdot v$.]

*Conservation of 4-momentum*

| *Example* | The low-velocity relation between 3-momentum and kinetic energy is given by |

$$E_{\text{KINETIC}} = \frac{p^2}{2m}. \tag{14.5}$$

What is the corresponding relativistic expression? The relativistic expression follows from

$$p \cdot p = m^2\lambda \cdot \lambda = -m^2. \tag{14.6}$$

[I hope you have noticed by now that the main way to get information out of 4-vectors is to compute all the dot products in sight.]

Written out this is

$$p^x p^x + p^y p^y + p^z p^z - p^t p^t = -m^2. \qquad (14.7)$$

We will write

$$E \equiv p^t \qquad (14.8)$$

and call $E$ the total energy, and also write

$$\boldsymbol{p} \equiv p^x \hat{x} + p^y \hat{y} + p^z \hat{z}, \qquad (14.9)$$

$$p^2 \equiv \boldsymbol{p} \cdot \boldsymbol{p}. \qquad (14.10)$$

The relativistic generalization of our law is thus

$$E^2 = p^2 + m^2. \qquad (14.11)$$

The low-velocity limit of the above expression is

$$E \simeq m + \frac{p^2}{2m}. \qquad (14.12)$$

*Photons*    We can define a 4-momentum even for a particle moving at the speed of light. We again take the 4-momentum parallel to the world line

$$p \propto \sigma, \qquad (14.13)$$

and since $\sigma$ is now a null vector, we have

[The letter "$p$" is getting overused here, but these are the expressions in common use.]

$$E^2 = p^2. \qquad (14.14)$$

Thus such particles are said to have zero rest mass. The normalization of $p$ should be chosen so that the $\hat{t}$ component is the total energy, and the spatial part the total momentum, defined by our rule that 4-momentum be conserved.

Think for a moment about all possible 4-momenta. The space of *Energy-momentum space* 4-momentum vectors is a vector space similar to spacetime, but whose axes are not $x$ and $t$ but $p^x$ and $E$. The geometry in such a diagram, which we will call a momentum-energy diagram, is governed not by the special-relativity time-interval rule, but by the momentum-energy rule,

$$E^2 - p^2 = m^2, \qquad (14.15)$$

which we found in the above example. This should be no surprise; it is just our old friend the hyperbola again.

*Example*    Suppose two particles of the same mass collide and form a single new particle. In Figure 14.1 we give both a spacetime diagram and also a momentum-energy diagram for the process. Note that the various 4-momenta are all parallel to the world lines at the collision event. We have drawn there the hyperbola

$$E^2 - p^2 = m^2 \qquad (14.16)$$

for m, the mass of the incoming particles. Their 4-momenta therefore lie on this hyperbola. A careful look shows the final particle moving off with rapidity of about 0.12 and having a mass of about 2.18 times the mass of an incoming particle. The latter number came from comparing the length of its 4-momentum vector to the given hyperbola, which provides momentum-energy space with a standard of mass much as the time-interval hyperbola provides spacetime with a standard of time. One would say in words that some of the incoming kinetic energy has been converted into mass. The non-conservation of mass that we see here is the analog in the momentum-energy diagram of the twin paradox in a spacetime diagram.

When physicists invent more sophisticated descriptions of particles, compatible with quantum mechanics, they always make sure that the particles of these theories behave in a Lorentz-invariant manner. No evidence for the failure of this idea has been found, and no theory that is not compatible with Lorentz invariance will be taken seriously until such evidence turns up.

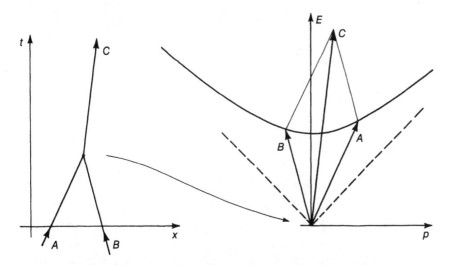

**Figure 14.1.** A spacetime diagram on the left and a momentum-energy diagram on the right for the same collision. Note origin of left diagram is arbitrary, but not so for the energy-momentum diagram. Note parallel lines between the two diagrams. The hyperbola is the curve $E^2 - p^2 = m^2$.

PROBLEMS

14.1 (18) Explain the units used in the equation

$$E^2 = p^2 + m^2.$$

14.2 (14) Give the explicit numerical calculations for the numbers assigned to Figure 14.1.

14.3 (27) Suppose that one has found a particle which, when it decays at rest, decays into two identical particles, which move off in opposite directions, each at twice the speed of light. Draw a spacetime diagram. Can one assign 4-momentum to the decay particles in a way that conserves 4-momentum? If one assumes that the process is Lorentz-invariant, then these particles can be sent back into one's own past (an anti-telephone). Show how to do it.

# CHAPTER TWO

# Geometry

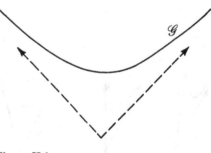

**Figure II.1.**

Our development of special relativity centered on clock structures represented by hyperbolas, as sketched in Figure II.1. In this chapter, we pause and develop the mathematics of such structures. They will be the building blocks of the theory of gravitation called general relativity.

Let me make the following analogy. Consider a complicated fluid-flow pattern. The vector describing the fluid velocity will be a complicated function of space and time. For short times and within small regions, the velocity will be approximately constant. This is schematically illustrated in Figure II.2. A general-relativistic spacetime similarly has a clock structure that is a complicated function of position. In small regions, the clock structure is uniform, and these local approximations all look like special relativity. This chapter is devoted to the tools by which one can make sense out of a situation like that given in Figure II.3. Whereas the local approximation to a vector field is a constant vector field (a field of free vectors), which is not a very deep structure, the local approximation to a general-relativistic spacetime looks like special relativity, which already has a complex geometric structure, as we saw in the last part.

The mathematical structures represented by these hyperbolas are called tensors. We are now going to discuss the properties of tensors and tensor fields. Just as no one would try to study electromagnetism without knowing vector analysis, so too we must learn a bit of tensor analysis in order to study gravitation. The simplest and most effective

*Special relativity as a local theory*

[In Section 51 we will see that Figure II.3 is a picture of a black hole.]

*Tensors*

75

**Figure II.2.** A magnified view of a flow field.

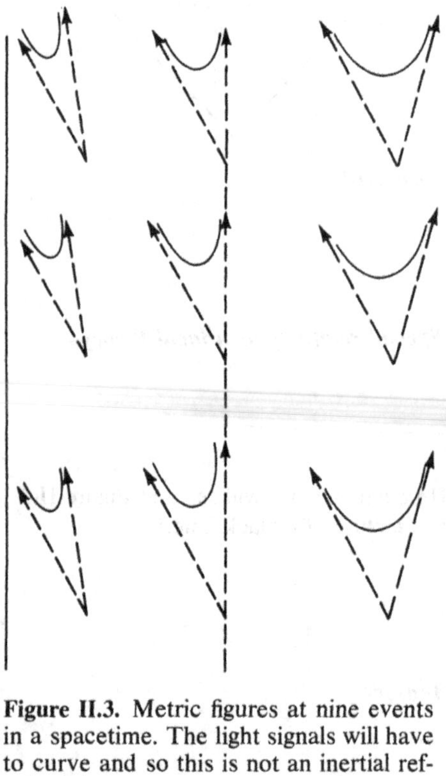

development of tensor analysis tries to stay as much in a covariant notation as possible, independent of any particular reference frame. The best example of such a calculation is the Doppler-shift calculation of Section 13. To emphasize the structures rather than the coordinates, I intentionally left off the axes in Figure II.1 and II.2.

In this chapter we will study the behavior of tensor fields in the immediate neighborhood of a single point, a neighborhood so small that the tensor fields are constant within the neighborhood. This is called tensor algebra. In Chapter III we will study the effects of tensor fields which are not constant, which is called tensor analysis. Using this tensor algebra, we will give a compact description of special relativity as the geometry of a simple tensor field.

**Figure II.3.** Metric figures at nine events in a spacetime. The light signals will have to curve and so this is not an inertial reference frame, nor can one be found.

## 15. Vectors and Covectors

Tensor algebra is the study of a vector space and the linear operators associated with it. A tensor algebra can be constructed for any vector space. The algebra of most interest to us is based on the tangent vectors in spacetime.

For any vector space $V$ the simplest and most important linear operators are those which act on elements of $V$ to produce real numbers. We will see in Section 16 that if the vectors $V$ represent velocity vectors. then these operators will represent gradients of functions. Let $\omega$ be an operator (or map; we use the word operator because these are special maps; you could even legitimately call them functions),

$$\omega: V \to \mathbb{R}; \; a \mapsto \omega \cdot a. \tag{15.1}$$

*Covectors*

[This notation for maps was explained on page 7.]

The operator $\omega$ is linear if for all vectors $a$ and $b$, and all numbers $k$, we have

$$\omega \cdot (ka + b) = k(\omega \cdot a) + \omega \cdot b. \tag{15.2}$$

Note that on the lefthand side we are adding vectors and on the righthand side we are adding numbers. Such linear operators will be called *covectors*, and we will see that they too form a vector space, usually written $V^*$. The dot used here for this operation is not the dot of either Euclidean or Minkowski geometry, which will be used only rarely in the rest of this book. I use the dot because the relation between $\omega$ and $a$ is symmetric. You can also think of $a$ as an operator on covectors.

These objects, vectors and covectors, have a numerical representation familiar to physicists in the row and column vectors of matrix algebra. We can also find a graphical representation of these objects that is both convenient and useful for one's intuition, since it makes it easy to see these objects in a frame-independent manner. Recall that we constructed a representation of our clock structure $\mathcal{G}$ from sets of elements of $V$ itself. We proceed here in a similar fashion.

*Graphical representations*

Look at the set of all vectors $v$ in $V$ satisfying

$$\omega \cdot v = 1. \tag{15.3}$$

Call this set "$\omega$." This set will be a representation for $\omega$; in fact, we will follow our usual casual notation, and drop the quotes after this section.

In a two-dimensional vector space, the set "$\omega$" must be a straight line. To see this, take two different vectors $a$ and $b$ such that

*Two dimensions*

$$\omega \cdot a = 1, \tag{15.4}$$

$$\omega \cdot b = 1. \tag{15.5}$$

Both $a$ and $b$ are elements of "$\omega$." Using our linearity condition 15.2, we have

$$\omega \cdot (a - b) = 0; \tag{15.6}$$

so for all $k$

$$\omega \cdot (k(a - b) + a) = 1. \tag{15.7}$$

Thus if $a$ and $b$ are in "$\omega$," then so is the straight line containing them. See Figure 15.1.

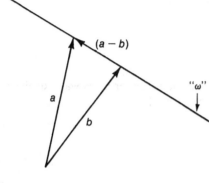

Figure 15.1.

In two dimensions there can be no other vectors in "$\omega$." Any other vector would generate a second line, and then between pairs of vectors on the two lines we would sweep out the entire two-dimensional vector space. But from 15.2 we have

$$\omega \cdot 0 = 0; \tag{15.8}$$

so that cannot be. In $n$ dimensions, the representation of a covector will be a linear subspace of $(n - 1)$ dimensions. In three dimensions, "$\omega$" will be a plane; and so on. If there are no vectors satisfying Equation 15.3, then we must have

$$\omega \cdot v = 0 \tag{15.9}$$

*Zero*  for all vectors. This operator we will call the zero operator.

We could draw an entire contour map of the operator. The contour lines would satisfy

[Being a confirmed two-dimension chauvinist, I say "lines" when I mean, of course, "$(n - 1)$-dimensional linear subspaces."]

$$\omega \cdot v = \ldots -1, 0, 1, 2, \ldots \tag{15.10}$$

*Contours*  You should be able to show that these are also straight lines parallel to "$\omega$" and equidistant. The zero contour passes through the zero vector.

[See Section 1 and also Figure 16.5 for free vectors.]

There is a notion of free covector corresponding to that of a free vector. For a bound vector, one point is a sufficient representation, and "$\omega$" consists of just such points. A free vector requires two points and a way to distinguish between the head and the tail for its representation; *Free covectors*  hence the usual arrow notation. For a free covector we must draw two contour lines, and somehow flag the upper one. My convention for this is shown in Figures 15.2 and 15.3 for vectors and covectors in two and three dimensions.

The operation of a covector on a vector is easily computed in this representation. One only has to count the number (and fraction) of contour lines of "$\omega$" crossed by the vector. The sign depends on whether the vector points uphill or down. This operation is sometimes called "evaluation."

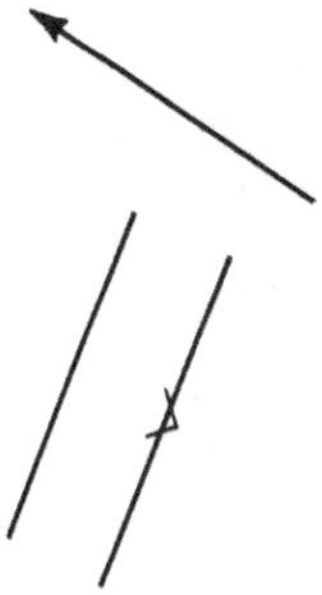

**Figure 15.2.**

*Example* | In Figure 15.4 we sketch two free covectors and three free vectors. The operations have the values

$$\omega \cdot a = +1, \qquad \nu \cdot a = -1,$$
$$\omega \cdot b = 0, \qquad \nu \cdot b = -1,$$
$$\omega \cdot c = -2, \qquad \nu \cdot c = +1.$$

[We are going to follow the notational convention in common use that reserves early-alphabet Latin letters for vectors and late alphabet Greek for covectors; $v$ for "vector" is an exception.]

These covectors themselves have a linear structure. We define the covector $k\omega$ to be the operator such that

*Linear structure for covectors*

$$(k\omega) \cdot v = k(\omega \cdot v). \tag{15.11}$$

Similarly the sum of two covectors $\omega$ and $\nu$ will be the covector such that

$$(\omega + \nu) \cdot v = \omega \cdot v + \nu \cdot v. \tag{15.12}$$

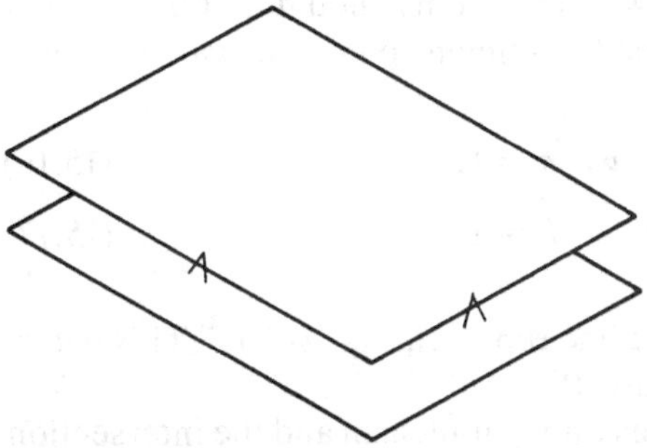

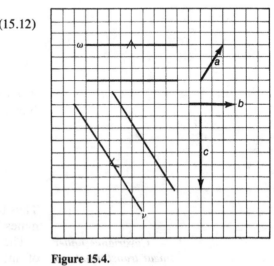

**Figure 15.3.**

**Figure 15.4.**

*Scaling*     These definitions render many of the parentheses used above unnecessary. What does "$k\omega$" look like? Suppose that $a$ is in "$\omega$," that is,

$$\omega \cdot a = 1. \tag{15.13}$$

Then, using linearity (Equation 15.2), we have

$$(k\omega) \cdot \left(\frac{a}{k}\right) = 1. \tag{15.14}$$

The contour lines of "$k\omega$" are parallel to those of "$\omega$" and closer together by a factor of $k$. This is sketched in Figure 15.5 for $k = 2$. Note the complementary behavior of vectors and covectors. This is necessary because there is no unique representation of a vector space. Just as with our inertial reference frames, there is the freedom of any linear transformation. In particular, our representation should be covariant under expansions, and it is. An expansion corresponds to a bigger vector but a small covector, and the operation $\omega \cdot v$ remains unchanged (see Figure 15.6).

*Addition*     What about the addition of two covectors? Given any two covectors "$\omega$" and "$v$," how do we construct their sum "$\omega + v$"? Again we can deduce this from linearity. Pick vectors $a$ and $b$ such that

$$\omega \cdot a = 1,$$
$$v \cdot a = 0,$$
$$\omega \cdot b = 0,$$
$$v \cdot b = 1. \tag{15.15}$$

The extension of this beyond two dimensions should be obvious, but it is much harder to draw. For the two-dimensional case, we must have

$$(\omega + v) \cdot a = 1, \tag{15.16}$$
$$(\omega + v) \cdot b = 1. \tag{15.17}$$

Thus both $a$ and b must lie on the line representing "$\omega + v$." This determines "$\omega + v$" as shown in Figure 15.7.

*Covariance under*     The addition operation involves only parallelism and the intersection
*linear transformations*  of lines. It is clearly covariant under any linear transformation. This

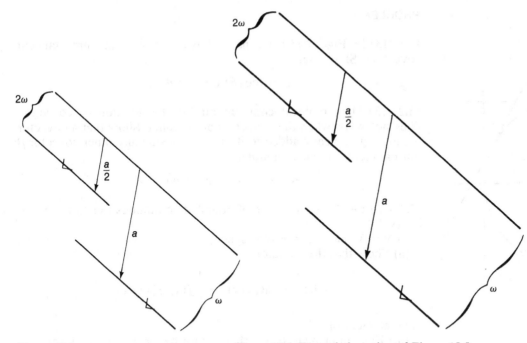

**Figure 15.5.**    **Figure 15.6.** Expanded version of Figure 15.5.

covariance will be very important when we come to discuss vector and tensor fields on manifolds.

A covector in two dimensions can be represented by the point on the unit contour that is closest to the origin. This representation is not covariant under linear transformations. This representation was used to generate the exotic addition rule for a vector space in $\mathbb{R}^2$ given in Section 1.

The vectors and the covectors are quite similar. The relation between them is called *duality*. Covectors are dual to vectors, we say. Just as the covectors are linear operators on vectors, clearly the vectors are linear operators on the covectors. Use the rule

$$a: V^* \to \mathbb{R}; \; \omega \mapsto \omega \cdot a.$$

Thus the vectors are dual also to the covectors. In fact, all linear operators mapping covectors to numbers are given by vectors. This is not really obvious, but for us not worth proving. (It is not true in some infinite-dimension function spaces.)

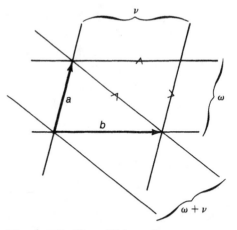

**Figure 15.7.** The addition of covectors $\omega$ and $\nu$ to give the covector $(\omega + \nu)$, using the auxiliary vectors $a$ and $b$.

*Duality*

## PROBLEMS

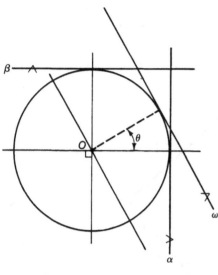

**Figure 15.8.**

15.1 (18) In Figure 15.8 we have drawn a circle and three tangent covectors. Show that

$$\omega = (\cos\theta)\alpha + (\sin\theta)\beta.$$

15.2 (13) The graphical construction for the addition of covectors does not work when the covectors are parallel. Show that a covector $\alpha$ can be graphically added to itself by choosing any other covector $\beta$ not parallel to it and computing

$$\alpha + \alpha = [(\alpha + \beta) + \alpha] - \beta.$$

15.3 (16) Let $V$ be the space of bounded continuous functions on the interval $0 \le x \le 1$.
   (i) Verify that $V$ is a vector space;
   (ii) Show that the operator

$$I_g: V \to \mathbb{R}; f(x) \to \int_0^1 f(x)\, g(x)\, dx$$

is an element of $V^*$.
   (iii) Find an element of $V^*$ that cannot be written in this form.

15.4 (22) Suppose that an unknown covector $\omega$ satisfies

$$\omega \cdot x = 0$$

for all vectors $x$ such that, for another given covector $\nu$, we have

$$\nu \cdot x = 0.$$

Sketch this and show that we must have

$$\omega = \lambda\nu,$$

where $\lambda$ is an undetermined numerical factor called a Lagrange multiplier.

15.5 (21) Continue Problem 15.4 to the case where the vector $x$ is constrained by a number of equations,

$$\alpha \cdot x = 0,$$
$$\beta \cdot x = 0,$$
$$\text{etc.}$$

15.6 (16) Extend Problems 15.4 and 15.5 to the dual case, where an unknown vector $x$ satisfies

$$\omega \cdot x = 0$$

for all $\omega$ such that

$$\omega \cdot a = 0,$$

$$\omega \cdot b = 0,$$

etc.

**15.7 (26)** Covectors in two dimensions can be represented by the two lengths $l$ and $m$, cut off on rectangular coordinate axes as shown in Figure 15.9.

(i) Find the change in this representation when the covector is multiplied by a factor $k$.

(ii) Show that the covector law of addition is given in this representation by

$$\frac{1}{l_3} = \frac{1}{l_1} + \frac{1}{l_2},$$

$$\frac{1}{m_3} = \frac{1}{m_1} + \frac{1}{m_2}.$$

(Hint: Use similar triangles, and don't be discouraged by having too few equations.)

(iii) What if the coordinate axes are not orthogonal?

**15.8 (24)** Go back and do Problem 1.11 again. (Note new rating!)

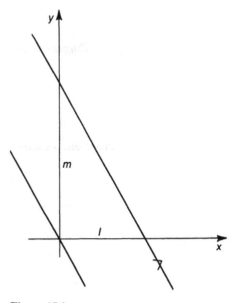

**Figure 15.9.**

# 16. Tangent Vectors and 1-Forms

*"It is the extreme of madness to learn what must then be unlearned."*

**ERASMUS**

The last section introduced covectors as the mathematically simplest operators on a vector space. The idea can also be given a geometric motivation. In this section we look at one realization of this mathematical idea. In the next two sections we will discuss concrete physical applications. One oversight of ordinary mathematical physics courses is their failure to distinguish between vectors and covectors. It is difficult to begin distinguishing between concepts that were introduced as a single idea. Thus one finds in the world many unreconstructed tensor-analysis types who do not see any usefulness in these distinctions. The physical world is full of covectors! I hope that the examples to follow show how much the idea of a covector simplifies our view of the world.

*Tangent vectors*        The vector space on which we operate will be the vector space of tangents to parametrized curves passing through some particular point $P$, fixed throughout this discussion. The tangent vector contains the information about the local behavior of a smooth curve passing through the point $P$. A free vector represents the local behavior of a smooth family of curves in the neighborhood of $P$.

*Parametrized curves*    Recall the definition of a tangent to a parametrized curve given on page 59. This was based on a limit process which involved magnifying the region around $P$ by some factor, and taking the limit as the magnification went to infinity. This limit process pushes off to infinity all of the information except that involved in the linear approximation to the curves near $P$.

*Example*    In the plane $\mathbb{R}^2$, consider the family of parametrized curves,

$$\gamma_a: \mathbb{R} \to \mathbb{R}^2; \ u \mapsto a\left(\cos\frac{u}{a^2}, \ \sin\frac{u}{a^2}\right), \qquad (16.1)$$

shown in Figure 16.1.

[We draw these figures on a grid to emphasize that they are quantitatively correct. Please note that any linear grid would do. Nothing here depends on the rectangular coordinate grid. This is our old friend, covariance under linear transformations.]

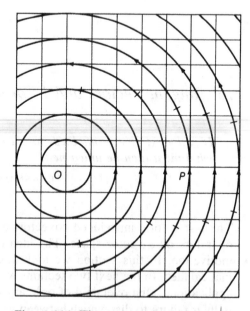

**Figure 16.1.** The parametrized curves described in the text. Arrows at unit parameter values, ticks at halves.

Here $a$ labels different curves, and the parameter $u$ identifies different points on the same curve. The curves are circles of radius $a$ around the origin; the velocity along these curves drops off with radius. Let us examine the behavior in the neighborhood of the point $P = (1,0)$ using the limit process described above. We will draw magnified pictures using stretched coordinates,

$$X \equiv \frac{x-1}{\varepsilon},$$

$$Y \equiv \frac{y}{\varepsilon}, \tag{16.2}$$

which magnify a region from a size $\varepsilon$ up to size unity. We scale the parameter $u$ so that we look only in the neighborhood of $u = 0$, corresponding to the point $p$, using a new parameter $U$:

$$U \equiv \frac{u}{\varepsilon}. \tag{16.3}$$

To look only at curves $\gamma_a$ near $(1,0)$, we introduce a new curve label $A$:

$$A \equiv \frac{a-1}{\varepsilon}. \tag{16.4}$$

Only the curve $\gamma_1$ passes through $(1,0)$. The curves in the magnified coordinates are given by the maps $\Gamma_A$:

$$\Gamma_A: U \mapsto (X,Y);$$

$$X = \frac{(1+\varepsilon A)}{\varepsilon}\left[\cos\frac{\varepsilon U}{(1+\varepsilon A)^2}\right] - \frac{1}{\varepsilon};$$

$$Y = \frac{(1+\varepsilon A)}{\varepsilon}\sin\frac{\varepsilon U}{(1+\varepsilon A)^2}.$$

Plots of these curves for $\varepsilon = \frac{1}{2}$ and for $\varepsilon = \frac{1}{5}$ are given in Figures 16.2 and 16.3. Since this is a family of smooth curves, if we take $\varepsilon$ small enough, the curves in $(X,Y)$ space do not change. We can find the limiting behavior using a Taylor's series expansion. We have

$$X = A,$$

$$Y = U. \tag{16.6}$$

These describe a family of parallel curves in the $y$-direction, as shown in Figure 16.4.

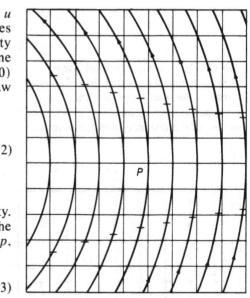

**Figure 16.2.** Previous figure has been doubled in size, $\epsilon = \frac{1}{2}$, more curves have been drawn halfway in between, and parameters are now marked at halves and quarters.

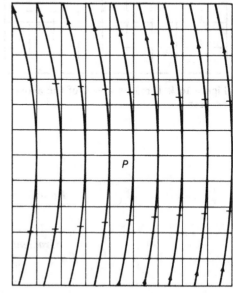

**Figure 16.3.** Stage corresponding to $\epsilon = \frac{1}{5}$, parameter marked at 0.2 and 0.1.

Having found the linear approximation to the family of curves, we can usefully transform it back into our original variables. We find

$$x = a,$$

$$y = u, \qquad (16.7)$$

and these describe a family of curves that form a local linear approximation to our given family around the point $P = (1,0)$.

The free vector field corresponding to this is sketched in Figure 16.5.

[To say that a curve is a local linear approximation means that the difference between the true curves and the approximate curves increases only as the square of the distance from the point $P$.]

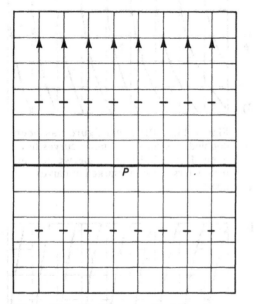

**Figure 16.4.** Limit as $\epsilon \to 0$ of the above process.

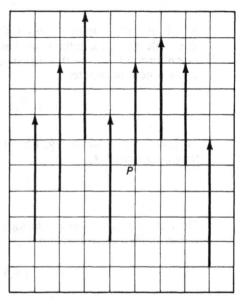

**Figure 16.5.** The field of free vectors corresponding to the limit in Figure 16.4. Note that a free vector can be drawn from any point, just as a curve can be drawn through any point in Figure 16.4.

*Functions*

*Gradient*

There is another spacetime object that is as fundamental as the idea of a curve. We can have a *function* on spacetime. A function has a value at each event. It is a map from spacetime to $\mathbb{R}$. A parametrized curve is a dual idea; it is a map from $\mathbb{R}$ to spacetime. The behavior of a smooth function in the neighborhood of a point is called its gradient at the point, just as the local behavior of a smooth curve is called its tangent. We can use the same limit process to find the local approximation to a function. Let us represent the function by a contour map; then expand the contour map by a factor of $1/\varepsilon$, and draw new contour lines at intervals of $\varepsilon$ rather than at unit intervals. If the function is smooth, then this process goes to a well-defined limit. This limit will be a family of parallel and equidistant contour lines. You should recognize this as our friend the free covector.

This gradient is one of a class of operators called differential forms. The gradient is the first member of the class, called a 1-form. Since "gradient" is often used to mean a vector, we will call this covector gradient a 1-form to avoid confusion. We will have no need here for the higher operators, 2-forms, etc.

*1-forms*

**Example**

Again in $\mathbb{R}^2$, look at a function

$$f(x,y) = x^2 + y^2 - 1 \qquad (16.8)$$

and find its local approximation near the point $P = (0,1)$. Introduce similar magnified coordinates,

$$X \equiv \frac{x}{\varepsilon},$$

$$Y \equiv \frac{y-1}{\varepsilon}, \qquad (16.9)$$

to find a new function (well, a function of new variables)

$$F(X,Y) = (1 + \varepsilon Y)^2 + \varepsilon^2 X^2 - 1. \qquad (16.10)$$

The magnified contour lines satisfy

$$(1 + \varepsilon Y)^2 + \varepsilon^2 X^2 - 1 = \varepsilon n, \qquad (16.11)$$

where $n = 0, 1, 2, \ldots$ These contours are shown in Figures 16.6 and 16.7 for $\varepsilon = 1$ and $\varepsilon = \frac{1}{4}$. In the limit $\varepsilon \to 0$, we have the contour lines given by

$$2Y = n, \qquad (16.12)$$

which are indeed straight, parallel, equidistant contour lines, shown in Figure 16.8. The free covector field corresponding to this is shown in Figure 16.9.

We can reverse our limit process once we have found the linear approximation. Writing in the original coordinates, we have

$$f \sim 2y - 2 \qquad (16.13)$$

for the linear approximation to the function

$$f = x^2 + y^2 - 1 \qquad (16.14)$$

in the neighborhood of the point $(0,1)$.

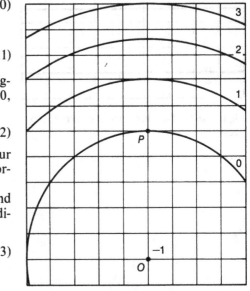

**Figure 16.6.** Contours of the function $x^2 + y^2 - 1$. Point $P$ has coordinates $(0,1)$.

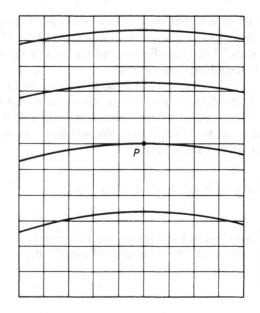

**Figure 16.7.** Figure 16.6 expanded by a factor of four.

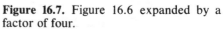

**Figure 16.8.** The limit as the expansion becomes large.

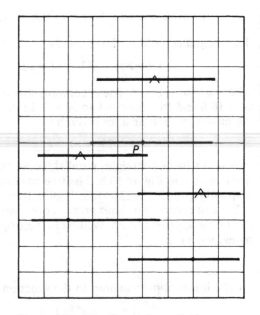

**Figure 16.9.** The free 1-form field corresponding to the limit in Figure 16.8.

The 1-forms at a point are linear approximations to functions at that point. Two such linear functions can be added, and a linear function can be multiplied by a constant factor. These operations turn the space of 1-forms into a linear vector space. Similarly, tangent vectors at a point are linear approximations to parametrized curves passing through that point. These linear approximations can be scaled and added, and also form a linear vector space.

*Linear structure of 1-forms*

*Linear structure of tangent vectors*

*Example*

The linear approximation to the curve

$$u \mapsto (x,y) = (u, \sin u) \qquad (16.15)$$

at $(0,0)$ is the linear curve

$$u \mapsto (x,y) = (u,u). \qquad (16.16)$$

Another linear curve is

$$u \mapsto (2u,-u), \qquad (16.17)$$

and the sum of these is the linear curve

$$u \mapsto (3u,0). \qquad (16.18)$$

Likewise, $k$ times the second curve is the curve

$$u \mapsto (2ku,-ku), \qquad (16.19)$$

another linear curve.

The space of all tangent vectors at a point is called the tangent space at that point. The space of all 1-forms at a point is called the cotangent space.

Covectors and 1-forms are linear operators, acting on vectors to produce real numbers. What is the geometric interpretation of this operation? Suppose we have a curve $\gamma$:

*Linear operators*

$$\gamma \colon \mathbb{R} \to M; \, u \mapsto \gamma(u), \qquad (16.20)$$

where we are going to call the general spacetime $M$ for Minkowski. Suppose further that we have a function $f$:

$$f \colon M \to \mathbb{R}; \, p \mapsto f(p). \qquad (16.21)$$

[None of these ideas will need modification when we discuss manifolds, and there you can think "M for manifold."]

If we evaluate the function $f$ at the points of the curve $\gamma$, then we have a map, written $f \circ \gamma$, and read "$f$ composed with $\gamma$," such that

$$(f \circ \gamma) \colon \mathbb{R} \to \mathbb{R}; \, u \mapsto f[\gamma(u)], \qquad (16.22)$$

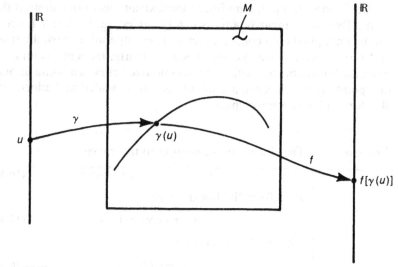

**Figure 16.10.**

as sketched in Figure 16.10. Now, $f \circ \gamma$ is a function in the sense of ordinary calculus. We will see in Section 17 that the operation of the covector representing the gradient of $f$ on the tangent vector to $\gamma$ is *Directional derivative* just the ordinary calculus derivative of $f \circ \gamma$: it is the rate of change of $f$ along $\gamma$ relative to the rate of change of the parameter $u$. There are two different ideas packaged in the notion of a tangent vector. One is the local linear approximation to a curve or a family of curves. The other is that of a directional derivative operator on functions. In spaces of finite dimension, these two ideas are represented by the same mathematical structure. This is also true in most spaces of infinite dimension. The vector $v$ used as a directional derivative is written $v \cdot \nabla$ in the common 3-vector notation. The directional derivative will be discussed further in Section 17.

Tangent vectors and 1-forms are called contravariant and covariant vectors in books on tensor analysis. I hope that the discussion of generalized force (a 1-form!) in Section 18 and the discussion of group velocity (tangent vector) and phase velocity (1-form) in Section 19 on wave propagation convince you that a proper geometric picture which distinguishes between tangent vectors and 1-forms is simple, intuitive, and natural.

PROBLEMS

16.1 (11) Show graphically and analytically that the gradient 1-form of the function

**Figure 16.11.**

$$f(x,y) = x^2 + y^2$$

vanishes at the origin.

16.2 (10) Figure 16.11 shows a portion of a topographic map. Sketch the 1-form gradient at a number of places.

16.3 (10) Let $\gamma$ be the helical curve

$$\gamma: s \mapsto (\cos s, \sin s, s),$$

and $f$ the function

$$f(x,y,z) = x^2 + y^2 + z^2.$$

Find the function $(f \circ \gamma): \mathbb{R} \to \mathbb{R}$.

# 17. Coordinate Basis Vectors

Although it is simple and elegant to leave coordinates out of our definitions, and although graphical algorithms are congenial to the intuition, the efficient manipulation of tensors does require numerical algorithms.

To get a numerical representation, we will use our coordinate system to generate a special set of basis vectors. We did this before for a canonical reference frame. Although we are still working with an image of spacetime in a linear vector space, the definitions introduced here and elsewhere will apply to the more general case of manifolds with little change. Even in a vector space we may elect to use curvilinear coordinates. Manifolds are spaces for which curvilinear coordinates are a necessity rather than a choice.

*Basis 1-forms*

In Section 16 we defined the gradient of a function to be the local linear approximation to the function. Since the sum of any two linear functions is another linear function, and since a linear function can be multiplied by a constant to give another linear function, these linear functions form a vector space. An efficient way to represent a vector space is to pick a basis, a standard set of vectors, and then to write all vectors as linear combinations of these basis vectors. Now, here the coordinates themselves are functions on spacetime; so they have gradients. We will use these gradients as a basis for the vector space of 1-forms.

[We discuss the case of covectors first because it is easier .]

*Gradient*

Let us go back to our discussion of gradients and repeat the argument given in the example on page 84, now for the general function $f(x,y)$. In the neighborhood of a point $(x_0, y_0)$ we have a Taylor's series expansion,

$$f(x,y) \approx f(x_0, y_0) + \frac{\partial f}{\partial x}(x - x_0) + \frac{\partial f}{\partial y}(y - y_0) + \ldots \quad (17.1)$$

The linear terms in this expression are what we call the gradient of $f$. We will follow the general practice and denote the covector gradient of $f$ by $df$. This $df$ is not an infinitesimal, but the 1-form defined above.

*df*

What about the coordinate $x$ itself? This is a function, and it can be written

[Throughout this book $d$ will be used in this technical sense. The physicist's small increment $dx$ will be written $\Delta x$.]

$$x = x_0 + (x - x_0). \quad (17.2)$$

*dx*

The second term here is what we call the gradient, and following our convention we call it $dx$. The linear functions $dx$, $dy$, and so on form a basis for the space of gradients. Our Taylor's series expansion above shows that

$$df = \frac{\partial f}{\partial x}dx + \frac{\partial f}{\partial y}dy, \quad (17.3)$$

and now you can see why we called these 1-forms $df$.

*Basis tangent vectors*

A coordinate system singles out a special set of curves as well as a special set of functions. These are the curves in which all the coordinates but one are held constant. The local linear approximation to a curve is called its tangent. The tangents to the special curves based on our coordinate system will be called $\partial/\partial x$, $\partial/\partial y$, and so on. In a few paragraphs we will see that these peculiar names are indeed appropriate. At the point $(x_0, y_0)$, $\partial/\partial x$ is the curve

$$u \mapsto (x_0 + u, y_0), \tag{17.4}$$

and $\dfrac{\partial}{\partial y}$ is the curve

$$u \mapsto (x_0, y_0 + u). \tag{17.5}$$

$\dfrac{\partial}{\partial x}$

The general curve $\gamma$,

$$\gamma: u \mapsto [X(u), Y(u)], \tag{17.6}$$

has a local approximation again given by a Taylor's series expansion

$$u \mapsto (x_0 + \frac{dX}{du} u, \; y_0 + \frac{dY}{du} u) \tag{17.7}$$

(we have taken $0 \mapsto (x_0, y_0)$ with no loss of generality). Again we can easily define addition and scaling on these linear curves. The linear approximation to the general curve $\gamma$, which we will call $\dot\gamma$, is given by

$$\dot\gamma = \frac{dX}{du} \frac{\partial}{\partial x} + \frac{dY}{du} \frac{\partial}{\partial y}. \tag{17.8}$$

We will use the tangents $\partial/\partial x$, $\partial/\partial y$, and so on as a basis for the tangent vectors. All the above can be easily extended from two to four or more dimensions.

The basis vectors $\partial/\partial x$ and $\partial/\partial y$ and the basis covectors $dx$ and $dy$ are dual to one another. That is, we have,

**Duality**

$$dx \cdot \frac{\partial}{\partial x} = 1,$$

$$dx \cdot \frac{\partial}{\partial y} = 0,$$

$$dy \cdot \frac{\partial}{\partial x} = 0, \tag{17.9}$$

$$dy \cdot \frac{\partial}{\partial y} = 1.$$

[Remember that we use the dot here for the operation of a covector on a vector. It is not the dot of Euclidean or Minkowski geometry.]

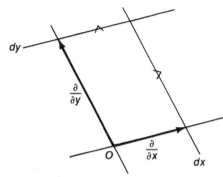

**Figure 17.1.**

Let us verify the first of these. We have defined $dx$ as the gradient of the function $x$, and $\partial/\partial y$ as the curve

$$u \mapsto (x_0 + u,\, y_0). \tag{17.10}$$

The value of the composite function is the $x$-value along the curve for each value of $u$. This is just $(x_0 + u)$. The rate of change of this with $u$ is clearly unity. A similar argument will verify the rest of them. In Figure 17.1 we sketch this result graphically, showing basis vectors and basis covectors for a skew coordinate system. All of this is covariant under linear transformations, as should be obvious.

The notation $\partial/\partial x$ for the $x$-basis vector emphasizes the idea of a directional derivative.

**Example 1** For any function $f$ we claimed in Section 16 that $df \cdot \partial/\partial x$ was the directional derivative in the $x$-direction. Using our above rules, we have

$$df = \frac{\partial f}{\partial x} dx + \frac{\partial f}{\partial y} dy + \ldots \tag{17.11}$$

and so

*Directional derivative*

$$df \cdot \frac{\partial}{\partial x} = \frac{\partial f}{\partial x}. \tag{17.12}$$

**Example 2** Suppose that a function $f$ satisfies the partial differential equation

$$\frac{\partial f}{\partial x} + 2\frac{\partial f}{\partial y} = 0 \tag{17.13}$$

and the boundary conditions

$$f(x,0) = e^{-x^2}. \tag{17.14}$$

This information determines $f$ in the entire plane.

The partial differential equation is equivalent to the expression

$$df \cdot \left( \frac{\partial}{\partial x} + 2\frac{\partial}{\partial y} \right) = 0, \tag{17.15}$$

which states that $f$ must be constant in the direction

$$\frac{\partial}{\partial x} + 2\frac{\partial}{\partial y}. \tag{17.16}$$

The value of $f$ can be found now from simple geometry. See Figure 17.2. We have

$$f(x, y) = e^{-(2x - y)^2/4}. \tag{17.17}$$

[This example is continued in Section 25.]

PROBLEMS

**17.1.** (08) Sketch the addition

$$dx + 3\,dy.$$

**17.2.** (12) Draw the vectors

$$\xi = \frac{\partial}{\partial x} + \frac{\partial}{\partial t},$$

$$\eta = \frac{\partial}{\partial x} - \frac{\partial}{\partial t},$$

and sketch 1-forms forming a basis dual to these. Write them in terms of $dx$ and $dt$.

**17.3.** (13) What is the tangent vector to the conical helix

$$s \mapsto (s\cos s,\, s\sin s,\, s)?$$

**17.4.** (13) What is the gradient of the function

$$f(x,y) = \sin(x+y)?$$

Sketch.

**17.5.** (16) Draw the coordinate basis and dual basis at some point in polar coordinates.

**17.6.** (22) Express the basis covectors of spherical polar coordinates in terms of $dx$, $dy$, and $dz$.

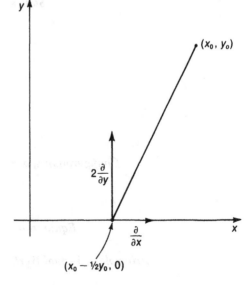

Figure 17.2.

# 18. Example: Static Equilibrium

We could now go on and discuss tensors, but it would be a mistake to do so without a thorough understanding of 1-forms and their relationship to tangent vectors. For that reason, let us pause for a concrete example. Generalized force is a simple example of a 1-form in physics. Although it has no direct application to cosmology, it should be familiar to everyone, but not as a 1-form. The standard books fail to recognize that generalized force is a 1-form; so they have to deal with it by means of abstract and *ad hoc* rules. The statements of the Principle of Virtual Work given in most books are confusing. Abstraction and jargon usually signal some basic defect in understanding.

*Statics*         The subject of our example is the static equilibrium of mechanical structures with a finite number of degrees of freedom. Electrical and hybrid systems could easily be included. The state of such a system can be specified by a finite number of parameters, called generalized coordinates. Specifying the system by giving the space locations of all its parts would be much less efficient. A fanciful example is sketched in Figure 18.1. If we neglect elastic waves in the springs, strings, beams, and masses, then the state of that system is specified by just two parameters, $x$ and $\theta$. The state of the system can be represented by a point in part of $\mathbb{R}^2$. The part of $\mathbb{R}^2$ accessible to the system we will call
*Configuration space*   configuration space. Configuration space has no natural notion of distance. The Euclidean metric of $\mathbb{R}^2$ has no physical significance here. Configuration space is a useful example of such spaces with no metrics.

If left to itself, a mechanical system will settle finally into some
*Equilibrium*          equilibrium configuration. The Principle of Virtual Work (PVW) is an efficient method for finding such equilibrium configurations. A typical
*Principle of Virtual Work*   statement of the principle is: "The given mechanical system will be in equilibrium if, and only if, the total virtual work of all the impressed forces vanishes. . . . The virtual work of the forces of reaction is always zero for any virtual displacement which is in harmony with the given kinematical constraints" (Lanczos 1949). The manipulations described are easily carried out. What is missing is an intuitive geometric picture. PVW asks us to compute the work done in a "virtual displacement" of the system. The generalized force relates this displacement to the "virtual work." Geometrically then, the generalized force is an operator taking displacements (tangent vectors) into the resulting work (number). The displacements are to be infinitesimal, and this process is

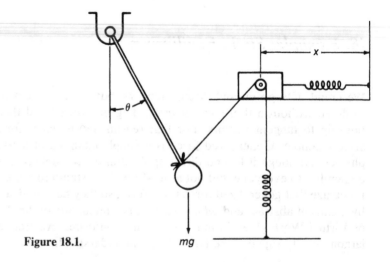

**Figure 18.1.**                              *mg*

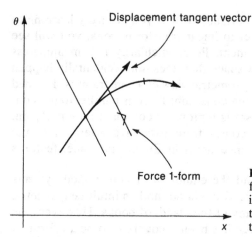

Figure 18.2. The geometry of generalized force as a 1-form. The situation pictured is one where the rate of doing work is about two units for motion with the tangent vector shown.

equivalent to the limit process that we introduced in the last section. This limit linearizes the operator. Generalized force, as a linear operator taking tangent vectors to numbers, is thus a 1-form. See Figure 18.2.

*Generalized force*

This recognition that force is a 1-form immediately dispels the mystery of the Principle of Virtual Work. The virtual displacements are being used just to compute the components of the force 1-form. Once we realize that generalized force is a 1-form, we can restate the Principle of Virtual Work: "The system is in equilibrium if the total generalized force including the constraint forces is zero."

Some forces are derived from potentials. Such forces are clearly gradients, that is, 1-forms. The amount of work done by a system in moving along a path in configuration space can be found by counting the net number of contour lines crossed. Any force field that can be derived from a potential does no work on a system which moves around any closed path. Not every force field can be written as the gradient of a potential. The force field sketched in Figure 18.3 cannot be derived from a potential. At any single point, however, any 1-form could have come from any number of functions.

*Potentials*

How is it that we have been able to think of force as a tangent vector? If we are dealing with a configuration space which has a natural Euclidean geometry, then we can describe force as a tangent vector. The Euclidean metric lets us associate a tangent vector to every 1-form. This will be shown when we discuss metric tensors. We all learn mechanics by first studying the mechanics of a particle moving in Euclidean space. We must realize that this is a special situation.

*Euclidean space*

A further payoff for having a clear geometric picture of force comes in situations involving constraints. A slippery constraint is one having no frictional forces. One usually says that the constraint force is per-

*Constraints*

pendicular to the constraint surface. Now if you are truly learning to think covariantly with respect to linear transformations, you will see that this is a nonsense statement. Perpendicularity is a meaningless concept in any configuration space that does not accidentally happen to have a metric. The correct geometric view of a constraint is sketched in Figure 18.4. As a 1-form, the constraint force is parallel to the constraint surface, and parallelism is a properly covariant notion. Again, the non-covariant language comes from excessive attention to the peculiar features of particle mechanics in Euclidean space, features which do not generalize.

*Lagrange multipliers*  The procedure used to find the equilibria of constrained systems involves Lagrange multipliers. You will not find an intuitive geometric picture of this procedure in any of the standard books. However, one can easily be given, once force has been recognized to be a 1-form. A constraint, by definition, can exert arbitrarily large forces to enforce the constraint condition. This force must be parallel to the constraint surface. For a constraint expressed by the condition

$$f(x,\theta) = 0, \tag{18.1}$$

a 1-form parallel to the surface is given by $df$. This should be obvious. The surface is just the zero contour line of $f$. Equilibrium demands zero total force. If $Q$ is the total force except for the constraints, and is thus known, then we must have

$$Q + \lambda\, df = 0. \tag{18.2}$$

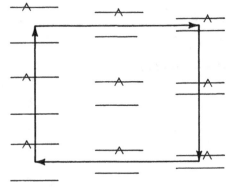

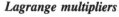

**Figure 18.3.** A nonconservative force field. Approximately 8 units of net work must be done in going once around the rectangular path shown.

The constraint force is given by $\lambda\, df$. The unknown multiplier $\lambda$ is needed because we know only the direction of the constraint force and not its magnitude. Equation 18.2 is a covector equation, to be solved for the $n$ coordinates of the equilibrium location and for $\lambda$. As a covector equation, it is really $n$ equations, one for each component. One further equation comes from the constraint itself, Equation 18.1, making $(n + 1)$ equations for $(n + 1)$ unknowns.

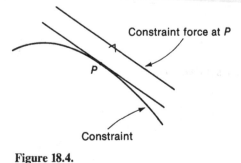

Constraint force at $P$

$P$

Constraint

**Figure 18.4.**

*Example*  Look at the pendulum sketched in Figure 18.5. The force of gravity (as a 1-form) is sketched, and also the unknown force provided by the rod.

If we describe the system by a single generalized coordinate $\theta$, then the gravitational potential energy is given by

$$V = mgl(1 - \cos\theta), \tag{18.3}$$

and the gravitational force 1-form $Q$,

$$Q = -dV, \tag{18.4}$$

is

$$Q = -mgl \sin \theta \, d\theta. \qquad (18.5)$$

The system will be in equilibrium when the force vanishes. This happens when

$$\sin \theta \, d\theta = 0, \qquad (18.6)$$

that is, when

$$\theta = 0, \pi. \qquad (18.7)$$

The configuration space here is a circle, and the force 1-form is a little hard to represent. Contour "lines" are really contour points. We indicate them by tick marks, and sketch the configuration space and the force 1-form in Figure 18.6.

We can, on the other hand, describe the system by two generalized coordinates $(x, y)$, provided we now take into account the elastic force in the bar. (The elastic force could be ignored in the $\theta$-space since it has no component in that direction.) The gravitational force $\gamma$ is given by

$$\gamma = -mg \, dy, \qquad (18.8)$$

and the elastic force due to the rod by

$$p = \lambda(\cos \theta \, dy - \sin \theta \, dx), \qquad (18.9)$$

where only the direction of $p$ is known and the magnitude $\lambda$ needs to be calculated. Equilibrium demands that the total force vanishes,

$$\gamma + p = 0; \qquad (18.10)$$

that is,

$$(\lambda \cos \theta - mg) \, dy - \lambda \sin \theta \, dx = 0. \qquad (18.11)$$

This is a covector equation, and each component must vanish separately. This will determine both the position of equilibrium,

$$\sin \theta = 0, \qquad (18.12)$$

and also the elastic force in the rod,

$$\lambda \cos \theta = mg; \qquad (18.13)$$

that is,

$$\theta = 0, \quad \lambda = mg, \qquad (18.14)$$

$$\theta = \pi, \quad \lambda = -mg. \qquad (18.15)$$

**Figure 18.5.**

[See Problem 15.1 for this.]

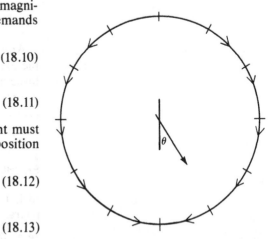

**Figure 18.6.** One dimensional configuration space for the pendulum. The gravitational force 1-form is indicated. Equilibrium occurs when the force vanishes, at points $\theta = 0$ and $\theta = \pi$.

We can also view this as a problem with a constraint and use Lagrange multipliers. Equilibrium occurs when the force is parallel to the constraint, as sketched in Figure 18.7. The details are left to Problems 18.1 and 18.2.

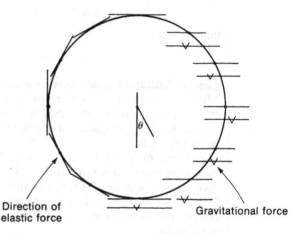

Direction of
elastic force

Gravitational force

**Figure 18.7.** The pendulum viewed as a two-dimensional problem with a constraint. On the right we sketch the gravitational force 1-form, on the left the direction of the elastic force constraining the pendulum bob. The magnitude of the constraint force adjusts as needed, but with no friction the direction must be that shown. The two forces can cancel only if their directions agree. This happens only at the top and bottom positions.

Any new formalism must prove its worth on problems. I hope that the examples and the problems will show you what a rational view of force and of all the $dx$'s is provided by the language of 1-forms. When we come to study the motion of particles, we will find that momentum is also a 1-form. Thus force and rate of change of momentum will have compatible geometric structures.

PROBLEMS

18.1. (20) Complete the example on page 98 using Lagrange multipliers. (See Problems 15.4 to 15.6 for Lagrange multipliers.)

18.2. (08) Suppose the pendulum of the example is acted on by a force

$$Q = 2x\,dx + y\,dy$$

instead of gravity. Find a potential for the above force.

18.3. (12) Find the equilibrium points for the system in Problem 18.2 by working directly in $\theta$-space. Sketch.

18.4. (15) Find the equilibria for Problem 18.2 by using constraints. Sketch and compare with 18.3.

18.5. (15) Find the force due to the bar in Problem 18.2.

18.6. (20) Suppose the force on the bar is that given in 18.2 plus gravity. Find the equilibria in any way you like. Sketch.

18.7. (20) Suppose the force on the pendulum comes from a potential

$$V = x^4 + y^4.$$

Again find the equilibria. First give a quick graphical argument, then actually calculate them using Lagrange multipliers.

18.8. (24) Do the same as in 18.7 for the potential

$$V = x^3 + y^3.$$

18.9. (20) Do the same as in 18.7 for the potential

$$V = |x|^3 + |y|^3.$$

18.10. (35) Catastrophe theory studies the changes in equilibrium positions caused by changes in the parameters of a problem. (See *Scientific American,* April 1976, and the book of Poston and Stewart, for example.) The standard example of such a system is shown in Figure 18.8. The energy in the springs is assumed to be given by

$$V = \frac{k}{2}(L - L_0)^2,$$

where $L_0$ is some constant and $L$ is the distance between the ends. Study the equilibria of the system as a function of the location of the point where the top spring is fastened for a system with

$$l = k = L_0 = 1.$$

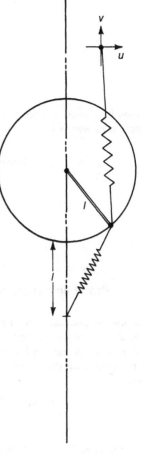

**Figure 18.8.**

# 19. Example: Dispersive Waves

Our second example of 1-forms that arise naturally in physics is the propagation of wave packets of dispersive waves, such as water waves, elastic waves, and the wave functions of quantum-mechanical particles.

[The book by Pierce is a good, simple introduction to waves.]

### Wave packets

### Partial differential equations

[Only a brief acquaintance with partial differential equations is needed for this discussion. The specialized knowledge found in a course on the subject will not be needed.]

### Semiclassical view

[See Figures 19.6, 19.7, and 19.8 for this limit.]

### Group velocity and phase "velocity"

[To aid the reader, I will carry along a particular example, setting it off with the usual vertical line. The same example will continue through the entire discussion.]

This example deserves careful study, not just to see the interplay between tangent vectors and 1-forms, but also because the motion of wave packets will be used as a model for the motion of particles in general relativity. We will use it to discuss the red shift, symmetries, tachyons, and black holes. The motion of water waves has a surprising symmetry, behaving like another physical realization of the mathematics of special relativity, and we will devote an entire section to this. This discussion of the geometry of wave-packet motion should be useful to solid-state physicists, who call such wave packets quasiparticles.

The motion of wave packets can be discussed without any mention of a metric structure on spacetime. We will later take an unusual view of metric structure, and will derive it from the motion of wave packets, instead of using it the other way around. Thus we will view the Lorentz metric of special relativity as the result of a particularly simple theory of waves.

A spacetime must have some structure in order to allow interesting physics. Lacking a metric, we must look to something else to generate structure. The structure of interest here will come from a linear partial differential equation. This partial differential equation should allow wavelike solutions, and we are going to look first at its plane-wave solutions. Later in this section we will form wave packets from these plane waves. For these wavepackets the wave amplitude is not constant. We will consider only high-frequency waves where the change in amplitude within a few wavelengths is small.

A coarse-grained view of such a wave packet would see only a concentrated bundle of energy and would not be able to discern the individual wave pattern. This will be our model for a particle. We will use the motion of these wave packets as a model for the motion of particles. We will see that these wave packets have naturally associated with them both a tangent vector and a 1-form. The tangent vector describes the direction of motion of the wave packet. The 1-form describes the configuration of the wave crests in spacetime. The first is called the group velocity; the second, the phase velocity. Conventional treatments try to deal with phase velocity as if it were a true tangent vector. Doing so has only limited success—it works in two dimensions, but fails beyond that. A 1-form in only two dimensions is described by parallel lines, and its orientation can be "sort of" considered to be a vector. The trouble caused by the "sort of" disappears once we recognize that phase velocity is naturally a 1-form. The treatment given here, especially the wave diagram to be developed in Section 27, is new. The conventional discussion of group velocity can be found in many books, the best being in Whitham (1974).

We start with a linear partial differential equation of perhaps quite high order.

*Example 1*  Let us consider the wave motion of spinless particles in spacetime, which is described by the Klein-Gordon equation:

$$\frac{\partial^2 \Psi}{\partial t^2} - \frac{\partial^2 \Psi}{\partial x^2} + \Psi = 0. \tag{19.1}$$

(You don't need to know where this equation comes from. Take it as the starting point. It also describes elastic waves on a supported beam.)

We want to study first of all plane-wave solutions to our partial differential equation, solutions of the general form

$$\Psi = A \cos \theta, \tag{19.2}$$

where the amplitude $A$ for now is taken to be constant, and $\theta$ describes the phase of the wave. Wave crests are given by

$$\theta = 2\pi n, \tag{19.3}$$

$n$ being an integer. A stationary observer sees a section of a plane wave along a vertical world line $W$ as sketched in Figure 19.1. The

*Plane waves*

[This factor of $2\pi$ will crop up throughout this discussion. It is related to the $2\pi$ in the Fourier transform and is also not removable.]

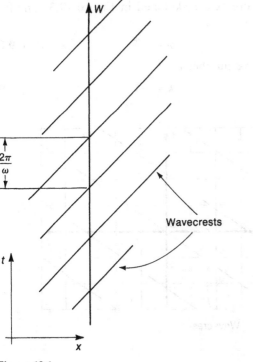

**Figure 19.1.**

rate at which he sees wave crests pass is called the frequency, and we use angular frequency $\omega$ defined by

$$\omega \equiv \frac{\partial \theta}{\partial t}. \tag{19.4}$$

At any fixed time, the spatial disposition of the wave is given by its behavior along the horizontal line in Figure 19.2. The rate at which wave crests occur in space is proportional to the wave number $k$, defined by

[In more dimensions, there would be three components to $k$, of course.]

$$k \equiv \frac{\partial \theta}{\partial x}. \tag{19.5}$$

[Example 1 continues]

The function

$$\Psi = A \cos (\sqrt{2}t - x) \tag{19.6}$$

is a particular plane-wave solution to Equation 19.2. The phase $\theta$ is

$$\theta = \sqrt{2}t - x. \tag{19.7}$$

The wave crests are sketched in Figure 19.3. The frequency $\omega$ is

$$\omega = \sqrt{2}, \tag{19.8}$$

and the wave number is

$$k = -1. \tag{19.9}$$

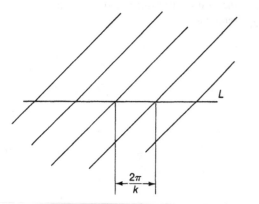

**Figure 19.2.**

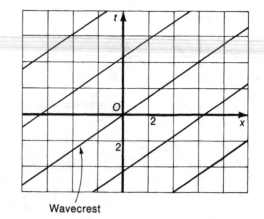

Wavecrest

**Figure 19.3.**

The frequency and wave number are clearly part of a 1-form, the phase gradient

$$d\theta = \frac{\partial\theta}{\partial x}\,dx + \frac{\partial\theta}{\partial t}\,dt, \qquad (19.10)$$

that is,

$$d\theta = k\,dx + \omega\,dt. \qquad (19.11)$$

The phase gradient describes the kinematics of the wave crests, their local arrangement in spacetime. To be precise, the wave crests have less structure than a 1-form. The negative of the 1-form describes the same wave crests. They also have more structure, since the phase gradient ignores a constant phase shift. Common sense handles both situations.

The plane wave of the example has a phase gradient

[Example 1 continues]

$$d\theta = \sqrt{2}\,dt - dx. \qquad (19.12)$$

The wave crests are moving to the right with a speed

$$\frac{\omega}{k} = \sqrt{2}, \qquad (19.13)$$

as can be seen from Figure 19.3.

What are the conditions under which a plane wave in the form of Equation 19.2 will satisfy our partial differential equation? Using the chain rule, the partial differential equation reduces to a single algebraic condition on the components of the phase gradient.

For

[Example 1 continues]

$$\Psi = A\cos\theta, \qquad (19.14)$$

we have

$$\frac{\partial^2\Psi}{\partial t^2} = -A\omega^2\cos\theta \qquad (19.15)$$

and

$$\frac{\partial^2\Psi}{\partial x^2} = -Ak^2\cos\theta, \qquad (19.16)$$

provided that $k$ and $\omega$ are constant, as they are in a plane wave. Thus our partial differential equation demands that

$$-\omega^2 + k^2 + 1 = 0. \qquad (19.17)$$

This algebraic condition on the components of the phase gradient is called the dispersion relation. It contains the same information as the partial differential equation itself. You should be able to see how to reconstruct the partial differential equation from the dispersion relation, one time derivative for each $\omega$, and so on. In fact, one often describes a wave problem by giving the dispersion relation rather than the partial differential equation. The experimental study of a wave system often measures the dispersion relation directly.

[If you have studied quantum mechanics, these rules will ring a bell.]

---

Some common dispersion relations:

ordinary wave equation,

$$\omega^2 = k^2; \tag{19.18}$$

Klein-Gordon equation,

$$\omega^2 = k^2 + 1; \tag{19.19}$$

elastic waves on a bar with tension and stiffness,

$$\omega^2 = k^2 - k^4; \tag{19.20}$$

deep water waves,

$$\omega^4 = k^2. \tag{19.21}$$

---

*Common dispersion relations*

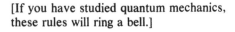

[I have left out the physical constants to simplify the relations. This is equivalent to choosing canonical coordinates for each problem.]

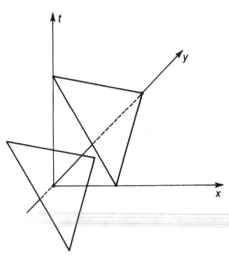

**Figure 19.4.** Wave crests in two space dimensions and time. Two wave crests are drawn, one through the origin, the other with vertices on the axes.

*Nearly plane waves*

What is the importance of the phase gradient being a 1-form rather than a tangent vector? Consider the phase gradient in three dimensions sketched in Figure 19.4. For this situation, one cannot point to any spatial direction as the direction of motion of the wave crests. It is possible to find the rate at which any observer sees the phase change along his world line. If his world line has a tangent vector $\lambda$, then the rate of change of phase is given by the operation $d\theta \cdot \lambda$. This is the natural use of a 1-form on a tangent vector.

Let us now consider solutions for which $k$ and $\omega$ are not exactly constant. Such waves are no longer plane. If $k$ and $\omega$ change only a little in any one wavelength and wave period, then the waves are nearly plane; so they will still be described approximately by our plane-wave formalism. A nearly plane wave is sketched in Figure 19.5. Calculations for such nearly plane waves are done by assuming that $k$ and $\omega$ depend not on $x$ or $t$, but on slowly changing variables $\varepsilon x$ and $\varepsilon t$, in the limit

$\varepsilon \to 0$. In this limit the frequency and wave number change slowly relative to the phase itself. For our solution

$$\Psi = A \cos \theta, \qquad (19.22)$$

we now have

$$\frac{\partial \Psi}{\partial t} = -A\omega \sin \theta \qquad (19.23)$$

and

$$\frac{\partial^2 \Psi}{\partial t^2} = -A\omega^2 \cos \theta - \varepsilon A \frac{\partial \omega}{\partial t} \sin \theta. \qquad (19.24)$$

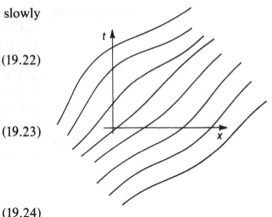

**Figure 19.5.**

The largest terms as $\varepsilon \to 0$ are the terms in the dispersion relation for plane waves. The changing frequency and wave number must still satisfy the dispersion relation at each point.

A nearly plane-wave solution for our example would be

$$\Psi = A \cos\left\{\frac{1}{\varepsilon}[1 + 2\sqrt{2}\,\varepsilon t - 2\varepsilon x - (\varepsilon t)^2 - (\varepsilon x)^2]^{1/2}\right\}. \qquad (19.25)$$

This can be used over long distances, where $x$ can be as large as $1/\varepsilon$, so that the $\varepsilon x$ and the $\varepsilon^2 x^2$ terms are not necessarily small and cannot be ignored. Near the origin we have

$$\Psi \sim A \cos(\sqrt{2}\,t - x + \text{constant}). \qquad (19.26)$$

This has a phase gradient

$$d\theta = \sqrt{2}\,dt - dx. \qquad (19.27)$$

Often when we are dealing with wave packets, the absolute phase is of little interest, so the constant in Equation 19.26 can be ignored.

For large $x$ and/or $t$, the terms in $\varepsilon x$ cannot be ignored. The frequency is given everywhere by

$$\omega = \frac{\sqrt{2} + (\varepsilon t)}{[1 + 2\sqrt{2}\,\varepsilon t - 2\,\varepsilon x + (\varepsilon t)^2 - (\varepsilon x)^2]^{1/2}}. \qquad (19.28)$$

Near any point it does change slowly with respect to the phase itself, and likewise for the wave number.

[Example 1 continues]

[It is not obvious how to find solutions such as the above. Look in the excellent book by Whitham for suitable methods.]

*Group velocity*   Once we go from a perfectly plane wave to a wave that can have a variable frequency and wave number, we can in effect mark the wavefront, and discuss the propagation of this mark through spacetime. The mark might be a bump in the frequency, for example. It does make sense to trace out in spacetime the path of such a mark, and a vector pointing in this direction is called the *group-velocity vector.* Later we will see that the energy in the wave, amplitude bumps, and frequency information all propagate in the same spacetime direction.

*Frequency*   The transport of frequency information is determined by the dispersion relation, as follows. We know that

$$\frac{\partial \omega}{\partial x} = \frac{\partial^2 \theta}{\partial x\, \partial t} = \frac{\partial^2 \theta}{\partial t\, \partial x} = \frac{\partial k}{\partial t}. \tag{19.29}$$

If we solve the dispersion relation for wave number

$$k = K(\omega) \tag{19.30}$$

and then use Equation 19.29, we find an equation describing the transport of frequency information.

*Example 2*   For water waves on deep water, we gave the dispersion relation above as,

$$\omega^4 = k^2 \tag{19.31}$$

and this has one branch of solutions

$$k = -\omega^2. \tag{19.32}$$

Using equation 19.29, we have

$$\frac{\partial \omega}{\partial x} + \frac{\partial}{\partial t} (\omega^2) = 0; \tag{19.33}$$

that is,

$$\frac{\partial \omega}{\partial x} + 2\omega \frac{\partial \omega}{\partial t} = 0. \tag{19.34}$$

This equation states that the frequency is constant in the spacetime direction given by the vector

[Note how natural our notation for a tangent vector is. See the discussion of directional derivatives in Section 17.]

$$v = \frac{\partial}{\partial x} + 2\omega \frac{\partial}{\partial t}. \tag{19.35}$$

This vector points in the group-velocity direction for our deep-water waves.

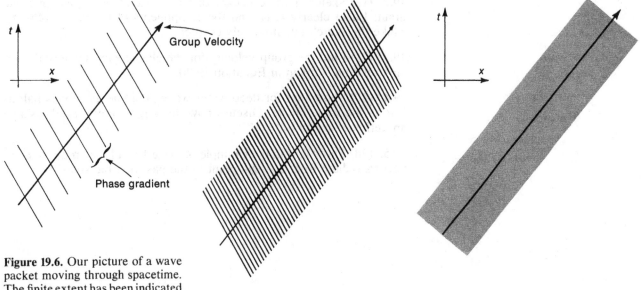

**Figure 19.6.** Our picture of a wave packet moving through spacetime. The finite extent has been indicated by drawing only short lengths of the wave crests. A long wavelength has been used for clarity. Figure 19.7 is a better high-frequency wave packet.

**Figure 19.7.** A wave packet with a frequency five times that of the preceding figure.

**Figure 19.8.** The limit.

We will see later that amplitude information is also transported in this same direction. Our view of a wave packet is the echelon diagram sketched in Figure 19.6. Figures 19.7 and 19.8 show higher-frequency wave packets. For an electron, the actual spacing is about $8 \times 10^{-21}$ sec. We have a wave whose amplitude has a lump somewhere in space, smoothly dying off away from the wave packet. The wave crests are described by a phase gradient whose components must satisfy the dispersion relation. The wave packet moves through spacetime in the direction of its group-velocity vector, which is found from the derivative of the dispersion relation. Later sections will extend this discussion to partial differential equations whose coefficients are also slow functions of position. We will use them to discuss the motion of wave packets in external fields.

*Amplitude*

PROBLEMS

19.1. (10) Calculate $k$ for the plane wave given in the example on page 107.

19.2. (12) Sketch a wave packet of water waves in a spacetime diagram. Label clearly $k$, $\omega$, and the components of the group velocity. Make the sketch quantitatively correct.

19.3. (21) Find the group velocity for the elastic waves whose dispersion relation is given in Equation 19.20.

19.4. (20) Show that for deep-water waves the energy moves half as fast as the wave crests. Discuss how this might be observed, using a spacetime diagram.

19.5. (28) Show that in the example on page 107 all the group-velocity vectors radiate from a single point in the past (the big splash).

## 20. Maps

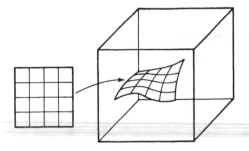

**Figure 20.1.** A map inserting a subspace into a larger space.

*Map*

*Function*

A fundamental element in mathematics is the idea of a map between two sets. If the sets have some mathematical structure, then it is natural to ask how this structure is carried back and forth by the map. We are interested here in sets which are vector spaces. No change will be needed in the results given here when we go on to discuss the more general case of manifolds in Section 26. As usual, there is a nice duality in the relations between vectors and covectors, a duality that is lost in the conventional coordinate-oriented treatments.

Let us start with a map $\psi$ between two vector spaces $A$ and $B$. The spaces need not have the same dimension. If $A = \mathbb{R}$, then this is a parametrized curve. If $B = \mathbb{R}$, then this is an ordinary function. We will use these results mostly for situations where $A$ has a smaller dimension than $B$, and we will draw our pictures for that case. The results hold without this restriction. For $A$ smaller than $B$, we can think of the map as representing a parametrized subset of $B$, a generalization of the idea of a parametrized curve, as sketched in Figure 20.1.

Suppose further that we have a function on $B$, a map

$$f: B \mapsto \mathbb{R}, \tag{20.1}$$

as in Figure 20.2. This defines a natural function, $f^*$, on the set A as well:

$$f^*: A \to R; \ a \mapsto f[\psi(a)]. \tag{20.2}$$

This is also shown in Figure 20.2. To evaluate $f^*$ on an element $a$ of $A$, first use $\psi$ to jump into the space $B$ and use the value of $f$ there. The function $f^*$ is called the *pullback* of $f$ by the map $\psi$.

*Pullback*

[Apologies for the use of this heavily used star symbol. It is the common convention. It has no relation to complex conjugation.]

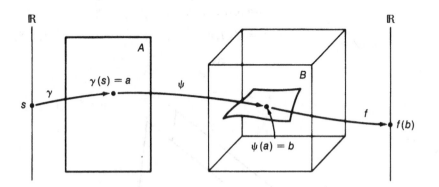

**Figure 20.2.** A parametrized curve into a two-dimensional space $A$, a map from $A$ into a three-dimensional space $B$, and a function on $B$.

The dual situation involves a parametrized curve in $A$,

$$\gamma: \mathbb{R} \to A \tag{20.3}$$

*Parametrized curve*

shown also in Figure 20.2. Again, this structure is carried along by the map $\psi$, this time in the same direction as the map. The curve $\gamma_*$,

$$\gamma_*: \mathbb{R} \to B; \ s \mapsto \psi[\gamma(s)], \tag{20.4}$$

*Push-forward*

is called the *push-forward* of the curve $\gamma$ by the map $\psi$.

Our tangent vectors and 1-forms are local linear approximations to curves and functions. It should not matter whether one approximates before or after the map $\psi$. This is really the definition of a smooth map. All of the following assumes that the maps are smooth.

For smooth maps one can also push-forward tangent vectors and pullback 1-forms, and the same star notation will be used. This cannot

[Your casual notion of smoothness will be adequate for our study. There is little physics in the distinctions that you will someday learn between $C^r$, $C^\infty$, and $C^\omega$.]

*Tangent vectors and 1-forms*

be done in the reverse direction. A tangent vector in $B$ cannot be pulled back to $A$. Think of $A$ as a subspace in $B$. There is no unique way to project a vector into this subspace, as we can see in Figure 20.3. Similarly, although a 1-form on $B$ naturally determines a 1-form on the subspace (Figure 20.4), the converse is not true.

*Calculation of pullback*     The calculation of a pullback is done using the chain rule. Let $(u, v, \ldots)$ be coordinates on $A$, and $(x, y, \ldots)$ be coordinates on $B$. These need not be linear coordinates. The map $\psi$ is represented by functions

$$x = X(u, v, \ldots),$$
$$y = Y(u, v, \ldots), \hspace{3cm} (20.5)$$
$$\text{etc.}$$

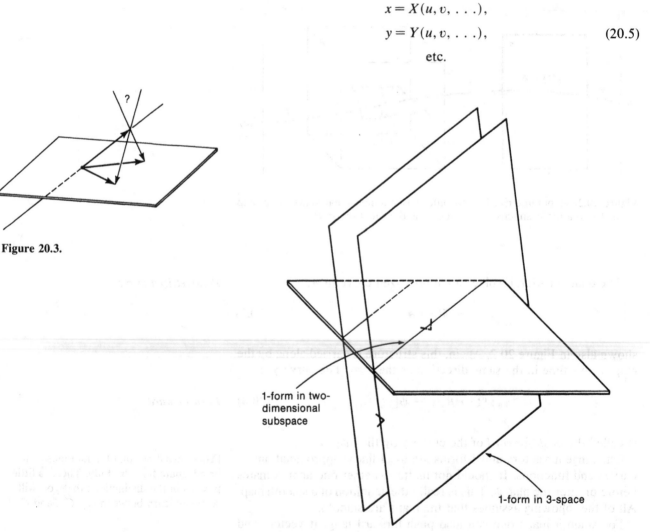

**Figure 20.3.**

1-form in two-dimensional subspace

1-form in 3-space

**Figure 20.4.**

The simplest function on $B$ is the function $f = x$. Its gradient on $B$ is the basis 1-form $dx$. The pullback of $x$ on $A$ is the function

$$x^* = X(u, v, \ldots), \qquad (20.6)$$

with gradient

$$dx^* = \frac{\partial X}{\partial u} du + \frac{\partial X}{\partial v} dv + \ldots . \qquad (20.7)$$

Since each basis 1-form can be pulled back, and since the gradient operator $d$ is linear and commutes with the map $\psi$, any 1-form can be written as a linear combination of basis 1-forms and so be pulled back.

*Example*

Let $A = \mathbb{R}^2$ and $B = \mathbb{R}^3$, and let $\psi$ be the map

$$\psi \colon (u, v) \mapsto (x, y, z) = (u, v, u^2 + v^2) \qquad (20.8)$$

shown in Figure 20.5. We have

$$\begin{aligned} x^* &= u, \\ y^* &= v, \\ z^* &= u^2 + v^2; \end{aligned} \qquad (20.9)$$

and so

$$\begin{aligned} dx^* &= du, \\ dy^* &= dv, \\ dz^* &= 2u\, du + 2v\, dv. \end{aligned} \qquad (20.10)$$

At $(0,0)$ we have $dz^* = 0$, which is reasonable, since the paraboloid is parallel to the 1-form $dz$ at that point.

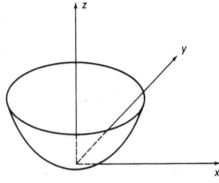

**Figure 20.5.**

The push-forward of tangent vectors is done similarly. The curve in $A$,

$$\gamma \colon s \mapsto (s, 0, \ldots), \qquad (20.11)$$

is called $\partial/\partial u$. The curve $\gamma_*$,

$$\gamma_* \colon \mathbb{R} \to B; \ s \mapsto [X(s, 0, \ldots), Y(s, 0, \ldots), \ldots], \qquad (20.12)$$

has a tangent vector which we write $(\partial/\partial u)_*$; given by the chain rule

$$\left( \frac{\partial}{\partial u} \right)_* = \frac{\partial X}{\partial u} \frac{\partial}{\partial x} + \frac{\partial Y}{\partial u} \frac{\partial}{\partial y} + \ldots . \qquad (20.13)$$

*Calculation of push-forward*

[Not to brag, but note again how natural the notation for tangent vector is.]

When $A$ and $B$ have the same dimension, the map $\psi$ can often be inverted, and vectors and 1-forms can then be carried both ways. This situation arises often in the transformation of coordinates.

[The first use of the pullback will be in the discussion of wave-packet motion in Section 29. It will also be used to find the metric on the 3-sphere in Section 39.]

## 21. Tensors

*Tenser, said the Tensor*
*Tenser, said the Tensor*
*Tension, apprehension,*
*And dissension have begun.*

ALFRED BESTER

We are now prepared to study linear operators that are more general than covectors. The covectors were maps $V \to \mathbb{R}$, and the set of all covectors was called $V^*$. What about operators $V^* \to \mathbb{R}$? There is nothing new with these. Every member of $V$, say, $v$, is such an operator, since we have

$$v: \omega \mapsto \omega \cdot v, \tag{21.1}$$

and every such operator is equivalent to some vector.

*Tensors*    The next level of complexity is to consider linear operators $V \to V^*$. Such linear operators we will call tensors. Some authors call these second-rank tensors, and use the term "tensors" to refer to any linear operator whatsoever. I will refer to this general class as general tensors. The key idea of tensor algebra is linearity. Whenever an operator acts between spaces $E$ and $F$ which have the structure of a linear vector space, we can single out the special class of linear operators. Both $V$ and $V^*$ are vector spaces. An operator $\mathscr{B}$ such that

[I will try to use the script type face for tensors. Here $v$ and $w$ are vectors, and $k$ is a real number.]

$$\mathscr{B}: V \to V^* \tag{21.2}$$

is linear if

*Linearity*
$$\mathscr{B}(kv) = k\,\mathscr{B}(v) \tag{21.3}$$

and

$$\mathscr{B}(v + w) = \mathscr{B}(v) + \mathscr{B}(w). \tag{21.4}$$

The addition on the lefthand side is the addition of vectors. The addition on the righthand side is the addition of covectors.

*Example*

The structure of Euclidean geometry is generated by a tensor. Our discussion in Section 3 showed that Euclidean geometry in canonical coordinates could be represented by a circle. The circle can be used as an operator, call it $\mathscr{E}$, that maps vectors into covectors. We do it as shown in Figure 21.1. Take any vector lying outside the unit circle. Draw tangents from the vector, call it $a$, to the circle. Draw the line through the two points of tangency. This line is to be the unit contour line of our covector.

Is this operation a tensor? Yes, if it is linear, and it is. We must check first that

$$\mathscr{E}(ka) = k\mathscr{E}(a) \tag{21.5}$$

and then also that

$$\mathscr{E}(a + b) = \mathscr{E}(a) + \mathscr{E}(b). \tag{21.6}$$

The first is easily shown by using the similar triangles in Figure 21.2. This scaling law can be used to allow our circle to act on vectors which lie inside the circle: first scale the vector up; operate; then scale the covector down.

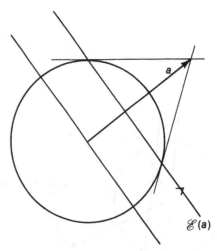

**Figure 21.1.**

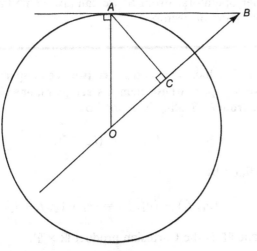

**Figure 21.2.** Triangles $OAC$ and $OBA$ are similar, hence the product $OC \times OB$ is constant.

The second linearity condition, Equation 21.6, is rather difficult to prove if one uses ordinary geometry, but the numerical techniques of the next section will handle it easily. You should appreciate the content of the equation even if the direct proof is not instructive. In Figure 21.3 we sketch the operations indicated by the lefthand side of the equation. In Figure 21.4 we sketch those on the righthand side. These should make it clear that Equation 21.6 is nontrivial.

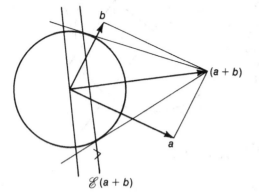

**Figure 21.3.**

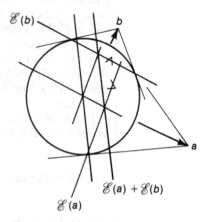

**Figure 21.4.**

Suppose that we have some tensor $\mathscr{B}$. Associated with it are a number of other operators. From $\mathscr{B}$ we can always find another operator that acts on pairs of vectors.

---

*Cartesian Product*

**Cartesian Product: If $E$ and $F$ are two vector spaces, then $E \times F$ is the vector space whose elements are pairs, one element from $E$ and one from $F$.** Scaling is defined by

[Here $e$ is a vector in $E$, $f$ a vector in $F$, and $k$ is a real number.]

$$k(e,f) = (ke, kf), \tag{21.7}$$

and addition by

$$(e_1, f_1) + (e_2, f_2) = (e_1 + e_2, f_1 + f_2). \tag{21.8}$$

The plane $\mathbb{R}^2$ is the Cartesian product $\mathbb{R} \times \mathbb{R}$.

---

Let us call this associated operator $\overline{\mathscr{B}}$

$$\overline{\mathscr{B}}: V \times V \to \mathbb{R} \qquad (21.9)$$

and define it by

$$\overline{\mathscr{B}}: (a, b) \mapsto \mathscr{B}(a) \cdot b. \qquad (21.10)$$

Here $\mathscr{B}(a)$ is a covector, which operates on the vector $b$ to give a number. Given an operator of the type of $\overline{\mathscr{B}}$, one can reverse the procedure. Define an operator $\mathscr{B}$ by

$$\mathscr{B}: V \to V^*; \; a \mapsto \overline{\mathscr{B}}(a, \_\_). \qquad (21.11)$$

The symbol $\overline{\mathscr{B}}(a, \_\_)$ represents a partially evaluated operator. $\overline{\mathscr{B}}(a, \_\_)$ is a covector! Why? Because it will accept a vector (fill in the blank) and produce a number. It is linear if $\overline{\mathscr{B}}$ is linear.

Linearity for operators acting on Cartesian products is easily defined. One just demands linearity term by term

$$\overline{\mathscr{B}}(a + b, c) = \overline{\mathscr{B}}(a, c) + \overline{\mathscr{B}}(b,c) \qquad (21.12)$$

and

$$\overline{\mathscr{B}}(a, b + c) = \overline{\mathscr{B}}(a, b) + \overline{\mathscr{B}}(a, c). \qquad (21.13)$$

Given an operator $\mathscr{B}: V \to V^*$ we can also define an inverse, a linear operator $\mathscr{B}^{-1}: V^* \to V$, such that

$$\mathscr{B}^{-1}[\mathscr{B}(a)] = a \qquad (21.14)$$

for all vectors $a$. Not every tensor has an inverse.

We will refer to linear operators of type $V \to V^*$ and all the associated operators $V \times V \to \mathbb{R}$, $V^* \to V$, $V^* \times V^* \to \mathbb{R}$, $V \to V$, and $V^* \to V^*$ as tensors. Most books call these second-rank tensors; they also call vectors and covectors first-rank tensors, and call numbers scalars. One can define more general tensors, but few of these will be of use to us here. We will make passing use of only one higher tensor, when we discuss curvature, the Riemann tensor, of type $V \times V \times V \to V$. This is why I simplify matters and use "tensor" to mean second-rank tensor.

Associated with our tensor $\mathscr{E}$ of Euclidean geometry, there is a tensor $\overline{\mathscr{E}}: V \times V \to \mathbb{R}$. This $\overline{\mathscr{E}}$ is special in that we have

$$\overline{\mathscr{E}}(a, b) = \overline{\mathscr{E}}(b, a). \qquad (21.15)$$

[You can find an easy proof of this by using Equations 21.5 and 21.6 and some geometry.]

***Symmetric tensors***    Such a tensor is called a symmetric tensor. The idea of distance and angle of conventional geometry can be added to a vector space if one introduces a symmetric nondegenerate tensor. Nondegenerate means that the inverse exists.

For the partial evaluation of a symmetric tensor, there is no need to indicate which argument is still to come; so we will use a shorthand notation for such partial evaluations:

$$\mathscr{E} \cdot a \equiv \mathscr{E} \cdot (a, \_\_). \tag{21.16}$$

***Anisotropic materials***

[Unfortunately for this discussion, anisotropic materials are usually ignored in classes precisely because they involve tensors.]

Symmetric tensors have other uses in addition to representing metric structures, primarily in the description of anisotropic (direction-dependent) situations. The main idea of a tensor is linearity, and tensors represent linear anisotropic situations. These discussions often take place in the framework of Euclidean geometry, where it is neither convenient nor necessary to use any but the usual orthonormal coordinates.

***Hooke's Law***    *Example 1*

Suppose a force ***F*** is applied to some point in a complicated mechanical structure. If the response does not depend on direction, then for small strains Hooke's law describes the relation between the force and the displacement ***d***:

$$F = kd. \tag{21.17}$$

For a general structure, such as the one fantasized in Figure 21.5, we expect the stiffness to be different in different directions. Remember also that force is more naturally represented by a 1-form. We can combine these ideas into a stiffness tensor $\mathscr{K}$. Here I give a graphical description of $\mathscr{K}$. In the next section you will learn how to deal with such tensors analytically and numerically.

The stiffness tensor $\mathscr{K}$ will be represented by an ellipse, and the force 1-form and the displacement vector will be related by the linear generalization of our rule for metric tensors. See Figure 21.6. It is not at all obvious that the operator $\mathscr{K}$ is properly linear. This is best shown by using the analytical techniques to be developed in Section 22.

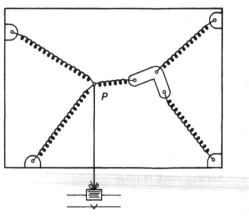

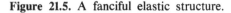

Applied force

**Figure 21.5.** A fanciful elastic structure.

***Tensor conductivity***    *Example 2*

Another natural tensor situation is the current flow (vector) in an anisotropic material like a crystal caused by an electrical potential gradient (1-form). By now you will not be surprised to find that the electrical properties of the crystal are described by a conductivity tensor $\mathscr{R}$.

The rule is similar to the above, and is sketched in Figure 21.7. Because this takes place in the framework of Euclidean geometry, the length of the current vector can be measured, and represents a current density of so many amperes per square meter in the direction of the vector.

Let me sketch some typical problems in which the conductivity tensor is used. There are two different simple situations. In one we cut out a thin sheet of the crystal and put conducting electrodes on each face, as shown in Figure 21.8. As long as the electrodes are much better conductors than the crystal, the equipotentials will be planes parallel to the electrodes. For a given conductivity tensor, the current can be found as shown in Figure 21.9. There are two components to this current. Part of it, labeled $J_\perp$ in the figure, carries a current between the plates. This current is measured by the meter shown in Figure 21.8. The component $J_\parallel$ travels up the plate. What happens to it depends on the details of the edges. The total amount of this current $J_\parallel$ is proportional not to the area of the sheet, but to its size times its thickness. For a thin sheet, this is a negligible amount of current.

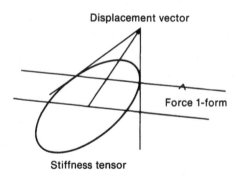

**Figure 21.6.** Use of the stiffness-tensor ellipse. For weak forces this construction does not work. Use linearity to scale up the force in that case and then scale down the displacement by the same factor.

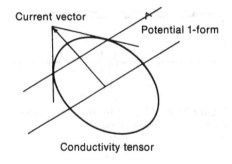

**Figure 21.7.**

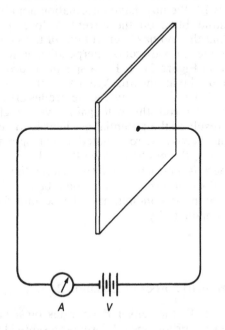

**Figure 21.8.** One way to measure the conductivity, using a thin sheet of material.

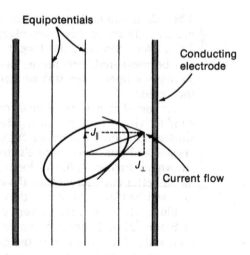

**Figure 21.9.** A detailed view of the conducting sheet. As sketched there are 4 units of potential across the sheet. The current density across the sheet is about $2\frac{3}{4}$ times the current density along the sheet.

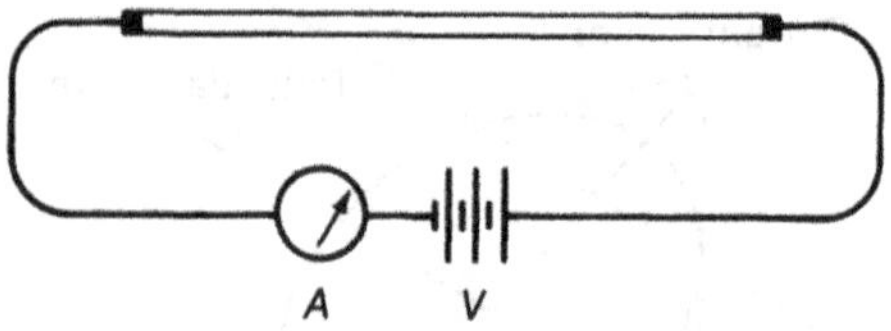

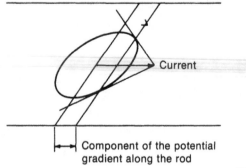

**Figure 21.10.** Measuring the conductivity using a thin rod of material.

**Figure 21.11.** Details of Figure 21.10. The voltage drop across the piece shown would be about 10 units as drawn.

A second way to proceed would be to use a long, thin rod of the material. We have the situation shown in Figure 21.10. We now have information not about the equipotentials but about the current. A close-up view, assuming that the rod has been cut out of the same material as the plate, in a direction perpendicular to the plate, is given in Figure 21.11. For a specified current the potential must be as shown. Again something happens at the ends, depending on how the electrodes are attached. For a long, thin rod, this will again have a negligible effect. The result of the potential gradient is to generate a net voltage across the rod related to its length and the component of the gradient along the rod.

There is a pleasing duality between these two experimental situations. The first would be practical for materials of poor conductivity, the second for ones with high conductivity.

PROBLEMS

21.1. (14) Give a geometric proof of Equation 21.5.

21.2. (26) In Figure 21.12 the tensor $\mathscr{E}$ operates on a vector whose head is at the point $(x,y)$, producing a 1-form represented by the lengths $l$ and $m$. Show that

$$lx = 1,$$
$$my = 1.$$

**21.3.** (16) Use the results of Problems 15.7 and 21.2 to give a proof that $\mathscr{E}$ is linear.

**21.4.** (34) Show that a procedure similar to that used for $\mathscr{E}$ but using instead a hyperbola also generates a tensor. The hyperbola

$$xy = 1$$

is a convenient one to use. Look carefully over Problems 21.1, 21.2, and 21.3.

**21.5.** (18) Many curves can be used to map vectors into covectors by using the construction with the two tangents. Show that the operator generated by the parabola

$$y = x^2$$

is not a tensor.

**21.6.** (20) Look at the map from vectors to covectors

$$\mathscr{D} : v \mapsto \mathscr{D}(v)$$

sketched in Figure 21.13. The $\mathscr{D}(v)$ are all parallel and the lengths $a$, $b$, $c$ satisfy

$$ab = c^2.$$

Show that this map is a tensor. Is it symmetric?

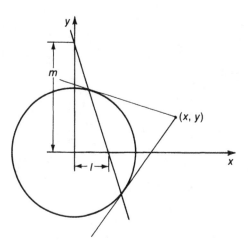

**Figure 21.12.**

[This is an important problem. Please do it carefully.]

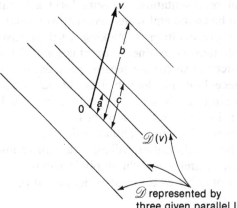

$\mathscr{D}$ represented by
three given parallel lines                       **Figure 21.13.**

**21.7.** (34) Compare the resistance, the ratio of total current to total voltage, for slabs cut horizontally and vertically from material whose conductivity tensor is given in Figure 21.7.

21.8. (17) What is the conductivity tensor for a material which consists of parallel, perfectly conducting rods in an insulating matrix?

21.19. (11) Why isn't the map

$$V \rightarrow V; \, a \mapsto a + b,$$

using a given vector $b$, a tensor?

## 22. Basis Tensors and the Tensor Product

*"Good fun," he used to say, "and
every bit as exciting as algebra."*

N. DOUGLAS

[The confusing language here is
unfortunately in universal use. Not only
are vectors tensors, but now you will
see that tensors are vectors.]

***Basis tensors***

***Tensor product***

[Read $\omega \otimes \nu$ as "omega tensor nu."]

[This centered dot is the natural
evaluation of a covector with a vector,
defined in Section 16. The cross is the
Cartesian product defined in the last
section.]

The idea behind all of tensor algebra is to exploit linearity. The linearity of the tangent vector comes from our limit process, which pushes off to infinity all the nonlinearity. The idea of a tensor operator as a linear operator should be clear to you from the definitions and the graphical examples of the last section. As with covectors, we need to develop both intuition and efficiency. The efficient manipulation of tensors requires a numerical representation. Tensors form a linear vector space; that is, they can be added and scaled. As with any vector space, we find a numerical representation by picking a set of basis elements and then writing all tensors as linear combinations of the basis tensors. We will construct a basis for our tensor space out of the basis of the original vector space. Not only does the basis of that vector space lead to a unique and natural basis for the covectors, but it will also provide basis tensors for the various tensor spaces.

We generate these basis tensors by defining a new rule of multiplication, the tensor product. This multiplication will allow us, for example, to take two covectors, say, $\omega$ and $\nu$, and "multiply" them to form a tensor $\omega \otimes \nu$. The operation of $\omega \otimes \nu$ as a tensor is defined to be

$$\omega \otimes \nu: V \times V \rightarrow \mathbb{R}; \, (a, b) \mapsto (\omega \cdot a)(\nu \cdot b) \qquad (22.1)$$

Here $a$ and $b$ are vectors, and $(\omega \cdot a)$ and $(\nu \cdot b)$ are numbers which we multiply to get our final answer. This operator is a tensor if it is linear. Since $\omega$ and $\nu$ are covectors, they are linear and therefore so is $\omega \otimes \nu$.

*Example 1* | Suppose

$$\omega = dx + dy,$$

$$v = dx - dy,$$

$$a = \frac{\partial}{\partial x},$$

[These basis vectors and covectors were defined in Section 18.]

$$b = \frac{\partial}{\partial y};$$

then

$$\omega \otimes v \cdot (a, b) = (\omega \cdot a)(v \cdot b) = -1. \quad (22.3)$$

[Here we are extending the centered dot to signify also the evaluation of a tensor on its arguments. This is a common notation.]

One can also evaluate $\omega \otimes v$ only partially. Thus $\omega \otimes v \cdot (a, \_\_)$ is a covector, and so is $\omega \otimes v \cdot (\_\_, a)$. They need not be equal.

$$\omega \otimes v \cdot (a, \_\_) = dx - dy,$$

$$\omega \otimes v \cdot (b, \_\_) = dx - dy,$$

$$\omega \otimes v \cdot (\_\_, a) = dx + dy,$$

$$\omega \otimes v \cdot (\_\_, b) = -dx - dy. \quad (22.4)$$

[Example 1 continues]

Other tensors can be constructed by multiplying together in a similar fashion two vectors or a vector and a covector. The rules should be obvious.

*Example* |

$$\frac{\partial}{\partial x} \otimes dx \cdot \left( dx + dy, 3\frac{\partial}{\partial x} - \frac{\partial}{\partial y} \right) = 3, \quad (22.5)$$

$$\frac{\partial}{\partial x} \otimes dy \cdot (3\,dx - dy, \_\_) = 3\,dy, \quad (22.6)$$

$$\frac{\partial}{\partial y} \otimes dx \cdot (3\,dx - dy, \_\_) = -dx, \quad (22.7)$$

$$\frac{\partial}{\partial y} \otimes dx \cdot (\_\_, dy) = \text{TILT!} \quad (22.8)$$

The last operation makes no sense. There is no way for one covector to act on another unless we introduce a special tensor operator.

The space of vectors such as $dx \otimes dx$ is called $V^* \otimes V^*$. The tensors in the above example were elements of the space $V \otimes V^*$.

The tensor-product space $V \otimes V^*$ is familiar to physicists in the multiplication of a column vector and a row vector to give a matrix, as in

$$\begin{pmatrix} 1 \\ 0 \\ 0 \end{pmatrix}(1 \ 1 \ 0) = \begin{pmatrix} 1 & 1 & 0 \\ 0 & 0 & 0 \\ 0 & 0 & 0 \end{pmatrix} \tag{22.9}$$

and also in the multiplication of quantum states to give an operator

$$|\psi> <\psi'|.$$

***Linear structure of tensors***    The tensors formed by the tensor product *generate* a vector space. They are not themselves a complete vector space. Not every tensor in, say, $V^* \circ V^*$ can be written as the product of two covectors. This is obvious for matrices. Not every matrix can be written as the outer product of two vectors. There are $n^2$ numbers in a matrix and only $2n$ numbers in a pair of vectors. However, every tensor can be written as a sum of such tensor products, and this is what we mean when we say that tensor products generate a vector space. The laws of scaling and addition for the tensor product differ from the laws of the Cartesian product, given in the last section. Scaling obeys

$$k(\omega \otimes \nu) = (k\omega) \otimes \nu = \omega \otimes (k\nu), \tag{22.10}$$

while addition obeys

$$\omega \otimes (\alpha + \beta) = \omega \otimes \alpha + \omega \otimes \beta. \tag{22.11}$$

One can derive these rules from linearity and the definitions.

***Example***    To derive the scaling law, note that linearity demands that

$$[k(\omega \otimes \nu) \cdot (a,b) = k[\omega \otimes \nu \cdot (a,b)], \tag{22.12}$$

which, by our definition,

$$= k[(\omega \cdot a)(\nu \cdot b)]$$
$$= (k\omega \cdot a)(\nu \cdot b) \tag{22.13}$$
$$= (\omega \cdot a)(k\nu \cdot b).$$

These rules make many of the above parentheses unnecessary.

***Coordinate basis***    We are now in a position to construct a basis for any of the tensor spaces. In two dimensions, with a basis for the covector space given by $dx$ and $dy$, the basis for $V^* \otimes V^*$ is given by four basis tensors:

$$dx \otimes dx,$$
$$dx \otimes dy,$$
$$dy \otimes dx,$$
$$dy \otimes dy.$$

(22.14)

A basis for the space $V \otimes V^*$ is

$$\frac{\partial}{\partial x} \otimes dx,$$

$$\frac{\partial}{\partial x} \otimes dy,$$

$$\frac{\partial}{\partial y} \otimes dx,$$

$$\frac{\partial}{\partial y} \otimes dy,$$

(22.15)

and so on. A basis for the symmetric tensors in $V^* \otimes V^*$ is

[See Equation 21.15 for the definition of a symmetric tensor.]

$$dx \otimes dx,$$
$$dx \otimes dy + dy \otimes dx,$$
$$dy \otimes dy,$$

(22.16)

and the space of antisymmetric tensors in $V^* \otimes V^*$ is one-dimensional with a single basis element,

$$dx \otimes dy - dy \otimes dx.$$

(22.17)

***Example***

Let us verify that the Euclidean metric tensor can be written in this basis:

[This example shows how to translate between graphical and algebraic arguments.]

$$\mathscr{E} = dx \otimes dx + dy \otimes dy.$$

(22.18)

This is clearly a linear operator, by construction. We must show that it is equivalent to the graphical construction given earlier.

First, note that the vectors

$$b = \cos \theta \, \frac{\partial}{\partial x} + \sin \theta \, \frac{\partial}{\partial y}$$

(22.19)

lie on the unit circle for all values of the parameter $\theta$. Also note that the vector

$$b_\perp \equiv -\sin \theta \, \frac{\partial}{\partial x} + \cos \theta \, \frac{\partial}{\partial y}$$

(22.20)

is always perpendicular to $b$. A 1-form $\beta$ will be tangent to the unit circle at $b$ if we have

$$\beta \cdot b = 1, \qquad (22.21)$$

$$\beta \cdot b_\perp = 0, \qquad (22.22)$$

as sketched in Figure 22.1. This determines

$$\beta = \cos \theta \, dx + \sin \theta \, dy. \qquad (22.23)$$

Now take any vector $a$ lying along the $x$-axis. Because of symmetry this can always be arranged. We now show that our graphical construction of a 1-form from $a$ as given in Section 21 and the operation $\mathscr{E} \cdot a$ agree. Let

$$a = k\frac{\partial}{\partial x}, \qquad (22.24)$$

where $k \geq 1$.

Our graphical construction asks us to draw tangents to the unit circle passing through $a$. These can be taken to be the unit contour lines of 1-forms $\beta_\pm$ in the form of Equation 22.23, which satisfy

$$\beta_\pm \cdot a = 1. \qquad (22.25)$$

If we pick $\theta$ such that

$$k \cos \theta = 1, \qquad (22.26)$$

then

$$\beta_\pm = \cos \theta \, dx \pm \sin \theta \, dy \qquad (22.27)$$

gives us the two 1-forms we need. They intersect the unit circle at the vectors

$$b_\pm = \cos \theta \, \frac{\partial}{\partial x} \pm \sin \theta \, \frac{\partial}{\partial y}. \qquad (22.28)$$

These are shown in Figure 22.2
Finally, we construct the 1-form $\alpha$ such that

$$\alpha \cdot b_\pm = 1, \qquad (22.29)$$

as shown in Figure 22.3. Clearly if we have

$$\alpha = k\, dx, \qquad (22.30)$$

then

$$\alpha \cdot b_\pm = k \cos \theta = 1. \qquad (22.31)$$

Thus our graphical construction maps

$$k\frac{\partial}{\partial x} \mapsto k\, dx. \qquad (22.32)$$

[See Problem 22.10 for the symmetry of $\mathscr{E}$ as defined in Equation 22.18.]

[The shorthand notation for partial evaluation was defined on page 118.]

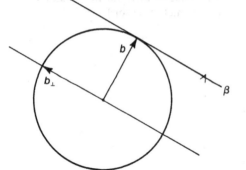

**Figure 22.1.**

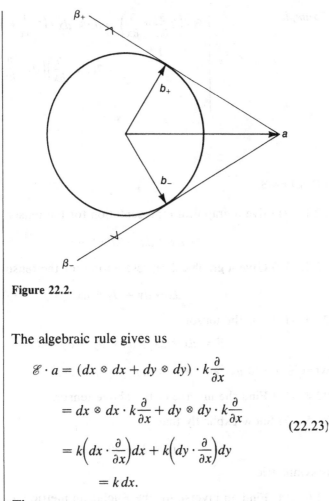

**Figure 22.2.**

The algebraic rule gives us

$$\mathscr{E} \cdot a = (dx \otimes dx + dy \otimes dy) \cdot k\frac{\partial}{\partial x}$$

$$= dx \otimes dx \cdot k\frac{\partial}{\partial x} + dy \otimes dy \cdot k\frac{\partial}{\partial x} \qquad (22.23)$$

$$= k\left(dx \cdot \frac{\partial}{\partial x}\right)dx + k\left(dy \cdot \frac{\partial}{\partial x}\right)dy$$

$$= k\,dx.$$

They agree.

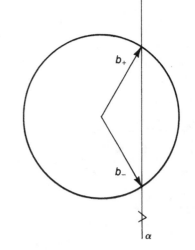

**Figure 22.3.**

A shorthand notation for metric tensors, symmetric tensors of the space $V^* \otimes V^*$, is generally used which omits the $\otimes$ symbol. One writes the Euclidean metric

$$\mathscr{E} = dx^2 + dy^2 \qquad (22.34)$$

and means by this

$$\mathscr{E} = dx \otimes dx + dy \otimes dy. \qquad (22.35)$$

The tensors defined here can also operate on other tensors as well as on pairs of vectors. An example will make this obvious.

$dx^2$

[In historical fact, this is really the coopting of the old notation by the new. Recall Christianity coopting the pagan holidays.]

***Example***

$$\mathscr{E} \cdot \left( 2 \frac{\partial}{\partial x} \otimes \frac{\partial}{\partial x} \right) = dx \otimes dx \cdot \left( 2 \frac{\partial}{\partial x} \otimes \frac{\partial}{\partial x} \right) + 0$$

$$= 2 \left( dx \cdot \frac{\partial}{\partial x} \right) \left( dx \cdot \frac{\partial}{\partial x} \right) \qquad (22.36)$$

$$= 2.$$

PROBLEMS

22.1. (7) Give a graphical representation for the tensor

$$dx \otimes dx + 2 \, dy \otimes dy.$$

22.2. (10) Give a graphical representation for the tensor

$$dx \otimes dy + dy \otimes dx.$$

22.3. (13) For the tensor

$$\mathscr{F} = dx \otimes dx + dx \otimes dy + dy \otimes dx,$$

what is $\mathscr{F} \cdot (\partial/\partial x)$ and $\mathscr{F} \cdot (\partial/\partial y)$?

22.4. (14) Find the inverse of the above tensor.

22.5. (7) Show explicitly that

$$dx \otimes dy + dy \otimes dx$$

is symmetric.

22.6. (14) Find an inverse for the Euclidean metric tensor.

22.7. (39) Invent a graphical representation for antisymmetric tensors in two, three, and four dimensions.

22.8. (13) Describe numerically the conductivity tensor shown in Figure 21.7.

22.9. (10) Show from Equation 22.18 that $\mathscr{E}$ is symmetric.

22.10. (16) Show that $\mathscr{E}$ has the same form when written in terms of rotated basis vectors

$$dx' = \cos \phi \, dx + \sin \phi \, dy,$$
$$dy' = -\sin \phi \, dx + \cos \phi \, dy.$$

## 23. Minkowski Spacetime

We can now reap the benefits of the tensor algebra that we have set up. We can finally translate the hyperbola of special relativity into mathematical language. It turns out to be a simple notion. The mathematical structure of special relativity is equivalent to that generated by the metric tensor $\mathcal{N}$:

*Special relativity metric tensor*

$$\mathcal{N} = -dt^2 + dx^2 = -dt \otimes dt + dx \otimes dx \qquad (23.1)$$

or, in four dimensions,

$$\mathcal{N} = -dt^2 + dx^2 + dy^2 + dz^2. \qquad (23.2)$$

Looking at it this way makes it easy for us to generalize special relativity to spacetimes that include gravity. We now show how this metric tensor is equivalent to our hyperbola, and along the way find a simple way to look at the Doppler-shift formula of Section 13.

In Section 22 we showed that the metric tensor of Euclidean geometry could be represented by its unit circle, the vectors $v$ such that

[I am not going to switch from the symbol $\mathcal{G}$ for a clock-rate structure to the $ds^2$ in common use. For us $d$ has a technical meaning that is not found in $ds^2$.]

*Metric figures*

$$\mathcal{E}(v, v) = 1. \qquad (23.3)$$

Such a representation works for any symmetric tensor. For the metric tensor $\mathcal{N}$ we have two separate families of unit vectors,

$$\mathcal{N} \cdot (v, v) = +1 \qquad (23.4)$$

and

$$\mathcal{N} \cdot (v, v) = -1. \qquad (23.5)$$

The negative branch will be our familiar hyperbola. The vector $v$ from the origin to the point $(t_0, x_0)$ has components

$$v = t_0 \frac{\partial}{\partial t} + x_0 \frac{\partial}{\partial x}, \qquad (23.6)$$

[Equation 23.6 says that $v$ is the tangent to the curve $x = x_0 s$, $t = t_0 s$, that for $s = 1$ passes through the point $(t_0, x_0)$.]

and we have

$$\mathcal{N} \cdot (v, v) = -t_0^2 + x_0^2. \qquad (23.7)$$

Our curve of vectors satisfying Equation 23.5, which we will call the timelike hyperbola, is given by

$$t_0^2 - x_0^2 = 1. \tag{23.8}$$

It is our familiar hyperbola.

The metric $\mathcal{N}$ also maps vectors into covectors. Let us define $\alpha$ by

$$\alpha \equiv \mathcal{N} \cdot a. \tag{23.9}$$

Then if $a$ is on the timelike hyperbola, we have

$$\alpha \cdot a = -1. \tag{23.10}$$

Look at a nearby point on the timelike hyperbola $a + \varepsilon$. Here $\varepsilon$ is a small vector and we will neglect its square, taking the limit $\varepsilon \to 0$. If $a + \varepsilon$ is also on the timelike hyperbola, then we have

$$\mathcal{N} \cdot (a + \varepsilon, a + \varepsilon) = -1, \tag{23.11}$$

and using linearity we have

$$-1 + \mathcal{N} \cdot (a, \varepsilon) + \mathcal{N} \cdot (\varepsilon, a) = -1. \tag{23.12}$$

Now $\mathcal{N}$ is symmetric,

$$\mathcal{N} \cdot (a, \varepsilon) = \mathcal{N} \cdot (\varepsilon, a); \tag{23.13}$$

hence we have

$$\mathcal{N} \cdot (a, \varepsilon) = 0. \tag{23.14}$$

That is,

$$\alpha \cdot \varepsilon = 0. \tag{23.15}$$

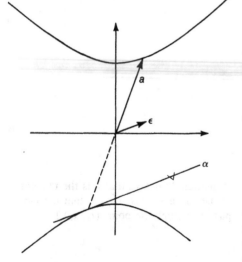

**Figure 23.1.**

By construction, the vector $\varepsilon$ is tangent to the hyperbola, and the above equation states that the contour lines of $\alpha$ are parallel to $\varepsilon$. Furthermore, Equation 23.10 states that the minus-one contour of $\alpha$ passes through $a$. Note the minus sign! The action of $\mathcal{N}$ on $a$ is thus given graphically by Figure 23.1. The action on vectors having the same direction as $a$ but different magnitudes can be found by scaling.

***Example***
$$\mathcal{N} \cdot \frac{\partial}{\partial t} = -dt. \qquad (23.16)$$

Spacelike vectors never intersect the timelike hyperbola. They do intersect the spacelike hyperbola.

$$x_0^2 - t_0^2 = 1. \qquad (23.17)$$

The same construction works but now with the other sign; see Figure 23.2.

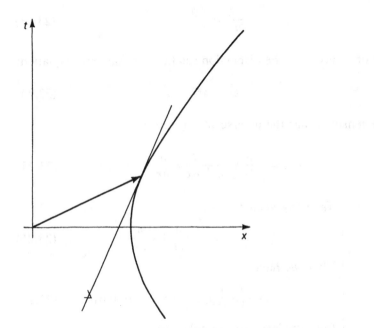

**Figure 23.2.**

***Example***
$$\mathcal{N} \cdot \frac{\partial}{\partial x} = dx. \qquad (23.18)$$

The 4-vector dot product can be easily written using the $\mathcal{N}$ metric:

$$\mathcal{N} \cdot (a,b) = a \cdot b \qquad (23.19)$$

[Note that the boldface dot on the righthand side stands for the 4-vector scalar product.]

***Proof*** Write the vectors

$$a = a^t \frac{\partial}{\partial t} + a^x \frac{\partial}{\partial x} + \ldots, \qquad (23.20)$$

$$b = b^t \frac{\partial}{\partial t} + \ldots; \qquad (23.21)$$

[Note the powerful mathematical symbol "$\ldots$"; it means "any fool can see how this goes."]

then

$$\mathcal{N} \cdot (a,b) = (-dt \circ dt + \ldots) \cdot (a^t \frac{\partial}{\partial t} + \ldots, b^t \frac{\partial}{\partial t} + \ldots)$$

(23.22)

$$= -a^t b^t + a^x b^x + \ldots.$$

*Wave equation*    The geometry of special relativity is intimately involved with the wave equation

$$\frac{\partial^2 \psi}{\partial t^2} = \frac{\partial^2 \psi}{\partial x^2}.$$

(23.23)

One use of $\mathcal{N}$ involves the dispersion relation for that wave equation:

$$\omega^2 = k^2.$$

(23.24)

Now the tensor $\mathcal{N}$ has the inverse $\mathcal{N}^{-1}$ given by

$$\mathcal{N}^{-1} = -\frac{\partial}{\partial t} \otimes \frac{\partial}{\partial t} + \frac{\partial}{\partial x} \otimes \frac{\partial}{\partial x}.$$

(23.25)

*Proof*    Take any vector

$$a = u \frac{\partial}{\partial t} + v \frac{\partial}{\partial x}.$$

(23.26)

Then we have

$$\mathcal{N} \cdot \left( u \frac{\partial}{\partial t} + v \frac{\partial}{\partial x} \right) = -u \, dt + v \, dx,$$

(23.27)

from the last two examples; so

$$\mathcal{N}^{-1} \cdot (-u \, dt + v \, dx)$$

$$= -\frac{\partial}{\partial t} \otimes \frac{\partial}{\partial t} \cdot (-u \, dt) + \frac{\partial}{\partial x} \otimes \frac{\partial}{\partial x} \cdot (v \, dx)$$

$$= u \frac{\partial}{\partial t} + v \frac{\partial}{\partial x}.$$

(23.28)

Thus $\mathcal{N}^{-1}$ is really the inverse of $\mathcal{N}$.

Using this inverse we can write the dispersion relation, Equation 23.24, as

*Dispersion relation*                     $$\mathcal{N}^{-1} \cdot (d\theta, d\theta) = 0,$$

(23.29)

which shows the natural relation between the special-relativity metric tensor $\mathcal{N}$ and the wave equation. One can also show that the group-velocity vector $\sigma$ associated with a phase gradient $d\theta$ is given by

$$\sigma = \mathcal{N}^{-1} \cdot d\theta. \tag{23.30}$$

[Wait for the discussion in Section 27 before trying to prove this.]

These relations are shown in Figures 23.3 and 23.4 for one and two spatial dimensions.

Electromagnetism obeys a wave equation which is the vector generalization of the scalar wave equation given above. It has the same dispersion relation. The above discussion applies as well to light waves. In the uniform spacetime of special relativity, a wave propagates without changing $k$ or $\omega$, and if you accept this you can see naturally how the Doppler-shift expression of Section 13 arises.

*Electromagnetism*

The geometric disposition of the wave crests is given by the phase gradient 1-form $d\theta$. The observed frequency is the rate at which an observer sees wave crests pass (see Figure 23.5). If the observer has a 4-velocity $\lambda$, then this rate is $(d\theta \cdot \lambda)/2\pi$. Two observers moving with different velocities see a relative rate

*Doppler shift*

[See Figure 13.5.]

$$\frac{d\theta \cdot \lambda}{d\theta \cdot \lambda'} \tag{23.31}$$

and using the inverse of Equation 23.30,

$$d\theta = \mathcal{N} \cdot \sigma, \tag{23.32}$$

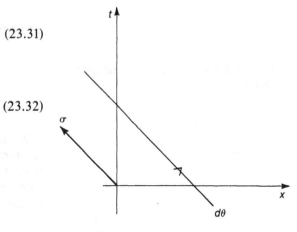

Figure 23.3.

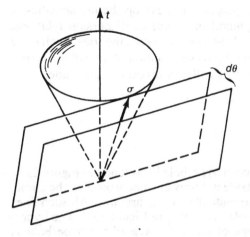

Figure 23.4.

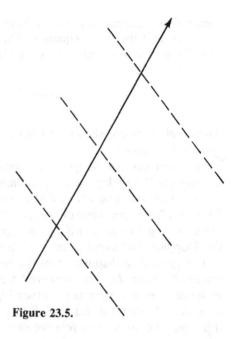

**Figure 23.5.**

we have a relative rate

[Note the 4-vector dot on the righthand side.]

$$\frac{\mathcal{N} \cdot (\sigma, \lambda)}{\mathcal{N} \cdot (\sigma, \lambda')} = \frac{\sigma \cdot \lambda}{\sigma \cdot \lambda'}. \tag{23.33}$$

Thus we have recovered our old expression.

We have succeeded in our program to develop the mathematics of clock-rate structures. The general ones will be dispersion relations. Simple ones define tensors and lead to very elegant theories. Such is the case with special relativity and the wave equation. A four-dimensional vector space with the metric $\mathcal{N}$ is called Minkowski spacetime.

## PROBLEM

[These envelope constructions are surprisingly ubiquitous.]

23.1. (31) One way to construct a hyperbola is sketched in Figure 23.6. Pick a circle and a point $P$ outside it. Draw any line from $P$ to the circle, and there draw the line $L$ perpendicular to the first line. All such lines $L$ will be tangent to a hyperbola, as shown in Figure 23.7. The hyperbola is said to be the envelope of the lines. Regard the hyperbola as representing the Minkowski metric $\mathcal{N}$, and the tangent lines $L$ as representing 1-forms. Show that the above construction is valid.

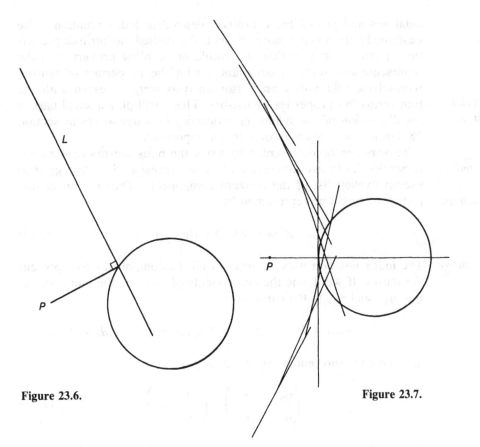

**Figure 23.6.**                          **Figure 23.7.**

## 24. Index Notation

> *"Ignorant people, like Faraday, naturally think*
> *in vectors. They may know nothing of their formal*
> *manipulations, but if they think about vectors,*
> *they think of them as vectors, that is, directed*
> *magnitudes. No ignorant man could or would*
> *think about the three components of a vector*
> *separately, and disconnected from one another."*
>
> O. HEAVISIDE

In the spirit of the epigraph, I have emphasized the intrinsic properties
of tensors. Both our definition of tensors as linear operators and
our graphical representations were coordinate independent. For cal-

culations and proofs the coordinate-dependent index notation to be described here is often more efficient. We studied the intrinsic properties of tensors first so that you would not confuse properties of the representation, such as covariance, with the properties of tensors themselves. Old books on tensor analysis were obsessed with the transformation properties of tensors. These will play a small part in the discussion of the symmetry properties of water waves in Section 28, but for us they are of only minor importance.

[The book of Synge and Schild is good and useful, in the index-shuffling style of tensor analysis.]

Tensors can be represented by using the basis tensors constructed in Section 23 in two different ways. One representation strategy is to use an explicit list of the nonzero components. Thus our spacetime metric $\mathcal{N}$ would be represented by

*Representations*

*Explicit list*

$$\mathcal{N} = -dt \otimes dt + dx \otimes dx. \tag{24.1}$$

*Array*   The index notation uses an array of all the components to represent the tensor. If we define the components of $\mathcal{N}$ to be the numbers $\eta_{tt}$, $\eta_{tx}$, $\eta_{xt}$, and $\eta_{xx}$ in the expansion

$$\mathcal{N} = \eta_{tt}dt \otimes dt + \eta_{tx}dt \otimes dx + \eta_{xt}dx \otimes dt + \eta_{xx}dx \otimes dx, \tag{24.2}$$

then we can represent $\mathcal{N}$ by the array

$$\begin{pmatrix} \eta_{tt} & \eta_{tx} \\ \eta_{xt} & \eta_{xx} \end{pmatrix} = \begin{pmatrix} -1 & 0 \\ 0 & 1 \end{pmatrix}. \tag{24.3}$$

The explicit list of nonzero components is more efficient for vectors and tensors that have only a few nonzero components. We will use it most of the time, and we could have used it exclusively. Since the index notation is in common use, and is more efficient for some problems, relativity students must learn it too.

The index notation represents the components of a vector or a tensor in an expansion that is based on our coordinate-derived basis vectors and that uses superscript and subscript indices for the summations. The placement of the indices will indicate the type of vector or tensor. Tangent vectors and their tensor products are represented by components with superscripts. Covectors and their products are represented with subscripts.

*Subscripts and superscripts*

*Example*   The expansion of a tangent vector $\lambda$ will be written

$$\lambda = \lambda^t \frac{\partial}{\partial t} + \lambda^x \frac{\partial}{\partial x} + \dots. \tag{24.4}$$

The covector $\alpha$ will be expanded

$$\alpha = \alpha_t\, dt + \alpha_x dx + \ldots . \qquad (24.5)$$

To handle these summations we use *dummy indices* to indicate the general superscript or subscript. We will use Greek letters for our spacetime indices, and will often not bother to indicate the index set explicitly. We slightly abuse the notation by writing the basis covectors in general as

*Dummy indices*

$$dx^\mu, \qquad \mu = \{x,y,z, \text{ and } t\},$$

and similarly write the general basis vector as $\partial/\partial x^\mu$. Our expansions are thus written

*Basis vectors and covectors*

$$\lambda = \sum_\mu \lambda^\mu \frac{\partial}{\partial x^\mu} \qquad (24.6)$$

for a tangent vector, and

$$\alpha = \sum_\mu \alpha_\mu\, dx^\mu \qquad (24.7)$$

for a 1-form. Some books are pedantic at this point, and demand that the coordinates be called $x^1$, $x^2$, and so on, so that the summation will be over numbers. This is no aid to the memory, and I prefer to follow a convention that sums the dummy index over whatever index set is convenient.

An equation with a free dummy index is to be true for all values of the index. Such a free index must occur in every term in the same upper or lower position; otherwise the linearity condition for tensors would fail.

*Free indices*

**Example 1**

The equation

$$a^\mu b^\nu = h^{\mu\nu} \qquad (24.8)$$

means that

$$a^t b^t = h^{tt}, \qquad (24.9)$$

and also

$$a^t b^x = h^{tx}, \qquad (24.10)$$

and so on, for a total of sixteen equations if we are in four dimensions. Please appreciate how compact this notation is.

*Example 2*

What is

$$dx^\mu \cdot \frac{\partial}{\partial x^\nu}?$$

These are covectors operating on vectors, so they must be numbers. The value of the number depends on the indices $\mu$ and $\nu$. Clearly we have

$$dx^\mu \cdot \frac{\partial}{\partial x^\nu} = \begin{cases} 1 \text{ if } \mu = \nu, \\ 0 \text{ if } \mu \neq \nu, \end{cases} \qquad (24.11)$$

since the $dx^\mu$ and the $(\partial/\partial x^\nu)$ are dual. For example

$$dx \cdot \frac{\partial}{\partial x} = 1 \qquad (24.12)$$

and

$$dx \cdot \frac{\partial}{\partial y} = 0. \qquad (24.13)$$

[A good strategy for approaching a tensor equation is to write out a few components.]

**Summation convention**

Finally, in tensor equations the summations will always occur over exactly two appearances of a dummy index, once up and once down. Great notational clarity results from a convention that a summation is automatically implied whenever an index is repeated. An equation in which a dummy index appears once in every term is to be true for every possible value of that dummy index. An equation in which a dummy index appears twice in every term is to be summed over that index. (Three or more appearances indicate that you have made an error.) Since any symbol can be used for a dummy index, one can freely change any dummy index to any convenient symbol, as long as that index is changed in every term in the equation. The most common error is to have four appearances of the same dummy index when one summation is substituted into another without proper relabeling. We will note the possibility of this error the first few times it comes up.

**Evaluation**

The index notation provides a compact notation for all the tensor operations that we have studied. The evaluation of a 1-form $\omega$ on a vector $v$ is defined by

$$\omega \cdot v \equiv (\omega_\mu dx^\mu) \cdot \left(v^\alpha \frac{\partial}{\partial x^\alpha}\right). \qquad (24.14)$$

Using linearity we can pull the summations through the evaluation to find

$$\omega \cdot v = \omega_\mu v^\alpha \left[ dx^\mu \cdot \frac{\partial}{\partial x^\alpha} \right]. \qquad (24.15)$$

The preceding example showed that the quantity in square brackets vanishes unless the indices $\mu$ and $\alpha$ are the same. This reduces our double sum to a single sum "down the diagonal":

$$\omega \cdot v = \omega_t v^t + \omega_t v^x + \cdots \tag{24.16}$$
$$\omega_x v^t + \omega_x v^x + \cdots$$
$$\cdots,$$

and this sum down the diagonal can be written

$$\omega \cdot v = \omega_\mu v^\mu. \tag{24.17}$$

Terms such as $dx^\mu \cdot (\partial/\partial x^\nu)$ come up quite often, and for simplicity we define a special array of numbers similar to the unit matrix, $\delta^\mu_\nu$, according to

*Kronecker delta*

$$\delta^\mu_\nu \equiv \begin{cases} 1 \text{ if } \mu = \nu \\ 0 \text{ if } \mu \neq \nu \end{cases}. \tag{24.18}$$

This array $\delta^\mu_\nu$ is called the Kronecker delta.

**Example** | We have

$$\omega_\mu \delta^\mu_\nu = \omega_\nu \tag{24.19}$$

as can be seen by writing out the $\nu = x$ component of this equation:

$$\omega_x \, \delta^x_x + \omega_t \, \delta^t_x + \cdots = \omega_x. \tag{24.20}$$

A tensor $\mathscr{G}$ of type $V^* \otimes V^*$ can be written

$$\mathscr{G} = g_{\mu\nu} dx^\mu \otimes dx^\nu. \tag{24.21}$$

[It is not necessary to use the same letter for the tensor and for its components, although doing so is usually convenient.]

The operation $\mathscr{G}: V \times V \to \mathbb{R}$ can be worked out using linearity as follows:

*Metric tensor*

$$\mathscr{G} \cdot (a, b) = (g_{\mu\nu} dx^\mu \otimes dx^\nu) \cdot \left( a^\alpha \frac{\partial}{\partial x^\alpha}, b^\beta \frac{\partial}{\partial x^\beta} \right)$$

$$= g_{\mu\nu} a^\alpha b^\beta \left( dx^\mu \cdot \frac{\partial}{\partial x^\alpha} \right) \left( dx^\nu \cdot \frac{\partial}{\partial x^\beta} \right) \tag{24.22}$$

$$= g_{\mu\nu} a^\alpha b^\beta \delta^\mu_\alpha \delta^\nu_\beta$$

$$= g_{\mu\nu} a^\mu b^\nu.$$

[Here is an example of the possibility of a labeling error. The dummy indices on the $a$ and $b$ expansions must be different and also different from $\mu$ and $\nu$.]

*Partial evaluation*    The operation of partial evaluation $\mathscr{G} \cdot a$ is represented by

[Recall that $\mathscr{G} \cdot a$ is short for $\mathscr{G} \cdot (a, \underline{\quad})$.]

$$\mathscr{G} \cdot a = g_{\mu\nu} a^{\mu} \, dx^{\nu}, \tag{24.23}$$

as can easily be shown.

*Tensor product*    The components of the tensor product are given by the product of the components

$$a \otimes b = a^{\mu} b^{\nu} \frac{\partial}{\partial x^{\mu}} \otimes \frac{\partial}{\partial x^{\nu}}. \tag{24.24}$$

The symmetry of a metric tensor $\mathscr{G} = g_{\mu\nu} dx^{\mu} \otimes dx^{\nu}$ is reflected in the symmetry of its components

$$g_{\mu\nu} = g_{\nu\mu}. \tag{24.25}$$

What about the inverse of a symmetric tensor? If we write the metric *Inverse* inverse

$$\mathscr{G}^{-1} \equiv g^{\mu\nu} \frac{\partial}{\partial x^{\mu}} \otimes \frac{\partial}{\partial x^{\nu}}, \tag{24.26}$$

then the relation

$$\mathscr{G}^{-1}(\mathscr{G} \cdot a) = a \tag{24.27}$$

is written

$$g^{\mu\alpha} g_{\alpha\beta} a^{\beta} = a^{\mu} = \delta^{\mu}_{\beta} a^{\beta} \tag{24.28}$$

Since this is true for all $a^{\beta}$, we must have

$$g^{\mu\alpha} g_{\alpha\beta} = \delta^{\mu}_{\beta}. \tag{24.29}$$

In four dimensions these are 16 algebraic equations for the 16 components of $\mathscr{G}^{-1}$, the $g^{\mu\nu}$.

*Tensors: operators or sets of numbers*    Relativists actually have a schizophrenic view of the index notation. Despite its formal definition, they usually treat the indexed symbols as just other names for the tensors. They are particularly convenient names, since the number and position of the indices tell us what type of tensor or vector it is. If some evaluations are to be made, these are shown by the repeated indices. Thus we will usually call $g_{\mu\nu}$ a tensor, rather than the components of a tensor. The failing of classical tensor analysis is not that it uses components, but that it considers $v^{\mu}$ and $v_{\mu}$

to be the same thing. Doing so leads to no errors when there is a metric given, but even then it is not particularly convenient.

**Example** | If you think of tensors as arrays of numbers, then you might be tempted to write

$$\delta_\nu^\mu = \delta_\mu^\nu. \qquad (24.30)$$

Why is this equation an abomination? Suppose we try to restore the underlying tensor equation by summing over a basis. For the $\mu$-index we have either

$$\delta_\nu^\mu \frac{\partial}{\partial x^\mu} = \delta_\mu^\nu \, dx^\mu, \qquad (24.31)$$

or

$$\delta_\nu^\mu \, dx^\mu = \delta_\mu^\nu \, dx^\mu, \qquad (24.32)$$

neither of which is an acceptable sentence in our language.

PROBLEM

24.1. (15) Consider the identity map

$$I: V^* \to V^*; \ \omega \mapsto \omega.$$

Show that it is a tensor, and find its components. (Use the $\delta_\nu^\mu$ symbol.)

# 25. *Vector Fields*

So far we have developed the algebra of vectors and tensors at a single point. The next step is to discuss vector and tensor fields. A vector field is a smooth assignment of a vector to every point in spacetime, and similarly for a tensor field. The analysis of vector and tensor fields is considerably more complicated in general spacetimes than it is in special relativity. Luckily, we will not have to delve deeply into these complexities. The intuitive notion of a vector field developed in a course on electromagnetism or fluid mechanics needs only a little modification to suffice for us. What is new here is the realization that there is a close relationship between vector fields and systems of ordinary differential equations and transformations.

*Vector field*
Once we have picked coordinates on spacetime, this coordinate system provides a set of basis vectors at every point. A smooth vector field is represented by giving the components of the vector at each point as smooth functions of position.

[We postpone a careful discussion of curvilinear coordinates to Section 26.]

**Example 1**  | A smooth vector field on the plane is given by

$$v = \frac{x}{2}\frac{\partial}{\partial x} - \frac{y}{2}\frac{\partial}{\partial y} \qquad (25.1)$$

and is sketched in Figure 25.1.

*Integral curves*
Given a vector field, one can try to find a family of curves such that at every point the vector field gives the local approximation to that family of curves. A single parametrized curve whose tangent vector at every point on the curve equals the vector field at that point is called an integral curve of the vector field.

[Example 1 continues. Here $a$ is a parameter identifying different curves. We take $a > 0$.]

The family of curves

$$\gamma_a: \mathbb{R} \to \mathbb{R}^2;\ \theta \mapsto (a\cos\frac{\theta}{2}, a\sin\frac{\theta}{2}) \qquad (25.2)$$

are all integral curves of the vector field $v$ defined above. The tangent vector to $\gamma_a$ at parameter value $\theta$, called $\gamma_a$, is given by

[See Section 17, and page 93 in particular, for the derivation of this.]

$$\gamma_a = -\frac{a}{2}\sin\theta\frac{\partial}{\partial x} + \frac{a}{2}\cos\theta\frac{\partial}{\partial y}, \qquad (25.3)$$

and along $\gamma_a$ we have

$$x^2 + y^2 = a^2,$$

$$\cos\theta = \frac{x}{a}, \qquad (25.4)$$

$$\sin\theta = \frac{y}{a},$$

and so

$$\dot\gamma_a = -\frac{y}{2}\frac{\partial}{\partial x} + \frac{x}{2}\frac{\partial}{\partial y}. \qquad (25.5)$$

This shows that the $\gamma_a$ are integral curves of $v$. They are shown in Figure 25.2.

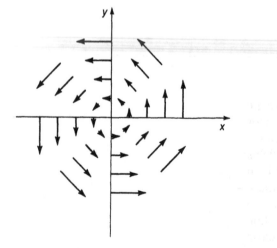

**Figure 25.1.**

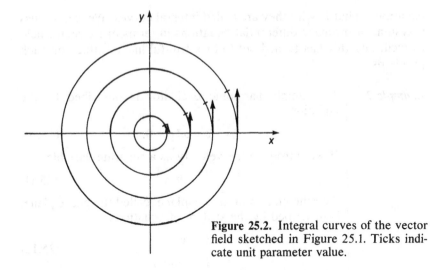

**Figure 25.2.** Integral curves of the vector field sketched in Figure 25.1. Ticks indicate unit parameter value.

How do we find integral curves? The general curve

$$\gamma: s \mapsto [X(s), Y(s), \ldots] \tag{25.6}$$

has a tangent vector

$$\dot{\gamma} = \frac{dX}{ds}\frac{\partial}{\partial x} + \frac{dY}{ds}\frac{\partial}{\partial y} + \ldots, \tag{25.7}$$

and for this to agree with a given vector field, say,

$$v = v^x\frac{\partial}{\partial x} + v^y\frac{\partial}{\partial y} + \ldots, \tag{25.8}$$

along the curve, we have

$$\frac{dX}{ds} = v^x(x, y', \ldots),$$

*Set of ordinary differential equations*

$$\frac{dY}{ds} = v^y(x, y, \ldots), \tag{25.9}$$

$$\ldots,$$

where each component $v^x$, $v^y$, $\ldots$ is a smooth function of the coordinates. We see that asking for the integral curves of a vector field is equivalent to asking for the solutions of a system of ordinary differential equations. That is why they are called integral curves. We can discuss a system of ordinary differential equations in terms of its vector field as well, and this has turned out to be a fruitful line of attack on such problems.

*Harmonic oscillator*     ***Example 2***     The simple harmonic oscillator is described by the equation

$$\ddot{x} + \omega^2 x = 0. \qquad (25.10)$$

If we introduce the velocity as a separate variable,

$$y \equiv \dot{x}, \qquad (25.11)$$

then the motion in the $x,y$-plane, called the phase plane, is described by the system of equations

$$\dot{x} = y, \qquad (25.12)$$
$$\dot{y} = -\omega^2 x.$$

This is equivalent to the vector field on the phase plane

$$v = y\frac{\partial}{\partial x} - \omega^2 x\frac{\partial}{\partial y}. \qquad (25.13)$$

Integral curves of $v$ are given by

$$\gamma_a: t \mapsto (a \cos \omega t, -a\omega \sin \omega t) \qquad (25.14)$$

as shown in Figure 25.3.

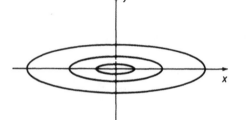

**Figure 25.3.** Integral curves drawn for $\omega = \sin 15° = 0.26$.

*Nonlinear oscillator*     ***Example 3***     A much studied nonlinear oscillator is the van der Pol oscillator, described by

$$\ddot{y} + ky - \mu\dot{y}[1 - \left(\frac{y}{d}\right)^2] = 0. \qquad (25.15)$$

This is equivalent to the vector field

$$v = \{ky - \mu x[1 - \left(\frac{y}{d}\right)^2]\}\frac{\partial}{\partial x} - x\frac{\partial}{\partial y}. \qquad (25.16)$$

*Hamiltonian mechanics*     The vector field is sketched in Figure 25.4, and a few integral curves are shown in Figure 25.5 for $k = 0.231$, $d = 0.465$, $\mu = 0.769$. The vectors are drawn one-fifth size to avoid clutter.

***Example 4***     Hamiltonian mechanics takes place on a phase space of $2n$ dimensions with coordinates

$$(q^1, q^2, \ldots, q^n, p_1, p_2, \ldots, p_n) = (q, p). \qquad (25.17)$$

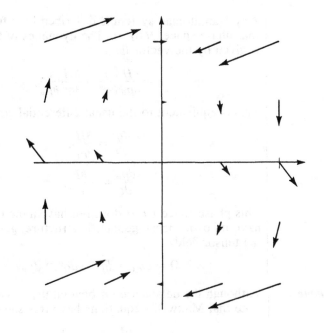

**Figure 25.4.** Van der Pol vector field.

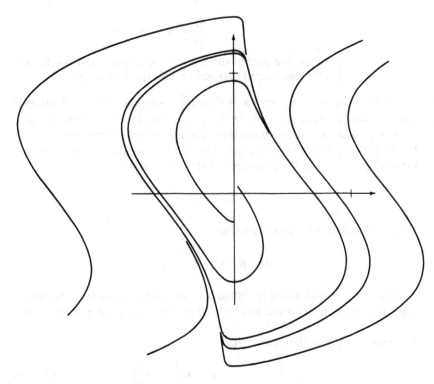

**Figure 25.5.** Integral curves for the Van der Pol vector field.

[For convenience, here we do not follow the usual practice of placing all the coordinate labels as superscripts. We use the obvious shorthand to refer to the coordinates collectively.]

Any Hamiltonian system is described by a function on that phase space, $H(q,p)$. The dynamics of the system is given by the vector field

$$v = \frac{\partial H}{\partial p_\mu} \frac{\partial}{\partial q^\mu} - \frac{\partial H}{\partial q^\mu} \frac{\partial}{\partial p_\mu}. \qquad (25.18)$$

[Summation convention!]

This is equivalent to the usual differential equations,

$$\frac{dq^\mu}{dt} = \frac{\partial H}{\partial p_\mu},$$
$$\frac{dp_\mu}{dt} = -\frac{\partial H}{\partial q^\mu}. \qquad (25.19)$$

[All the $dp_\mu$ and the $dq^\mu$ are covectors; we are not following the usual index placement.]

This phase space $(q,p)$ does not have a metric. It does have its own special geometric structure, generated by the tensor field

$$\Omega = dp_\mu \otimes dq^\mu - dq^\mu \otimes dp_\mu. \qquad (25.20)$$

*Electromagnetism*    *Example 5*

Although the details are far beyond us, it is amusing to note that Maxwell's equations have this same form:

$$\frac{\partial B}{\partial t} = -\nabla \times E,$$
$$\frac{\partial E}{\partial t} = \nabla \times B - 4\pi J. \qquad (25.21)$$

These are partial differential equations, and the vector field must be in a space of infinite dimension.

*Transformations*

[Transformations were discussed in Sections 4 and 6.]

Let us suppose that we have found a complete family of integral curves, one passing through every point. We can use these integral curves to generate a one-parameter family of transformations, $\Phi_w$, of the space as follows. For any point $p$, find the integral curve $\gamma$ passing through $p$. Let $u$ be the parameter value corresponding to $p$:

$$\gamma(u) = p. \qquad (25.22)$$

Then we define the transformation

$$\Phi_w: p \mapsto \gamma(u + w). \qquad (25.23)$$

Each point $p$ is slid along its integral curve for a parameter change w. This family of transformations is called the flow of the vector field.

*Example*

For the vector field

$$v = x\frac{\partial}{\partial y} - y\frac{\partial}{\partial x}. \qquad (25.24)$$

the flow $\Phi_w$ corresponds to a rotation of the $x,y$-plane through an angle $w$.

The vector field represents the infinitesimal version of the transformation.

*Example* | The vector field

$$v = x\frac{\partial}{\partial t} + t\frac{\partial}{\partial x} \qquad (25.25)$$

represents an infinitesimal Lorentz transformation:

$$(x,t) \mapsto (x\cosh\psi + t\sinh\psi,\ x\sinh\psi + t\cosh\psi). \qquad (25.26)$$

See Figures 25.6 and 25.7.

*Infinitesimal transformation*

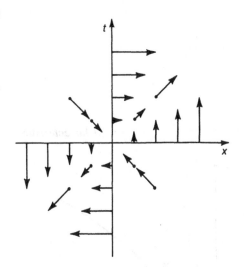

**Figure 25.6.** Vectors are drawn half-size to avoid clutter.

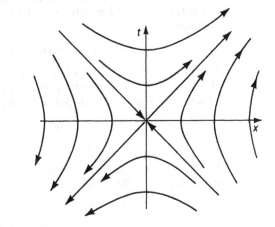

**Figure 25.7.**

An important property of a spacetime is its symmetry. We describe this symmetry by giving the transformations which leave it unchanged. It is sufficient and less complicated to give the infinitesimal transformations, that is, the vector fields.

*Symmetry*

[See Rosen for more on symmetry.]

A smooth covector field is defined similarly, and would be used to represent, for example, the phase of a wave packet in spacetime. Tensor fields are important as well. The geometry of a general spacetime will be specified by a smooth tensor field representing the metric.

The relationship between tangent vectors and ordinary differential equations comes from the role of the tangent vector as the local linear approximation to a curve. A tangent vector also acts as a linear operator on functions, and in this role it has applications to partial differential equations. We will need these in our study of dispersive wave propagation.

*Differential operators*

[This is the continuation of Example 2 in Section 17.]

*Example*

Suppose that a function $f(x, y)$ satisfies both

$$2y\frac{\partial f}{\partial x} + \frac{\partial f}{\partial y} = 0 \tag{25.27}$$

and the boundary conditions

$$f(x, 0) = e^{-x^2}. \tag{25.28}$$

Do these conditions determine $f$? How do we calculate it?

The geometric significance of Equation 25.27 is that the function $f$ is constant in the direction

$$2y\frac{\partial}{\partial x} + \frac{\partial}{\partial y}.$$

*Characteristics*

Therefore it will be constant along the integral curves of this vector field. Such integral curves are called *characteristics*. The integral curves here are the parabolas

$$\gamma_a: u \mapsto (x, y) = (u^2 + a, u), \tag{25.29}$$

where $a$ is a parameter which labels the different curves.

To find $f$ at an arbitrary point $(x_0, y_0)$ is straightforward (see Figure 25.8). Solve for $u$ and $a$:

$$u = y_0, \tag{25.30}$$

$$a = x_0 - y_0^2. \tag{25.31}$$

The curve $\gamma_a$ crosses the $x$-axis at $x = a$. The value of $f$ there is equal to the value at $(x_0, y_0)$, and so

$$f(x_0, y_0) = e^{-(x_0 - y_0^2)^2}. \tag{25.32}$$

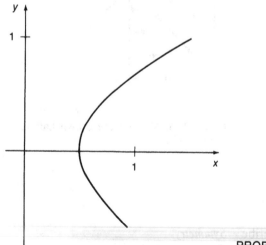

**Figure 25.8.** Line along which $f$ is constant.

## PROBLEMS

25.1. (8) Describe the transformation generated by the vector field

$$y\frac{\partial}{\partial x}.$$

25.2. (16) Find the vector field representing a uniform expansion.

25.3. (14) Find the circles that are invariant as sets of points under infinitesimal rotations given by the vector field

$$(x - 1)\frac{\partial}{\partial y} - y\frac{\partial}{\partial x}.$$

24.3. (19) Show that the hyperbola

$$t^2 - x^2 = 1$$

is invariant under the Lorentz transformation

$$x \frac{\partial}{\partial t} + t \frac{\partial}{\partial x}.$$

25.5. (38) If you have access to a programmable calculator, you can find some numerical solutions to the van der Pol equation.

25.6. (33) What special problems are posed by the partial differential equation

$$\frac{\partial f}{\partial x} + 2x \frac{\partial f}{\partial y}$$

with boundary conditions given on the *x*-axis?

# 26. Manifolds

> He had bought a large map representing the sea,
>   Without the least vestige of land:
> And the crew were much pleased when they found it to be
>   A map they could all understand.
>
> "What's the good of Mercator's North Poles and Equators,
>   Tropics, Zones, and Meridian Lines?"
> So the Bellman would cry: and the crew would reply
>   "They are merely conventional signs!"
>
> "Other maps are such shapes, with their islands and capes!
>   But we've got our brave Captain to thank"
> (So the crew would protest) "that he's bought us the best—
>   A perfect and absolute blank!"
>
> LEWIS CARROLL

We have worked so far as if spacetime were a vector space. Our definitions, however, were framed to accommodate curvilinear coordinates, such as polar coordinates, and the assumption of an underlying vector space has not really been used. The general spacetimes of Einstein's general relativity are not modeled in a vector space as special relativity is. We will now discuss these more general spaces, which look like a vector space only within small regions, and look nothing at all like a vector space in the large. Such spaces, suitably defined, are called manifolds.

|  |  |
|---|---|
| *Example* | The most familiar example of a manifold is the surface of a sphere. Near any point it can be approximated by a plane, but over the whole globe it is qualitatively quite different. You should refer to this example often for concrete instances of the properties of manifolds. |

[You might want to glance ahead at Section 38 to see why we need these ideas and how we are going to use them.]

*Manifolds*   A manifold differs from a vector space in many ways. It does not have the linear structure of a vector space. There is no point called zero, nor can points be added or scaled. There is no way to define any global idea of distant parallelism. The idea of a constant vector field is missing. Nor can one always find a single coordinate system that smoothly describes the whole manifold. Despite these losses, the geometric ideas developed here all apply to manifolds without any change. The last use of the vector-space structure of spacetime was in the definition of acceleration back in Section 12. It was with an eye to the future that we gave there such a careful definition of a tangent vector.

The manifold is the proper setting for a general-relativity description of the universe. Why, then, do so many books gloss over the idea and pretend that it is so obvious that it needs no discussion? Most introductory books are purely descriptive, and one can easily fast-talk one's way around this idea (or any idea). Here we intend to do actual calculations, and cannot be content with vague and unchecked ideas on the subject. A further reason why even many introductory mathematics books skip over the idea is that a full mathematical treatment is abstract and quite technical, and yet does not contribute much to one's ability to calculate on manifolds. We take here an informal approach. We admit that manifolds are necessary, here and in much of physics. Every serious physicist should know a little bit about them. But we will not attempt a precise definition of a manifold. That would not be a good way to start, nor would the fancy definition do you much good. We rely instead on the remarkable ability of the human mind to spot the pattern in a few examples. After you have absorbed the material in this book, then you can study carefully the ideas of calculus on manifolds. The examples given here will provide you with a concrete framework on which to anchor the abstractions of that mathematics.

For the mathematically handicapped, who are not able to proceed without formal definitions, I will mention that all the manifolds which play an important role in cosmology can be easily studied as subspaces of vector spaces. In such cases the tangent spaces are truly tangent hyperplanes, and the vector-space definitions can be legally used. We will, in fact, use such embeddings to define some of our manifolds.

In going from a Euclidean metric space to a vector space, we lose some geometric structure: length, angle, and perpendicularity no longer have any meaning. This was reflected in the representations that we made of such spaces. We went from representations that are covariant under orthogonal transformations to ones covariant under general linear transformations.

When we go to a manifold we lose the linear structure of our space. Our representations now should be covariant under arbitrary curvilinear transformations. In fact, all through this discussion of geometry, we have been casting our concepts in a form that has this covariance. With such preparations, we can describe manifolds without making much change beyond realizing the above-listed differences. Our manifolds no longer have a preferred class of linear coordinates, but, then, we haven't used that idea anywhere in the development.

The two important questions are "What exactly is a manifold?" and "How do we represent it?" The two questions have a common answer. *Manifolds* are sets which can be represented in the following manner. Pieces of manifolds will be represented by charts. A *chart* is a one-to-one, invertible map from a part of the manifold into $\mathbb{R}^n$. More precisely, a chart maps an open region of the manifold into an open set in $\mathbb{R}^n$ (see Figure 26.1). A manifold is represented by a collection of these charts. It is not usually possible to find a single chart for an entire manifold. Each chart furnishes us with coordinates for part of the manifold. Where two charts overlap, the transformations from one set of coordinates to the other should be smooth and have smooth inverses.

*Lost structure*

[You might want to go back and review the discussions of covariance given in Sections 3 and 4.]

*Representation and definition*

*Charts*

[For us the ordinary ideas of open sets and closed sets will suffice. Other definitions of open sets consistent with the rules of topology could be used, but there is little use for these exotic possibilities.]

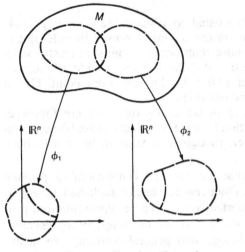

**Figure 26.1.** Two overlapping charts for a manifold $M$.

*Dimension*   Every point must appear in some chart. All charts map into the same dimension $\mathbb{R}^n$, and this number $n$ is called the dimension of the manifold. These restrictions ensure that, in the neighborhood of any point, a manifold looks like ordinary $\mathbb{R}^n$.

*The circle*   ***Example 2***

The circle, the set of points in $\mathbb{R}^2$ such that

$$x^2 + y^2 = 1, \qquad (26.1)$$

is a 1-manifold. Charts for the circle must be maps into $\mathbb{R}$. The following two maps define suitable charts for the circle:

$$(x,y) \mapsto u = 2\sqrt{\frac{1-y}{1+y}}, \qquad (26.2)$$

for all points of the circle except $(0,-1)$; and

$$(x,y) \mapsto v = 2\sqrt{\frac{1+y}{1-y}} \qquad (26.3)$$

for all points except $(0,1)$. See Figures 26.2 and 26.3. Each map is onto the entire real line, which is an open set.

For $x \neq 0$ these two charts overlap, and the coordinates $u$ and $v$ are smoothly related by

$$u = \frac{4}{v}. \qquad (26.4)$$

Since every point of the circle appears in one or the other of these charts, the circle is a manifold.

***Example 3***

Why doesn't the usual coordinate, the angle $\theta$, provide us with a chart in the technical sense? In order not to cover points more than once, we must at least restrict $\theta$ to the range $0 \leq \theta \leq 2\pi$. But this is a closed interval. The open interval $0 < \theta < 2\pi$ misses one point. Thus it does not follow our rules.

Defects creep in when the rules are not followed. Here the function $f = \theta$ is not a single-valued function on the circle, even though it appears to be in a $\theta$-chart.

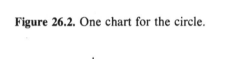

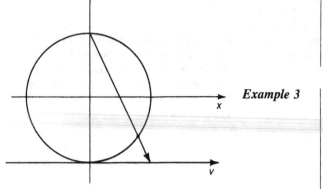

**Figure 26.2.** One chart for the circle.

**Figure 26.3.** Another chart for the circle.

*Defective charts*

[A glance at Section 38 will show how such defective charts are used.]

A defective chart, such as that in Example 3, is not useful for proving that a given set is a manifold. Once that has been established, however, such a defective chart is a perfectly serviceable representation, provided one remembers that some points are not properly represented. The manifolds used in cosmology and general relativity are nearly always represented in such technically defective charts.

| | |
|---|---|
| **Example 4** | The set in $\mathbb{R}^2$ defined by |

$$xy = 0 \qquad (26.5)$$

(Figure 26.4) is not a manifold. The point $(0,0)$ is the culprit. One cannot smoothly map an open set containing this point in a one-to-one manner onto an open set of the real line. Open sets containing $(0,0)$ are given by the conditions

$$a < x < b, \qquad a < 0 < b, \qquad (26.6)$$
$$c < y < d, \qquad c < 0 < d.$$

If the $x$-axis is faithfully represented, then there is no place for the $y$-axis' points and vice versa.

Alternatively, one cannot define the sets

$$a < x < b, \qquad (26.7)$$
$$y = 0,$$

and

$$c < y < d, \qquad (26.8)$$
$$x = 0,$$

to be open, since the intersection of two open sets, here the point $(0,0)$ must also be open. If we call the point $(0,0)$ open, then we cannot map it reversibly and one-to-one onto a conventional open set in $\mathbb{R}$. There single points are closed sets.

The important ideas of curves, functions, tangent vectors, and 1-forms were all defined in such a way that they make perfectly good sense on manifolds.

| | |
|---|---|
| **Example 5** | Suppose we try to do calculus on the cross set given in Example 4. For curves through $(0, 0)$, we can find local linear approximations. It will not be possible, however, to add them. The tangent vectors will not form a vector space. |

Our definition of the tangent vector as the local linear approximation to a curve was framed in such a manner that it made no reference to any special linear coordinate system. The linearity came from the limit process. Over infinitesimal regions, all smooth transformations are linear. There will be different representations of any tangent vector in different coordinate systems, but these turn out to be representations of the same geometric concept.

*Counterexample*

[These remarks should be skipped if you have no acquaintance with the topological idea that open sets are whatever you define them to be, subject to certain postulates of consistency.]

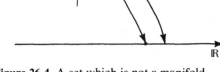

**Figure 26.4.** A set which is not a manifold. The attempt to find a one-to-one map fails.

[Recall the discussion about the addition of tangent curves on page 89.]

*Tangent vectors*

*Linear structure locally*   *Details*

[The $X$ here are $n$ functions of one variable, referred to collectively as just $X$.]

We can explicitly verify that the linear structure of the tangent space is preserved by arbitrary curvilinear coordinate transformations. Suppose that in one set of coordinates $(x^1, x^2, \ldots, x^n)$, which we will abbreviate as just $x$, we have a curve $\gamma$ represented by the map $\gamma_x$

$$\gamma_x: s \mapsto X(s). \tag{26.9}$$

This curve has a tangent vector represented in this coordinate basis by

$$\dot{\gamma}_x = \frac{dX^\mu}{ds} \frac{\partial}{\partial x^\mu}. \tag{26.10}$$

Consider a second system of coordinates $y$, related to the $x$ coordinates by $n$ given functions of $n$ variables,

$$y = Y(x). \tag{26.11}$$

The curve $\gamma$ in the new coordinate system is given by the map

$$\gamma_y: s \mapsto Y[X(s)], \tag{26.12}$$

and has a tangent represented by

$$\dot{\gamma}_y = \frac{d}{ds}\left\{ Y^\mu[X(s)] \right\} \frac{\partial}{\partial y^\mu}. \tag{26.13}$$

This can be simplified, by using the chain rule for partial derivatives, to find

$$\dot{\gamma}_y = \frac{\partial Y^\mu}{\partial x^\nu} \frac{dX^\nu}{ds} \frac{\partial}{\partial y^\mu}. \tag{26.14}$$

This result shows that the components in one coordinate system are linear functions of the components in any other system. The numbers $\partial Y^\mu/\partial x^\nu$ are just the components of an $n$-by-$n$ matrix. The nonlinear transformations $Y(x)$ are linearized by the differentiation.

Since the components are related by linear equations, all the linear structure is covariant under these transformations. Two curves which have equal tangents in one system have equal tangents in all systems. Scaling and addition done in either of two systems agree.

It is possible to find definitions of tangent vectors which are manifestly covariant, ones that make no mention at all of the coordinates used to represent the manifold. Such definitions eliminate the need to verify that one's definitions are properly covariant. Such coordinate-free definitions tend to be a little abstract and to lack motivation for physicists. One such definition defines a tangent vector as an equivalence class of parametrized curves that are tangent at the given point. Another defines a tangent vector as a linear differential operator on functions. Although they are efficient and elegant, I think that such definitions are a poor way to introduce the subject. Since the definitions are all equivalent, it really doesn't matter that we haven't taken the most elegant route.

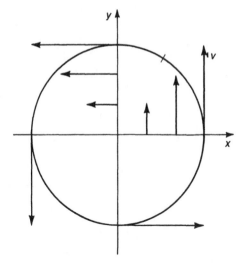

**Example 6** | The vector field

$$v = x\frac{\partial}{\partial y} - y\frac{\partial}{\partial x} \qquad (26.15)$$

has integral curves

$$\gamma_x: s \mapsto (a\cos s, a\sin s) \qquad (26.16)$$

sketched in Figure 26.5. It can also be represented in polar coordinates. The curves are then given by

$$\gamma_y: s \mapsto [R(s), \Theta(s)] = (a, s), \qquad (26.17)$$

and the vector field is given by

$$v = \frac{\partial}{\partial\theta}. \qquad (26.18)$$

This is sketched in Figure 26.6. There we use polar coordinates without prejudice, drawing them at right angles just like any other coordinates. Remember the covariance of manifolds. We need not give any coordinates preferential treatment.

**Figure 26.5.** A vector field and integral curves. Tick marks at unit parameter intervals.

The covariance of our representations of the tangent space shows up in our graphical representations as well. We have been drawing our tangent spaces directly onto the chart of the manifold. This is somewhat an abuse of notation. Tangent vectors are the result of a limit process, and live in a vector space of their own. Drawing them on the chart of the manifold does provide us with a faithful representation of their linear structure. There is no significance, however, to the coincidence of the head of a tangent vector and any particular point on the manifold. There is only significance to it if you take $\varepsilon$ times the vector and let $\varepsilon$ go to zero.

*Covariance*

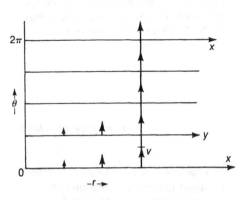

**Example 7** | Look again at Figures 26.5 and 26.6. The head of the vector *a* lies on the point $(x, y) = (1, 1)$ in the first drawing, and on the different point $(r, \theta) = (1, 1)$ in the second.

**Figure 26.6.** The same vectors, integral curves, and axes as drawn in Figure 26.1. Why are there five vectors along the curve here?

The representation of the tangent space right on the chart of the manifold is a useful device, and you should learn how to use it. The idea is

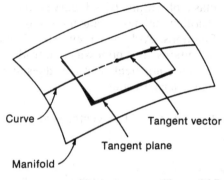

Curve

Tangent vector

Tangent plane

Manifold

**Figure 26.7.**

sketched in Figure 26.7. Imbedded surfaces are examples of manifolds, and for these the tangent spaces are really tangent planes. Our representation above projects these tangent planes down on the manifold. There is clearly no unique way to do this.

*Vectors at different points*

The most important structure lost in going from a vector space to a manifold is the idea of distant parallelism. Even in a manifold with a metric one cannot compare the directions of vectors at two different points. In Figure 26.6 our vector field appears constant in direction, and in Figure 26.5 it doesn't. The concept of a constant vector field just doesn't exist on a manifold.

*Several charts may be needed*

Another important property of manifolds is that we may not be able to cover them with a single smooth coordinate chart. The charts may have the kind of difficulties we are familiar with from polar coordinates. Such defects are generally unavoidable when we are dealing with manifolds. They give us no more trouble than polar coordinates do. Mathematicians formally handle this difficulty by using several different charts, so that every point is an interior point in at least one chart. We will easily be able to handle the difficulties by using common sense.

[This will be discussed further and used when we come to discuss the topology of the 3-sphere in Section 38.]

*Example 8*

The set of all possible rigid-body rotations in three dimensions is an important set that has the structure of a manifold. We can find charts for this manifold as follows. Any rotation leaves an axis fixed. This is not geometrically obvious, but can be shown algebraically. For the present argument just accept it. To represent any rotation, first find this fixed axis. Then find the amount of rotation about this axis. For this we need a sign convention, and we will use a right-hand rule as shown in Figure 26.8. We will represent rotations by vectors lying along the fixed axis whose magnitude is the angle of rotation $(<\pi)$, and whose sense is given by

*The space of rotations*

[To show it you need to recognize that rotations are linear transformations and so can be represented by $3 \times 3$ matrices. Every such matrix has at least one real eigenvector. This eigenvector defines the fixed axis.]

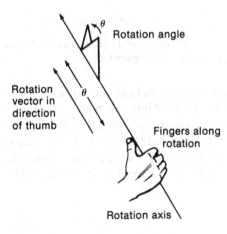

**Figure 26.8.** Example showing the representation of a rotation by a vector.

the right-hand rule. An easily visualized rotation is shown in Figure 26.9, along with its vector representation. These vectors are points in $\mathbb{R}^3$, and so can be used to generate charts for this manifold.

The first chart maps all rotations through angles less than $\pi$ into the solid ball

$$x^2 + y^2 + z^2 < \pi^2. \tag{26.19}$$

This chart contains all rotations except for those represented by vectors of length $\pi$. Call these rotations *flips*. The behavior around these flips is a bit more complicated. Different points on the surface $x^2 + y^2 + z^2 = \pi^2$ do not necessarily represent different rotations. Diametrically opposite points represent the same rotation. To find a chart around any flip $F$, we can take any open pillbox around the point, as sketched in Figure 26.10. The points beyond the flips correspond to rotations by an angle $\alpha$ of more than $\pi$. These are equal to rotations of the opposite sense with a magnitude of $(2\pi - \alpha)$. These rotations are represented also by points just inside the flips on the opposite side. This pillbox is open, each point inside it represents a unique rotation, and no rotation appears inside it more than once. Thus it is a legitimate chart.

Such a chart can be constructed around any flip. Thus the set of all rotations in three dimensions is a 3-manifold. To deal with this manifold, we find it more convenient to use only the first chart, let it extend beyond

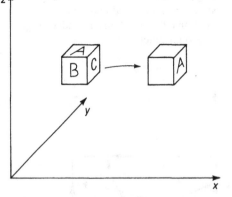

**Figure 26.9.** The rotation represented by the vector $(2\pi/3\sqrt{3})[(\partial/\partial x) + (\partial/\partial y) + (\partial/\partial z)]$.

*Flips*

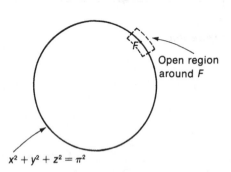

**Figure 26.10.** The space of all rotation vectors and an open chart around the flip $F$, drawn in only two dimensions.

$$x^2 + y^2 + z^2 = \pi^2 \qquad (26.20)$$

to cover the flips, and remember that some rotations are being doubly represented.

*Manifolds in general relativity*    The model of spacetime used in general relativity is a manifold with a metric tensor. We are forced into this position by the observed failure of the free-particle postulates of special relativity, as we will soon discuss. When gravitational fields are present, one cannot find inertial reference frames for distances that are more than infinitesimal. That is why we had to set up all the machinery of manifolds.

[Only after you have done several computations with them will tangent spaces seem natural and familiar. Sections 27 and 28 are intended to give you such practice.]

PROBLEMS

26.1. (20) A cylinder can be represented by a strip of the $x,y$ plane such that $0 \le x \le 1$, with the convention that for all $y$, $(0, y)$ and $(1, y)$ are really the same point. Which of the following vector fields are continuous on the cylinder:

(i) $\dfrac{\partial}{\partial x}$,

(ii) $\dfrac{\partial}{\partial y}$,

(iii) $\sin 2\pi x \dfrac{\partial}{\partial y}$,

(iv) $x\dfrac{\partial}{\partial y} - y\dfrac{\partial}{\partial x}$?

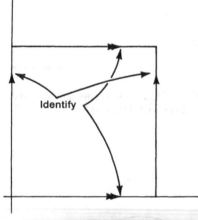

**Figure 26.11.**

26.2. (12) A torus can be represented by the portion of the plane such that $0 \le x \le 1$ and $0 \le y \le 1$, identified as shown in Figure 26.11. Is the vector field

$$y\frac{\partial}{\partial x}$$

continuous?

26.3. (20) How many different coordinate charts does it take to cover the torus in such a way that every point appears in at least one such chart as an interior point?

26.4. (28) Show that on a vector space, the manifold definition of a tangent vector given in Section 17 and the vector space definition given in Section 12 agree.

26.5. (30) Look at the set of all great circles on a sphere. Show that this set is a manifold (it is called real projective 2-space, $RP^2$), and find a useful representation of it.

26.6. (28) In the space of rotations given in the last example of this section, show that any plane through the origin in that representation cuts out a subset that is the above manifold $RP^2$.

26.7. (38) Read the example of page 2 of Porteous, *Topological Geometry*, and discuss it. See also *Am. J. Phys.* **47**, 379 (1979).

# CHAPTER THREE

# Gravitation

We have now assembled the mathematical tools needed to set up a theory of gravitation compatible with special relativity. Before we attack the problem of gravitation, we will first practice a bit with these tools on problems that will be needed later. Our mathematical model for a particle will be a high-frequency wave packet as described in Section 19. This is called a semiclassical model, since it is halfway between classical mechanics and quantum mechanics. To show that a composite system composed of many particles such as the Earth follows the same rules is far beyond our abilities, and in fact is still a subject of current research.

*Semiclassical models*

The study of these wave packets will provide us with a concrete realization of the idea of an intrinsic clock taken as a primitive notion in Chapter I. We will also find a model for the idea of Lorentz invariance, and such a simple system as the motion of waves on deep water will turn out to be another physical realization of a relativity symmetry.

We finally come to gravitation, and look first at uniform gravitational fields. We will find that special relativity can still be used in such situations, provided that we redefine a free particle as a particle with no forces acting on it except possibly gravity. Since over small regions all smooth gravitational fields look uniform, this extension means that locally the geometry of spacetime will be approximately the geometry of special relativity, Minkowski geometry. The extension of this to nonuniform gravitational fields is now straightforward. We need only let the metric tensor be a function of position rather than a constant.

*Free particles*

We will then end this chapter with a few simple applications of gravitation. We will first develop some more tools for the study of the motion of wave packets in nonuniform fields. We then use them to show that the model usually taken in general relativity for the spacetime near a massive body does describe the familiar effects of gravitation. We will briefly mention the idea of spacetime curvature and the Einstein field equations, but a careful and quantitative treatment of them is beyond the scope of this book.

## 27. The Motion of Wave Packets

Our model for the dynamics of particles will use the idea of a wave packet. We can put the central idea of such motion, the idea of a group velocity or energy velocity, into a simple and useful form if we pay attention to its geometric aspects. The discussion in Section 19 was marred by a dependence on a special role for the time coordinate. For this and the following two sections, we will restrict our attention to the straight-line motion of wave packets propagating through uniform conditions. We will study general motion in Section 30.

*Dispersion relation*    We start from a description of a wave equation in terms of a dispersion relation. In two dimensions, this dispersion relation is a function

[For simplicity we discuss only the case of two dimensions. The extension should be obvious.]

$$W(k, \omega, x, t) = 0, \tag{27.1}$$

and for uniform conditions, $W$ will not depend explicitly on either $x$ or $t$. A given high-frequency wave packet has everywhere a wave number $k(x, t)$ and a frequency $\omega(x, t)$ which together satisfy

$$W[k(x, t), \omega(x, t)] = 0. \tag{27.2}$$

The partial derivatives of this equation with respect to $x$ and $t$ must also vanish. They can be found from the chain rule:

$$\frac{\partial W}{\partial k} \frac{\partial k}{\partial x} + \frac{\partial W}{\partial \omega} \frac{\partial \omega}{\partial x} = 0, \tag{27.3}$$

$$\frac{\partial W}{\partial k} \frac{\partial k}{\partial t} + \frac{\partial W}{\partial \omega} \frac{\partial \omega}{\partial t} = 0. \tag{27.4}$$

Recall also (see page 108) that since $k$ and $\omega$ were already phase gradients, we have

$$\frac{\partial k}{\partial t} = \frac{\partial \omega}{\partial x}. \tag{27.5}$$

Thus we can write the above equations

$$\frac{\partial W}{\partial k}\frac{\partial k}{\partial x} + \frac{\partial W}{\partial \omega}\frac{\partial k}{\partial t} = 0, \tag{27.6}$$

$$\frac{\partial W}{\partial k}\frac{\partial \omega}{\partial x} + \frac{\partial W}{\partial \omega}\frac{\partial \omega}{\partial t} = 0. \tag{27.7}$$

These equations state that both $k$ and $\omega$ are constant in the direction of the vector, called the group-velocity vector,

*Group velocity*

$$\frac{\partial W}{\partial k}\frac{\partial}{\partial x} + \frac{\partial W}{\partial \omega}\frac{\partial}{\partial t}. \tag{27.8}$$

[Recall the discussion of characteristics in Section 25.]

*Example 1* | For deep-water waves, we had

$$W(k, \omega) = \omega^4 - k^2 = 0, \tag{27.9}$$ *Water waves*

so we have $k$ and $\omega$ constant in the direction

$$4\omega^3 \frac{\partial}{\partial t} - 2k\frac{\partial}{\partial x}. \tag{27.10}$$

Previously we calculated a group-velocity vector for deep-water waves

$$\frac{\partial}{\partial x} - 2\omega\frac{\partial}{\partial t}, \tag{27.11}$$

on page 108. Note that these vectors both point in the same spacetime direction.

Group velocity, as defined so far, is only a direction in spacetime, a direction along which wave number and frequency must be constant. The length of the group-velocity vector does not yet have any particular significance. There are several convenient normalizations for the length of the group-velocity vector. One is to make the $\partial/(\partial t)$ component unity, but doing so clearly gives some special importance to the coordinate $t$. A unique coordinate-free normalization is also possible, using the condition

*Normalization*

$$d\theta \cdot v = 1. \tag{27.12}$$

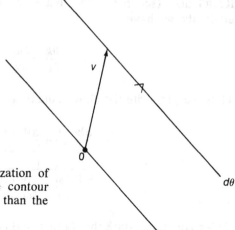

**Figure 27.1.** A natural normalization of the group-velocity vector. The contour lines of $d\theta$ are closer together than the wave crests by a factor of $2\pi$.

Here $d\theta$ is the phase gradient and $v$ the group velocity, sketched in Figure 27.1. If we use Equation 27.8 for the direction of the group-velocity vector, and 27.12 for normalization, we find that the group-velocity vector $v$ is given by

$$v = \frac{\dfrac{\partial W}{\partial k}\dfrac{\partial}{\partial x} + \dfrac{\partial W}{\partial \omega}\dfrac{\partial}{\partial t}}{k\dfrac{\partial W}{\partial k} + \omega\dfrac{\partial W}{\partial \omega}}. \tag{27.13}$$

*Asymptotes*  Situations where the denominator vanishes are important. For such waves there is no dispersion, and waves of different frequencies propagate together. This happens for light waves, for example. It will not happen for the wave packets representing our particles.

*Wave diagram*  For each $k$ and $\omega$ which satisfy our dispersion relation, there is a unique group-velocity vector which describes the motion of the wave packet through spacetime. For each point on the dispersion relation, there is a unique tangent vector. We have, therefore, a curve in the tangent space which corresponds point by point to the dispersion relation.

[Example 1 continues]

For deep-water waves we can parametrize points on the dispersion relation by the wave number $k$:

$$d\theta = \sqrt{|k|}\, dt + k\, dx. \tag{27.14}$$

Each value of $k$ leads to a group velocity vector

$$v = \frac{4\omega^3\dfrac{\partial}{\partial t} - 2k\dfrac{\partial}{\partial x}}{4\omega^4 - 2k^2}. \tag{27.15}$$

We can use the dispersion relation to simplify this to

$$v = \frac{2}{\sqrt{|k|}} \frac{\partial}{\partial t} - \frac{1}{k} \frac{\partial}{\partial x}, \qquad (27.16)$$

which expresses the 1-parameter family of tangent vectors "dual" to the dispersion relation.

When we are representing sets of vectors, it is most convenient to put all their tails at the origin, and to represent their heads by single points. Each vector is represented by a point in the tangent space. We can use coordinates in this space called $\dot{x}$ and $\dot{t}$. The relation

$$v = \dot{x} \frac{\partial}{\partial x} + \dot{t} \frac{\partial}{\partial t} \qquad (27.17)$$

shows how to go from coordinates $(\dot{x}, \dot{t})$ to a vector $v$, just as the relation

$$d\theta = k\,dx + \omega\,dt \qquad (27.18)$$

shows how to go from wave number and frequency coordinates to the phase gradient 1-form.

Our 1-parameter family of group velocity vectors is represented by

$$\dot{x} = -\frac{1}{k}, \qquad (27.19)$$

$$\dot{t} = \frac{2}{\sqrt{|k|}}.$$

We can eliminate the parameter $k$ to find an implicit equation for the set of group-velocity vectors:

$$\dot{t}^2 + 4\dot{x} = 0 \qquad (27.20)$$

and for the other branch of the dispersion relation

$$\dot{t}^2 - 4\dot{x} = 0. \qquad (27.21)$$

These sets of vectors are shown in Figure 27.2.

The set of all normalized group-velocity vectors will turn out to be a very useful tool for us. Although this construction has been used by mathematicians studying the calculus of variations, it has not been used before by physicists studying dispersive waves. I call it the *wave diagram*. We now demonstrate an important geometric relation which is responsible for the usefulness of the wave diagram. The phase

*$\dot{x}$ and $\dot{t}$*

[A glance at the equations for integral curves in Section 25 will show why this notation was chosen. Those familiar with Lagrangian mechanics will recognize this to be the same as taking $q$ and $\dot{q}$ to be independent coordinates.]

[Example 1 continues]

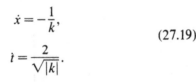

**Figure 27.2.** Wave diagram for deep-water waves.

[See Arnold's mechanics book, page 349, for example.]

*Duality*

[Example 1 continues]

[Do not forget that the tangent space has a unique zero vector. It is an important part of the diagram.]

gradient 1-form corresponding to any group velocity is tangent to the wave diagram at that point. The general construction is sketched in Figure 27.3.

In Figure 27.4 we use the above construction on the wave diagram for deep-water waves. You should be able to see immediately that the wave packet travels more slowly than the wave crests, in fact, half as fast. In Figure 27.5 we draw a spacetime diagram showing the wave packet described in Figure 27.4.

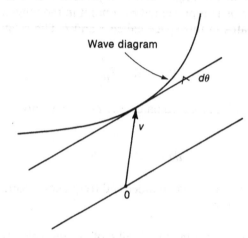

**Figure 27.3.** Use of the wave diagram to relate the group velocity with the phase gradient.

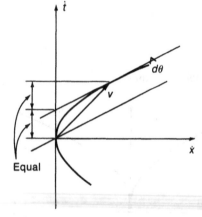

**Figure 27.4.** Phase gradient and group velocity for water waves.

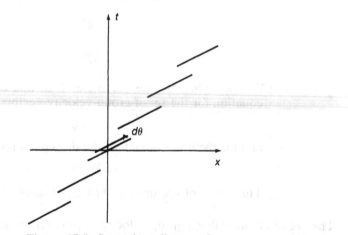

**Figure 27.5.** Spacetime diagram for a wave packet corresponding to Figure 27.4. Scale is 1/10 that of the previous figure. Note phase gradient at the center. Lines are wave crests. The finite size of the wave packet is indicated by the finite length of these wave-crest world lines. See Figures 20.6 and 20.7.

To demonstrate that the 1-form $d\theta$ is indeed tangent to the wave diagram, consider a group-velocity vector $v$ and its associated phase gradient $d\theta$. Because of our normalization, we have

$$d\theta \cdot v = 1, \tag{27.22}$$

which forces the unit contour of $d\theta$ to pass through the head of $v$. Look at a nearby group-velocity vector, $v'$, and its phase gradient $d\theta'$. Write the nearby phase gradient

$$d\theta' = (k + \Delta k)\,dx + (\omega + \Delta\omega)\,dt, \tag{27.23}$$

and the new group velocity

$$v' = v + \Delta v. \tag{27.24}$$

We must have

$$d\theta' \cdot v' = 1. \tag{27.25}$$

and taking the limit $\Delta k$, $\Delta\omega$, and $\Delta v \to 0$, and discarding all square terms, we have

$$(\Delta k\,dx + \Delta\omega\,dt) \cdot v + d\theta \cdot \Delta v = 0. \tag{27.26}$$

Look now at the first term here, and write out $v$ explicitly:

$$\frac{(\Delta k\,dx + \Delta\omega\,dt) \cdot \left(\dfrac{\partial W}{\partial k}\dfrac{\partial}{\partial x} + \dfrac{\partial W}{\partial \omega}\dfrac{\partial}{\partial t}\right)}{\left(k\dfrac{\partial W}{\partial k} + \omega\dfrac{\partial W}{\partial \omega}\right)} = \frac{\Delta k\dfrac{\partial W}{\partial k} + \Delta\omega\dfrac{\partial W}{\partial \omega}}{\left(k\dfrac{\partial W}{\partial k} + \omega\dfrac{\partial W}{\partial \omega}\right)}. \tag{27.27}$$

Now the phase gradient $d\theta'$ must satisfy the dispersion relation

$$W(k + \Delta k,\ \omega + \Delta\omega) = 0, \tag{27.28}$$

and a Taylor's expansion shows that

$$\frac{\partial W}{\partial k}\Delta k + \frac{\partial W}{\partial \omega}\Delta\omega = 0. \tag{27.29}$$

Thus the first term of Equation 27.26 vanishes, and then so too must the second, and we have

$$d\theta \cdot \Delta v = 0, \tag{27.30}$$

[This proof may be skipped without serious loss.]

You can see from Figure 27.6 that this condition says precisely that the unit contour of $d\theta$ is tangent to the wave diagram.

The wave diagram is a useful tool in dispersive-wave problems. It is the generalization of the hyperbola of special relativity. The wave-diagram construction is the way to see the hyperbola in the wave equation.

*Special-relativity hyperbola*     *Example 2*

Let us find the wave diagram for the equation

$$\frac{\partial^2 \psi}{\partial t^2} - \frac{\partial^2 \psi}{\partial x^2} + \psi = 0. \tag{27.31}$$

The dispersion relation is

$$W = \omega^2 - k^2 - 1 = 0, \tag{27.32}$$

and the normalized group-velocity vector is

$$v = \frac{\dfrac{\partial W}{\partial \omega}\dfrac{\partial}{\partial t} + \dfrac{\partial W}{\partial k}\dfrac{\partial}{\partial x}}{\omega\dfrac{\partial W}{\partial \omega} + k\dfrac{\partial W}{\partial k}}$$

$$= \frac{\omega\dfrac{\partial}{\partial t} - k\dfrac{\partial}{\partial x}}{\omega^2 - k^2} \tag{27.33}$$

$$= \omega\frac{\partial}{\partial t} - k\frac{\partial}{\partial x}.$$

Writing $v$ in terms of its components $\dot{t}$ and $\dot{x}$, we have

$$v = \dot{t}\frac{\partial}{\partial t} + \dot{x}\frac{\partial}{\partial x}, \tag{27.34}$$

and so

$$\begin{aligned} \dot{t} &= \omega, \\ \dot{x} &= -k; \end{aligned} \tag{27.35}$$

hence the wave diagram is given by

$$(\dot{t})^2 - (\dot{x})^2 = 1, \tag{27.36}$$

which is indeed the hyperbola.

We sketch the phase gradient and the group velocity for a moving wave packet in Figure 27.7. Note the peculiar way the wavefronts tilt relative to the group-velocity vector. This is the same strange tilt that occurred in our definition of simultaneity. Thus our definition turns out to be useful. Had we started with Maxwell's equations and similarly constructed a wave diagram, we would also have found a hyperbola. The

Figure 27.6.

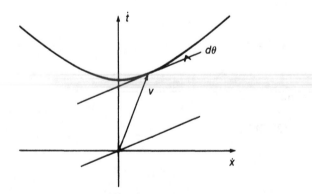

**Figure 27.7.** Phase gradient and group velocity for a wave-packet solution of the ordinary wave equation.

definition of simultaneity would then have appeared quite natural. That would have been a difficult way to start the theory, however.

For the straight-line motion of a wave packet under uniform conditions, there is little difference between spacetime $(x, t)$ and the tangent space $(\dot{x}, \dot{t})$ except for the location of the origin. When we go to nonuniform fields, where the dispersion relation depends explicitly on $x$ and $t$, we will have to construct the wave diagram in a linear space, and for this we need the tangent space. The tangent space is easier to work with than the dual space of phase gradients, because the tangent space is a linear approximation to the manifold itself. This makes the wave diagram easier to think about than the dispersion relation.

*Straight-line motion*
[Remember, the tangent space is the space of tangent vectors at a point.]

The role of the metric tensor in spacetime is to describe the wave diagram. Group-velocity vectors for the above wave equation, which has the Lorentz symmetry of special relativity, must satisfy

*Metric tensor*

$$\dot{t}^2 - \dot{x}^2 = 1, \qquad (27.37)$$

and this can be written

$$\mathcal{N} \cdot (v, v) = -1, \qquad (27.38)$$

where $\mathcal{N}$ is the Minkowski metric tensor,

$$\mathcal{N} = dx^2 - dt^2. \qquad (27.39)$$

Also, the phase gradient $d\theta$ associated with a velocity $v$ is given by

$$d\theta = -\mathcal{N} \cdot v. \qquad (27.40)$$

[This dot notation is explained on page 118.]

Let me now sketch briefly why it is that amplitude information also propagates with the group velocity. We discovered the dispersion relation from the largest terms in an expansion using nearly plane waves, such as

*Wave amplitude*

$$\psi \sim A \cos \theta. \qquad (27.41)$$

We now consider the case where not only $k$ and $\omega$, but also the amplitude $A$, depend slowly on $x$ and $t$. A term such as $\omega^4$ in the dispersion relation reflects a term $(\partial^4\psi)/(\partial t^4)$ in the partial differential equation. Such a term leads to the $\omega^4$ in the dispersion relation if all four differentiations hit $\theta$, the fastest changing thing around. The next rank of terms result from one of the four differentiations hitting either $k$, $\omega$, or $A$. Here the result would be a term $4\omega^3[(\partial A)/(\partial t)]$, the $\omega^3$ coming from

[We are continuing the argument of Section 19.]

**Figure 27.8.**

the three remaining differentiations of $\theta$. The factor of four comes from the fact that any of four differentiations can hit $A$. The $A$ differentiations will thus lead to an equation

$$\frac{\partial W}{\partial \omega}\frac{\partial A}{\partial t} + \frac{\partial W}{\partial k}\frac{\partial A}{\partial x} = \text{(something known)}. \qquad (27.42)$$

Again the behavior of $A$ is determined in the group-velocity direction. Because the waves can spread out, the amplitude is not necessarily constant. One can construct a careful argument based on the above sketch. Doing so is not essential for us, however, and we will not pursue it beyond this intuitive discussion.

The basic idea here is that every tangent space has its wave diagram. Sometimes we can find coordinates for which these wave diagrams are the same in all tangent spaces. Most of our examples are like this. What is important is that there must be some rule for constructing a wave diagram in every tangent space. A wave diagram is a well-defined geometric object, but it is not a tensor. It is not necessarily a linear operator. General relativity is special, and its wave diagram can be specified in terms of a tensor.

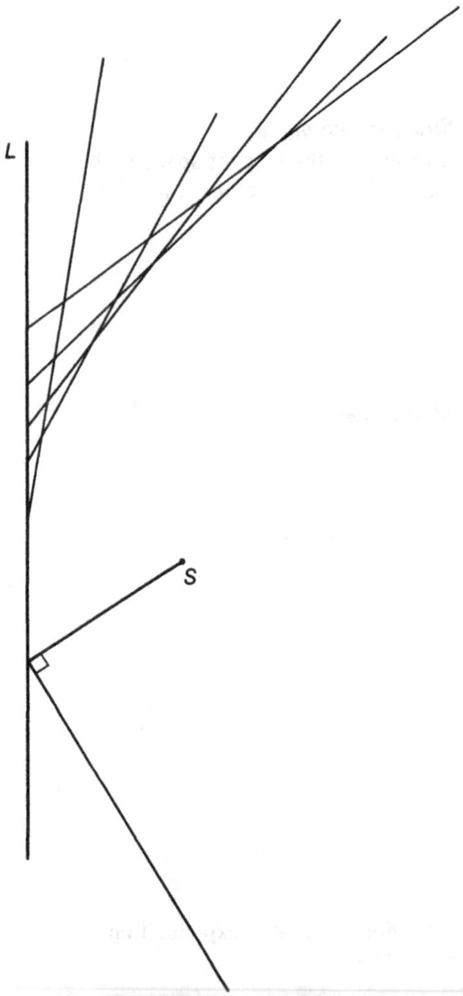

PROBLEMS

27.1. (18) Introduce suitable units and give numerical estimates for the situation sketched in Figure 27.4.

27.2. (32) A parabola can be constructed as the envelope of a family of lines constructed from a fixed point $S$ and a fixed line $L$ as shown in Figure 27.8. Express this construction in our geometric language by treating the lines as covectors. The envelope will be a wave diagram. Show that it is in fact a parabola. See Problem 23.1.

27.3. (21) Draw the wave diagram for the dispersion relation

$$\omega = k - k^3.$$

27.4. (21) For capillary waves on deep water, we have a dispersion relation

$$\omega^2 = k^3.$$

Draw the wave diagram.

27.5. (21) Find the wave diagram for the dispersion relation

$$\omega = k(1 + \alpha k^2).$$

27.6. (19) Show that in a two-dimensional spacetime diagram, the speed of the wave packet is given by $\partial \omega / \partial k$, and the speed of the wave crests by $\omega / k$.

## 28. Water-Wave Relativity

The motion of waves on the surface of water exhibits surprising similarities with special relativity. Before we discuss gravitation, we will explore the physics of water waves a little bit to practice with our new tool, the wave diagram. We will find that deep-water waves have a symmetry remarkably similar to Lorentz invariance. This second view of a relativity symmetry will give us a new outlook on special relativity as well as practical experience with actual calculations. Whereas most books on tensors for physicists spend considerable time on the transformation of tensors from one coordinate system to another, we will find here that the explicit notation we use makes such transformations obvious and routine.

The water-wave theory described by the dispersion relation

$$\omega^4 = k^2 \tag{28.1}$$

neglects many features of real water waves. It describes only the interaction of gravity and inertia. No influence of the bottom is included, nor is surface tension. It is also a linearization, and so no interaction between waves appears, nor any breaking of the waves. Including surface tension changes the dispersion relation to

$$\omega^4 = k^2 \left( g + \frac{T}{\rho} k^2 \right)^2, \tag{28.2}$$

where $T$ is the coefficient of surface tension and $\rho$ the density. Ignoring uninteresting constants, we look at the dispersion relation

$$\omega^4 = k^2 (1 + \alpha k^2)^2, \tag{28.3}$$

which describes waves which include the effects of surface tension. For large $k$ the surface-tension term replaces gravity as the balance to inertial forces. We will call this new mode "ripples." The ripples move much slower than the deep-water waves, and have short wavelengths. We will forego any detailed description of the ripples, and represent them on the wave diagram as a single vertical line, as shown in Figure 28.1. This represents the limit, as the surface tension becomes negligible, of a more complicated wave diagram. The exact dispersion relation is shown in Figure 28.2, and the wave diagram in Figure 28.3. We want to use a slightly improved theory of water waves that includes the ripples, but does not contain any specific details of their motion except their approximate group velocity of zero.

[This section and the next discuss interesting topics that should give you practice with vectors, 1-forms, and wave diagrams. None of the results, but only the skills, will be specifically used later.]

[A brief introduction to water waves can be found in the Feynman Lectures. More can be found in Lighthill and in the paper by Synge.]

### Dispersion relations

[Really, $\omega^4 = g^2 k^2$ in ordinary units.]

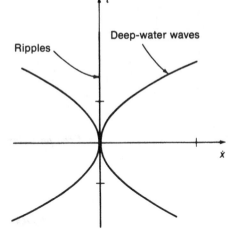

Figure 28.1.

### Ripples

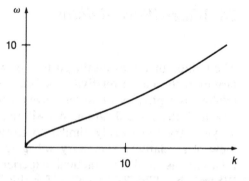

**Figure 28.2.** Dispersion relation for water waves, including surface tension but not the effects of finite depth. Drawn for $\alpha = 0.01$.

*Observations*    We now explore this theory. You should imagine that you are flying above the surface of the water in a blimp, say, and that you can observe the surface of the water and the waves on it, but nothing else. The blimp may be moving with some uniform velocity over the water. A number of questions come to mind. For example, can you deduce the speed of the blimp from observations of the water waves alone? This water-wave theory will turn out to be remarkably similar to special relativity. To bring this out we discuss this blimp observer in the same manner that we discussed spacetime observers in the first chapter.

*Free particles*    We start by considering wave packets of these waves. These will be the "free particles" of our theory. (When discussing this situation, Synge could not resist calling these particles "hydrons.") Of course, these "particles" are clearly not of zero extent, but then neither are the usual "particles" of Newtonian mechanics. As observers in our blimp above the waters, we can define intrinsic clocks. These are not

*Clocks*    the ordinary physical clocks, whose use would be cheating, but clocks made up only out of water waves. To make a clock out of water waves, a given observer need only look at a wave packet of waves whose group velocity is the same as the velocity of the observer. Such a wave packet as a whole will move along with the observer. A packet could be easily made by dropping something into the water and waiting. After a while the waves whose group velocities differ from your velocity will have gone off, and the only waves that you will have with you will be those whose group velocity matches your velocity. Because water waves are dispersive, the wave crests will travel at a different speed, here faster, and so you will see the water surface at any point directly under your position oscillate up and down. Such oscillations are precisely the idea of an intrinsic clock.

*Example*

Consider an observer moving with speed $v$ to the right. The wave diagram for right-going waves is given by

$$\dot{t}^2 = 4\dot{x}, \tag{28.4}$$

or, in parametric form,

$$\dot{x} = -\frac{1}{k}, \tag{28.5}$$

$$\dot{t} = \frac{2}{\sqrt{-k}},$$

where $k$, here negative, is the wave number. The point on this curve with wave number $k$ corresponds to a velocity

$$v = \frac{\dot{x}}{\dot{t}} = \frac{1}{2\sqrt{-k}}. \tag{28.6}$$

Thus we can express the group-velocity vector corresponding to velocity $v$ by

$$\dot{x} = 4v^2, \tag{28.7}$$

$$\dot{t} = 4v.$$

An observer moving along with such waves will see new wave crests at time intervals

$$\Delta t = 8\pi v. \tag{28.8}$$

See Figure 28.4. The phase gradient for this wave motion,

$$d\theta = \frac{1}{2v}dt - \frac{1}{4v^2}dx, \tag{28.9}$$

is also shown in the figure.

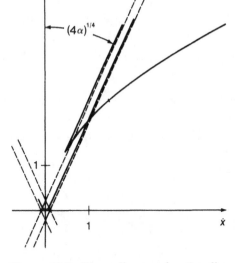

**Figure 28.3.** Wave diagram for the dispersion relation of Figure 28.2. Note how much more structure appears in the wave diagram.

[That pesky $2\pi$ again; see Section 19.]

To complete the list of primitive elements, we need a special class of waves to be called "light signals." For these we take the ripples. We now have a list of primitive elements with structure similar to that of the elements in the fundamental structure of spacetime. The motion of water waves provides us with another realization of that abstract structure.

*Light signals*

---

**Translation from water waves to spacetime language:** wave packet of water waves → free particle; vibrations of water waves moving along with a given observer → clock; usual $(x,t)$-coordinate system → inertial reference frame; ripples → light signals.

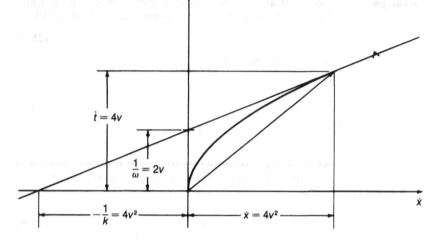

**Figure 28.4.** The details of a water-wave packet moving with a group velocity $v$. Note that the wave crests move twice as fast as the wave packet itself.

As long as our observer is moving with nonzero speed, these primitive elements satisfy all our postulates about spacetime structure except that instead of having both left-going and right-going light signals, we have only a single set of ripples.

*Inertial reference frames*

The usual $(x, t)$ coordinates form an inertial reference frame. What about a canonical reference frame? How would a given observer make linear transformations to simplify his description of the wave diagram? Let us assume that the observer has made enough observations to construct the entire wave diagram. There are four degrees of freedom in a linear transformation in two dimensions. We will describe the transformation to a canonical reference frame by giving four successive simple transformations which simplify the wave diagram for a given observer.

*Canonical reference frames*

[You might want to review the discussion of a canonical reference frame in special relativity given in Section 6.]

The initial situation, as seen not by our moving observer, but by an observer at rest with respect to the ripples, is shown in Figure 28.5. We will transform this diagram into canonical form. The first requirement on a canonical form for the wave diagram would be to put the moving observer "at rest," i.e., make his world line vertical, as in Figure 28.6. A second requirement on a canonical representation will be to adjust the time axis to the rate of the intrinsic clock. Thus in Figure 28.7 we have put the intersection of the wave diagram with the vertical axis at unit time. A third simplification would be to make the wave diagram cross the vertical axis at right angles, as in Figure 28.8. Finally, one

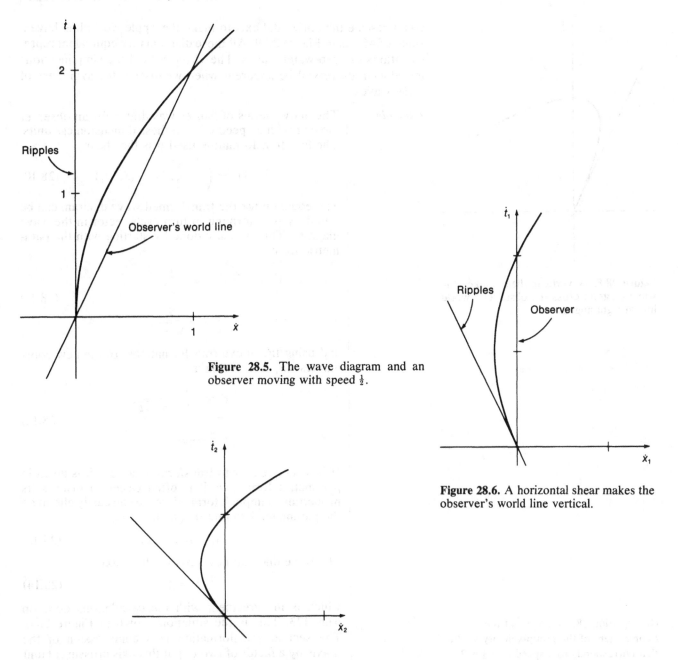

**Figure 28.5.** The wave diagram and an observer moving with speed $\frac{1}{2}$.

**Figure 28.6.** A horizontal shear makes the observer's world line vertical.

**Figure 28.7.** Stretching the vertical scale so that a unit space interval is equal to the interval between the origin and the intersection of the wave diagram and the observer's world line.

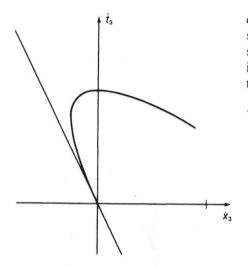

**Figure 28.8.** A vertical shear makes the wave diagram cross the observer's world line at right angles.

could rescale the horizontal axis to make the ripple world line have a slope of 45°, as in Figure 28.9. All these diagrams are equivalent representations of deep-water waves. The virtue behind the simplifications introduced above will be apparent when we discuss the symmetry of these waves.

*Example*

The above series of figures was drawn for an observer moving with a speed $v = \frac{1}{2}$ in these dimensionless units. The first transformation used was the shear

$$(\dot{x}, \dot{t}) \mapsto \left( \dot{x} - \frac{\dot{t}}{2}, \dot{t} \right) \equiv (\dot{x}_1, \dot{t}_1). \qquad (28.10)$$

The equation for the transformed wave diagram can be found by transforming in turn each vector in the wave diagram. This is easily done by starting with the parametric form,

$$\dot{x} = -\frac{1}{k},$$
$$\dot{t} = \frac{2}{\sqrt{-k}}, \qquad (28.11)$$

and using the above transformations. In the new coordinates, $(\dot{x}_1, \dot{t}_1)$, we have

$$\dot{x}_1 = -\frac{1}{k} - \frac{1}{\sqrt{-k}},$$
$$\dot{t}_1 = \frac{2}{\sqrt{-k}}. \qquad (28.12)$$

It is usually easier to transform a curve if it is given in parametric form, but it is often easier to discuss its properties in implicit form. Here we can easily eliminate the parameter $k$ to find the implicit form

$$4\dot{x}_1 = \dot{t}_1(\dot{t}_1 - 2). \qquad (28.13)$$

The wave diagram now crosses the $\dot{t}_1$ axis at

$$\dot{t}_1 = 2, \qquad (28.14)$$

which is in agreement with the calculations done on page 173. This is the situation shown in Figure 28.6. The second transformation is a compression of the $\dot{t}_1$ axis by a factor of two to put this axis crossing at unit height. We use the transformation

$$(\dot{x}_1, \dot{t}_1) \mapsto \left( \dot{x}_1, \frac{\dot{t}_1}{2} \right) \equiv (\dot{x}_2, \dot{t}_2), \qquad (28.15)$$

and now the wave diagram is given by

[In Equation 28.7 we see that the $t$-component of the group-velocity vector that corresponds to a speed $v$ of $\frac{1}{2}$ is 2, which agrees with Equation 28.14.]

$$\dot{x}_2 = \dot{t}_2(\dot{t}_2 - 1) \qquad (28.16)$$

This is Figure 28.7.

The third transformation is a shear to make the wave diagram horizontal where it crosses the $\dot{t}_2$ axis. Near $\dot{x}_2 = 0$, we have

$$\dot{x}_2 \approx (\dot{t}_2 - 1), \qquad (28.17)$$

and from this we can see what shear is needed to straighten this out:

$$(\dot{x}_2, \dot{t}_2) \mapsto (\dot{x}_2, \dot{t}_2 - \dot{x}_2) \equiv (\dot{x}_3, \dot{t}_3). \qquad (28.18)$$

The new wave diagram is shown in Figure 28.8, and was found by writing Equation 28.16 parametrically, in the form

$$\dot{x}_2 = \alpha(\alpha - 1),$$
$$\dot{t}_2 = \alpha, \qquad (28.19)$$

from which we have the parametric form in $(\dot{x}_3, \dot{t}_3)$ coordinates

$$\dot{x}_3 = \alpha(\alpha = 1),$$
$$\dot{t}_3 = \alpha(2 - \alpha). \qquad (28.20)$$

The final transformation will put the ripple world line at 45°. We have not been keeping track of the ripple line. It can easily be found since it passes through the origin, an invariant point of all these linear transformations, and is tangent there to the wave diagram, and tangency is preserved under linear transformations. In terms of our parameter $\alpha$, the origin is the point

$$\alpha = 0, \qquad (28.21)$$

and we can find the tangent from the small $\alpha$ expansion:

$$\dot{x}_3 \approx -\alpha,$$
$$\dot{t}_3 \approx 2\alpha. \qquad (28.22)$$

We need to expand the $\dot{x}_3$ axis by a factor of two to put this line at 45°; so we have the final transformation,

$$(\dot{x}_3, \dot{t}_3) \mapsto (2\dot{x}_3, \dot{t}_3) \equiv (\dot{\xi}, \dot{\eta}). \qquad (28.23)$$

We will use these final coordinates quite a bit and so we have given them more convenient names than $\dot{x}_4$ and $\dot{t}_4$. The final form of the wave diagram is given parametrically by

$$\dot{\xi} = 2\alpha(\alpha - 1),$$
$$\dot{\eta} = \alpha(2 - \alpha). \qquad (28.24)$$

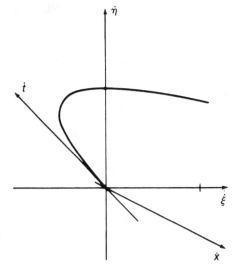

**Figure 28.9.** Horizontal compression makes the ripple's world line at a 45° angle.

*Shortcut*    The successive transformations used in the above example can be used to construct the canonical wave diagram for an observer moving with any speed. Rather than redo this argument for the general observer, we can usefully consider another, quicker way to canonize the wave diagram. The canonical reference frame is determined by its two basis vectors, call them $\partial/\partial\xi$ and $\partial/\partial\eta$. In our original frame, at rest with respect to the water, they are situated as shown in Figure 28.10. All four conditions on our wave diagram are needed to construct Figure 28.10. The vector $\partial/\partial\eta$ is by definition the group-velocity vector for waves moving along with the observer. The vector $\partial/\partial\xi$ must be parallel to the wave diagram at the tip of $\partial/\partial\eta$, and its length is chosen to make the ripple's world line appear at 45°.

Let us now translate these ideas into our precise geometric language. Equation 28.7 showed how to express the group-velocity vector as a function of the speed $v$. Thus we have

$$\frac{\partial}{\partial\eta} = 4v\left(\frac{\partial}{\partial t} + v\frac{\partial}{\partial x}\right),\tag{28.25}$$

and the phase gradient was given there as well, as

$$d\theta = -\frac{1}{2v^2}dx + \frac{1}{v}dt.\tag{28.26}$$

[This dot represents the natural operation of a covector on a vector. No metric operator, Euclidean or Minkowski, is needed. See Section 15.]

The condition that $\partial/\partial\xi$ be parallel to $d\theta$ is just

$$d\theta \cdot \frac{\partial}{\partial\xi} = 0.\tag{28.27}$$

The condition that the ripple's world line be at 45° states that the vector $\partial/\partial t$, parallel to the ripple's world line, and the vector $(\partial/\partial\eta - \partial/\partial\xi)$ are parallel. This can be written

[Often equations with unknown constants of proportionality can be translated into explicit equations involving 1-forms.]

$$dx \cdot \left(\frac{\partial}{\partial\eta} - \frac{\partial}{\partial\xi}\right) = 0\tag{28.28}$$

We can find the components of $\partial/\partial\xi$ from Equations 28.27 and 28.28. From 28.28, we have

$$\xi^x = \eta^x = 4v^2,\tag{28.29}$$

and from Equation 28.27, we have

$$-4v^2 + 2v\xi^t = 0,\tag{28.30}$$

and so

$$\xi^t = 2v. \qquad (28.31)$$

Our new basis is thus

$$\frac{\partial}{\partial \xi} = 2v\left(\frac{\partial}{\partial t} + 2v\frac{\partial}{\partial x}\right), \qquad (28.32)$$

$$\frac{\partial}{\partial \eta} = 4v\left(\frac{\partial}{\partial t} + v\frac{\partial}{\partial x}\right). \qquad (28.33)$$

To find the wave diagram in the new frame, we have only to write the vectors of the wave diagram in terms of their new components. We are calling these components $\dot{\xi}$ and $\dot{\eta}$, defined by

$$u = \dot{\xi}\frac{\partial}{\partial \xi} + \dot{\eta}\frac{\partial}{\partial \eta}, \qquad (28.34)$$

for any vector $u$. Using our relation between the new basis and the old, we can write this

$$u = \dot{\xi}2v\left(\frac{\partial}{\partial t} + 2v\frac{\partial}{\partial x}\right) + \dot{\eta}4v\left(\frac{\partial}{\partial t} + v\frac{\partial}{\partial x}\right), \qquad (28.35)$$

and collecting terms we have

$$u = (2v\dot{\xi} + 4v\dot{\eta})\frac{\partial}{\partial t} + (4v^2\dot{\xi} + 4v^2\dot{\eta})\frac{\partial}{\partial x}.$$

The components of $u$ in $(x, t)$ coordinates are defined by

$$u = \dot{x}\frac{\partial}{\partial x} + \dot{t}\frac{\partial}{\partial t}, \qquad (28.37)$$

and we can thus read off from Equation 28.36 the relations

$$\dot{x} = 4v^2(\dot{\xi} + \dot{\eta}), \qquad (28.38)$$

$$\dot{t} = 2v(\dot{\xi} + 2\dot{\eta}).$$

From these the wave diagram

$$\dot{t}^2 = \pm 4\dot{x} \qquad (28.39)$$

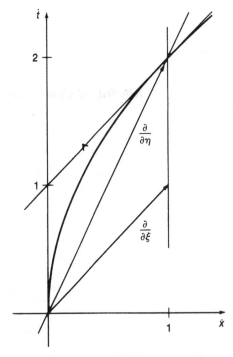

**Figure 28.10.** The basis vectors of a canonical reference frame represented in a frame at rest with the ripples and the water.

[The textbooks on tensors for physicists are full of complicated expressions for these coordinate transformations. In my experience these are rarely useful. It is usually easier and safer to work them out explicitly, as we did here.]

can be written

$$(\dot{\xi} + 2\dot{\eta})^2 = \pm(\dot{\xi} + \dot{\eta}). \qquad (28.40)$$

This is what was sketched in Figure 28.9.

*Relativity of all observers*   The wave diagram in Figure 28.9 is the water-wave counterpart of the hyperbola for special relativity. Note in particular the surprising result that the canonical wave diagrams seen by different observers are identical! The wave diagram given by Equation 28.40 does not depend on $v$. Similarly, in special relativity, every observer sees the same hyperbola. Here everyone sees the same parabola. Thus, if an observer moving over the water observes only the wave patterns, and interprets them using only a water-wave clock, such an observer will be unable to deduce his speed over the water from these observations. All observers are equivalent—relativity! Recall that our Newtonian absolute-time clocks also had a relativity symmetry, the Galilean symmetry. Clearly such symmetries are neither rare nor special.

We could now repeat all the calculations of the special-relativity chapter. There would be a transformation between canonical reference frames similar to the Lorentz transformation, a Doppler-shift formula, and a velocity-addition formula. Since the ripples are at 45° in every canonical reference frame, and since there is a canonical reference frame for every observer, this would seem to imply that one cannot ever hope to move as fast as the ripples. Although you probably accepted this argument in special relativity, the conclusion here that one cannot catch the ripples seems paradoxical. Yet it is correct in some narrow sense. One cannot slow a wave packet of water waves down to complete rest. One cannot catch the ripples, because they are not moving. But in fact our theory is only approximate, and does not correctly describe slow water waves. The more precise theory shows that if you slow down a wave packet of water waves, it *becomes* a ripple.

The above model for special-relativity symmetries should be taken as just good fun. The problem of the ripples, however, should lead us to be a bit humble about flat statements that faster-than-light travel is impossible. Very little in special relativity would need to be changed to accommodate such behavior, except, of course, the textbooks.

PROBLEMS

28.1. (08) In the example on page 173, why is $k$ negative?

28.2. (16) Draw some wave packets in the coordinates of Figure 28.9.

28.3. (18) Show that in a canonical reference frame for water waves (Figure 28.9), the wave crests and the energy flow in opposite directions.

28.4. (22) The parabola in Figure 28.9 extends back across the $\xi$-axis. Thus it appears possible to travel backward in time. Explain why this is not so.

28.5. (28) Describe the twin paradox for water-wave clocks.

28.6. (30) Find a matrix representation of the water-wave symmetry.

28.7. (25) Give an algorithm for constructing Figure 28.9 as an envelope.

28.8. (29) Give an algorithm for constructing Figure 28.2 as an envelope.

28.9. (40) The dispersion relation for water waves on water of finite depth $D$ is

$$\omega^2 = gk \tanh (kD).$$

Explore this.

28.10. (23) Show that the asymptotes in Figure 28.3 are given by

$$\frac{d\omega}{dk} = \frac{\omega}{k}.$$

28.11. (20) What is the meaning of the cusp in Figure 28.3? Where does it appear in Figure 28.2?

28.12. (34) Find a vector field which generates the symmetry of deep-water waves.

28.13. (36) Use the symmetry of water waves to show that the angle in the wake behind a boat does not depend on its speed. You do not need to actually calculate the angle.

28.14. (36) Find a relativity-like symmetry for the dispersion relation

$$\omega = k - k^3$$

of Problem 27.3.

28.15. (36) Find a relativity-like symmetry for the dispersion relation

$$\omega^2 = k^3$$

of Problem 27.4.

28.16. (36) Find a relativity-like symmetry, if there is one, for the dispersion relation

$$\omega = k(1 + \alpha k^2)$$

of Problem 27.5.

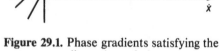

**Figure 29.1.** Phase gradients satisfying the water-wave dispersion relation.

*Initial-value problem*

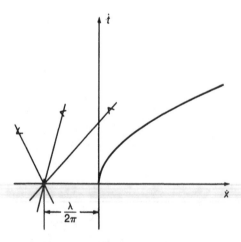

**Figure 29.2.** Phase gradients having a specified wavelength.

*Coordinate-free approach*

## 29. The Interaction of Wave Packets

The dispersion relation and its associated linear partial differential equation describe the motion of an already given wave packet. There will usually be some interaction between wave packets as well. There will also be sources for the waves. These interactions will be represented by other terms in the partial differential equation, linear terms for sources and nonlinear terms for the interactions. We can learn a surprisingly large amount from considering only the geometry of these interactions, without any discussion of the details. The analogous situation in optics is the difference between Snell's Law for refraction, and the more detailed description of the amplitude and polarization given by the Fresnel equations. From the basic geometry of the situation we will find a general form of Snell's Law valid for all wave fields, a generalization of the conservation law of 4-momentum of special relativity, and a nice view of the difficulties with tachyons.

The easiest approach to the generation of a wave packet is to put the generation so far into the past that we can ignore it. Instead we take given information about the wave at some initial time and discuss only how it propagates from then on. This is called an initial-value problem, and its geometry is common to all the discussions in this section. To describe the state of a wave packet, we need its phase gradient. This phase gradient 1-form in two dimensions must satisfy two conditions:

(i) its unit contour must be tangent to the wave diagram (that is, the phase gradient must satisfy the dispersion relation);
(ii) its contours must have the correct spacing to agree with the given initial conditions.

Figure 29.1 sketches phase gradients satisfying only the first condition for the water-wave diagram. Figure 29.2 sketches phase gradients satisfying only the second condition. In Figure 29.3 we sketch the solution of both conditions simultaneously, and in Figure 29.4 we show in a spacetime diagram the future behavior of such a wave packet. Most waves have both a right-going and a left-going mode, and for such waves one also needs information like $\partial\psi/\partial t$ to separate these modes.

These intial-value conditions can be written without using special coordinates if we use the pullback operation of Section 20. The given initial data will specify the wave structure along some surface, and we must have the pullback of the phase gradient of the wave agree with the gradient of the given phase. The gradients should have the same magnitude and direction, but need not have the same sign.

Another way to generate waves is to have some system that interacts with the field, which I will call a transmitter. We might specify, for

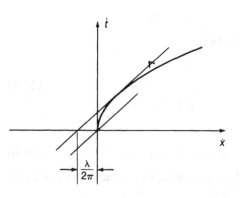

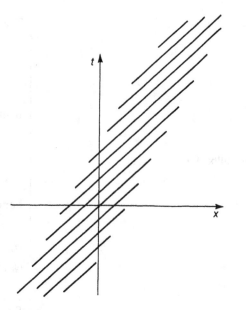

**Figure 29.3.** Phase gradient which satisfies the dispersion relation and has a specified wavelength.

**Figure 29.4.** Spacetime diagram of the particular wave packet given in Figure 29.3. (Reduced scale.)

example, the frequency of the transmitter, and we can distinguish several different cases. A small transmitter (small compared with a wavelength) will be nondirectional and represented by a single world line. Any transmitter will interact only with waves that stay in phase with the oscillations of the transmitter. The transmitter will couple only with waves whose phase gradients, when pulled back to the transmitter world line, match the transmitter frequency. Such a coordinate-independent statement lets us work problems with moving sources with little additional effort.

*Antennae*

[Remember that the dispersion relation or its dual, the wave diagram, is a complete theory of waves. It is not just a theory valid in the coordinates in which it is given.]

*Example 1*

A source of water waves moves along the path

$$s \mapsto (t, x, y) = (s, vs, 0), \qquad (29.1)$$

and oscillates with a frequency $\omega_0$ relative to the parameter $s$. The dispersion relation for the water waves is

$$\omega^4 = k_x{}^2 + k_y{}^2, \qquad (29.2)$$

and *this* tells us that these are the familiar $(t,x,y)$ coordinates at rest with the water. Write

$$d\theta = \omega \, dt + k_x dx + k_y dy. \qquad (29.3)$$

*Moving source*

[This situation is similar to one that comes up quite often in general relativity, in which the meaning of one's coordinates is implicitly contained in the geometric objects defined on the space, such as the metric.]

The pullback onto the path is found from

$$t^* = s,$$
$$x^* = vs,$$     (29.4)
$$y^* = 0,$$

from which we have

$$dt^* = ds,$$
$$dx^* = v\,ds,$$     (29.5)
$$dy^* = 0;$$

hence we have

$$d\theta^* = (\omega + vk_x)\,ds.$$     (29.6)

We are given that the frequency relative to $s$ is to be $\omega_0$; so we must have

$$\omega_0 ds = (\omega + vk_x)\,ds.$$     (29.7)

See Figure 29.5. We have two equations, 29.2 and 29.7, for three variables, so there is a one-parameter family of possible waves, corresponding to waves being sent out over a range of possible directions.

[See Section 20 for pullback.]

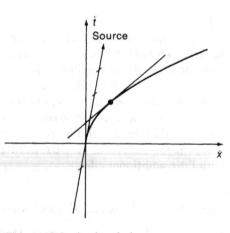

**Figure 29.5.** A sketch in one space and one time dimension of a water wave sent out by a moving periodic source.

It is also possible for the transmitter to extend over many wavelengths in one or two dimensions. Such a system is usually called an antenna, and it beams the radiation. Radio antennas are often long in one dimension, in order not to waste energy on waves sent upward,

*Beaming*

and a radar antenna is two-dimensional, to beam the waves in a particular direction. We will now have the pullback onto a two- or three-dimensional surface. The condition that our pullback agree up to a sign with the given phase will be two or three equations in addition to the dispersion relation. It is even possible to have a source extend over many wavelengths in all three space dimensions. Then only for special values of the given transmitter phasing will any waves at all be possible. Such a situation occurs in x-ray diffraction, for example. The conditions are then called the Laue equations.

One way to arrange a distributed source with the correct phase relations is to excite the sources with an incident wave packet. The simplest example of this is the reflection of a wave from a mirror. The incident wave excites currents in the conducting surface of the mirror, and these currents act as sources for the reflected wave. The same argument also applies to the refraction of a wave at an interface where the index of refraction changes abruptly. For both of these cases we can find important information from the requirement that the phases of the various waves should agree on the surface. If $\alpha$ is the phase gradient of the incident wave, and $\beta$ that of the reflected wave, then we must have

*Reflection and refraction*

$$\alpha^* = \beta^*; \tag{29.8}$$

that is, the pullbacks of $\alpha$ and $\beta$ onto the surface must be equal.

**Example 2** | A wave packet of water waves

*Moving reflector*

$$\omega^4 = k^2 \tag{29.9}$$

is incident on a moving reflector, traveling along the world line $x = vt$. Find the reflected wave. The geometry is sketched in Figure 29.6.

Let the incident wave have a frequency $\omega_0$ and a wavenumber $k_0$, satisfying the dispersion relation. Let the reflector world line be given parametrically by

$$u \mapsto (t, x) = (u, vu). \tag{29.10}$$

The pullbacks of the basis 1-forms are

$$dt^* = du,$$
$$dx^* = v\,du, \tag{29.11}$$

and so the incident wave pulls back to

$$\alpha^* = (\omega_0 + vk_0)\,du. \tag{29.12}$$

The reflected wave will have a phase gradient

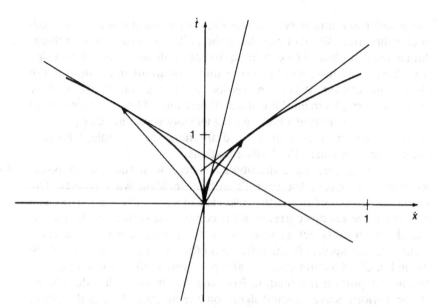

**Figure 29.6.** The geometry in the tangent space for a water wave reflecting from a moving boundary. The case shown has $v = 0.10$, $\omega = 1.40$, $k_0 = 1.97$, $\omega_1 = 2$, and $k_1 = 4$. Note unequal scales on the axes.

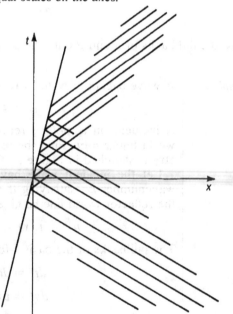

**Figure 29.7.** A spacetime diagram showing the incident and reflected waves of Figure 29.6. (Drawn to half scale.)

$$\beta = \omega_1 \, dt + k_1 \, dx, \qquad (29.13)$$

and we have equations

$$\omega_0 + v k_0 = \omega_1 + v k_1, \qquad (29.14)$$

$$k_1 = -\omega_1{}^2. \qquad (29.15)$$

We have taken $k_1$ to be negative so that the wave will move to the right. Figure 29.7 shows one solution to this problem. These geometric considerations say nothing about the phase relation between the waves, beyond matching their gradients. The situation in Figure 29.8 shows a fixed phase shift between the waves. Only a detailed study of the actual partial differential equations can tell us which situation is correct. Many uses ignore the phase of the wave, however.

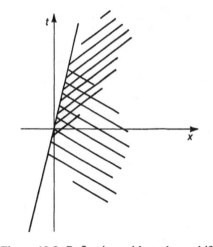

**Figure 29.8.** Reflection with a phase shift of $\pi$.

*Example 3*

Light waves in an isotropic material have a dispersion relation

$$k_x{}^2 + k_y{}^2 = \left(\frac{n}{\lambda}\right)^2. \qquad (29.16)$$

*Snell's Law*

[See the example in Section 33 also.]

The index of refraction $n$ depends on position. If it changes slowly compared to the wavelength, then a wave packet follows a path given by Huygens' construction, which we will discuss in Section 33. If the index changes rapidly compared with a wavelength, new waves are generated at the interface by reflection, and the direction of the transmitted wave changes abruptly. Look, for example, at an interface along the surface $x = 0$, sketched in Figure 29.9.

Write the phase gradient of the incident wave

$$\alpha = k_x \, dx + k_y \, dy. \qquad (29.17)$$

In terms of a direction of the wave given by the angle $\phi_1$ shown in Figure 29.10, we have

$$\alpha = \frac{n_1 \cos \phi_1}{\lambda} \, dx + \frac{n_1 \sin \phi_1}{\lambda} \, dy. \qquad (29.18)$$

The pullback of $dx$ here is zero, and the incident and transmitted waves must have the same value of $k_y$. Thus we have

$$n_1 \sin \phi_1 = n_2 \sin \phi_2, \qquad (29.19)$$

which is the familiar Snell's Law for refraction.

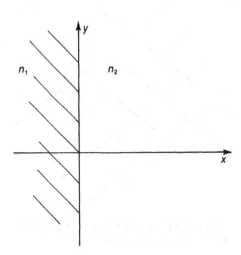

**Figure 29.9.**

[The full advantages of the pullback rule appear in situations in three or four dimensions.]

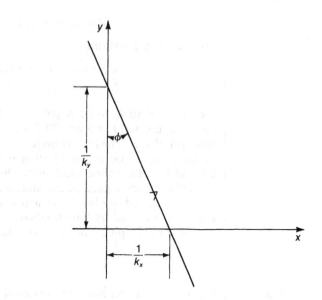

**Figure 29.10.**

*Interactions*    Wave packets may also interact with other wave packets. In Figure 29.11 I sketch two wave packets passing through the same region of spacetime. Let us suppose that there is some weak mutual interaction, not enough to destroy the waves, but enough to create some additional waves. If the source of these waves depends on the product of the fields of these two waves, then this source will be largest where both of these waves are large. These are the points on the crests of both waves, marked in Figure 29.12. The interaction of these two waves will generate a new wave only if these source points remain in phase with the

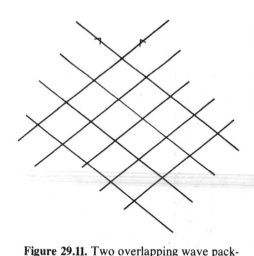

**Figure 29.11.** Two overlapping wave packets. Lines are wave crests.

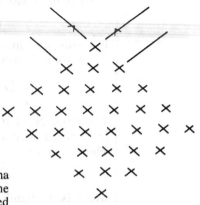

**Figure 29.12.** Crosses mark the maxima of the product of the two wave fields. The 1-forms shown are phase gradients divided by $2\pi$.

new wave over the volume of spacetime where the two incident waves overlap. A wave in phase with the above sources is sketched in Figure 29.13. This wave is simply related to the incident waves. Its phase gradient is the sum of the phase gradients of the incident waves.

*Example 4*

Wave packets of particles in spacetime satisfy the dispersion relation

$$\omega^2 = k^2 + \frac{m^2}{\hbar^2}, \qquad (29.20)$$

where $m$ is the mass of the particle. The wave diagram is given by

$$\dot{x} = \frac{2\omega\hbar^2}{m^2}, \qquad (29.21)$$

$$\dot{t} = \frac{2k\hbar^2}{m^2},$$

and the speed of the wave packet by

$$v = -\frac{\omega}{k}. \qquad (29.22)$$

This phase gradient is related to the 4-momentum of special relativity by the definitions

$$E = \hbar\omega, \qquad (29.23)$$

$$p = -\hbar k. \qquad (29.24)$$

In an interaction, our condition on the sum of the phase gradients becomes

$$E_1 + E_2 = E_3, \qquad (29.25)$$

$$p_1 + p_2 = p_3. \qquad (29.26)$$

This is the law of total 4-momentum conservation that was used in Section 14. Similar conservation laws can be written for any wave fields.

*Example 5*

Tachyons are hypothetical particles which have a dispersion relation that is Lorentz-invariant, but which have a group velocity that exceeds the speed of light. Such a dispersion relation must be of the form

$$\omega^2 = k^2 - \frac{m^2}{\hbar^2}, \qquad (29.27)$$

where $m$ is some constant. The wave diagram for such particles is given by

**4-momentum conservation**

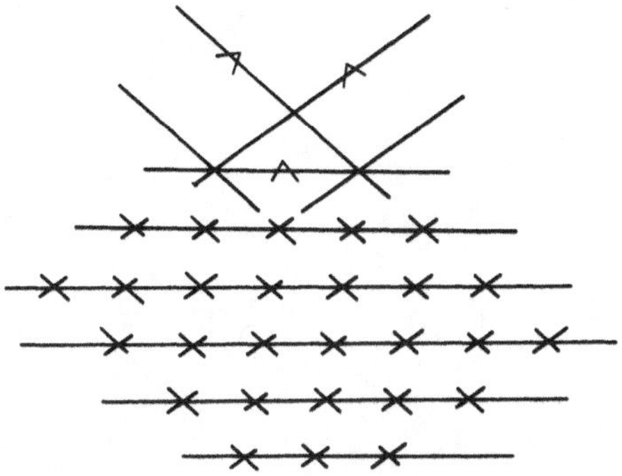

**Figure 29.13.** A wave packet that stays in phase with the above source. Its phase gradient is the sum of the phase gradient 1-forms of the incoming waves.

[Note how force and momentum always come up as 1-forms.]

*Tachyons*

$$(\dot{x})^2 = (\dot{t})^2 + \left(\frac{\hbar}{m}\right)^2 \qquad (29.28)$$

and this is drawn in Figure 29.14. A corresponding wave packet appears in Figure 29.15.

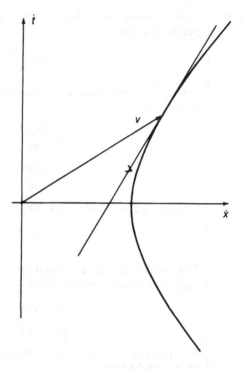

Figure 29.14. A wave diagram for a tachyon wave packet, drawn in the tangent space.

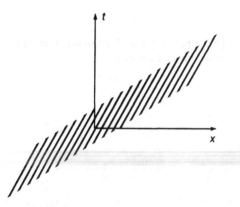

Figure 29.15. A spacetime diagram showing the wave packet described in Figure 29.14. (Reduced scale.)

Now, this looks like a perfectly consistent theory. It is just the usual theory of particles with time and space interchanged. Since time and space enter symmetrically, can anything be wrong? Is there any difference between space and time? There is, but it lies not in the equations, but in the boundary conditions. The questions that science answers are of the form: given such and such conditions, what will happen? Such questions single out the time direction as the direction of prediction. There is a direction in spacetime provided by the questions that we ask and the situations that we can arrange.

These tachyons will not be of any interest unless we can interact with them. Suppose we have some field at our disposal that interacts with the tachyon field. One simple situation would be to have a periodic arrangement of the source in space at time zero. This source will couple to a particular pair of tachyons as shown in Figures 29.16 and 29.17. So far, so good. But what if the wavelength of the source is increased? This should be easy to arrange. For the situation in Figure 29.18, note that the two modes have disappeared. The dispersion relation is a quadratic equation, and the frequency is given by the dispersion relation 29.27. For small $k$ there are no *real* solutions. The frequency becomes imaginary. Our oscillatory solutions turn into exponentials. There will be a pair of these, and one of them will be a growing exponential. The tachyon wave is unstable if perturbed with long-wavelength perturbations. No one has found a cure for this instability. It is a fatal flaw in the theory, and this has been known for some time. People still discuss tachyons, hoping, I suspect, that some elegant cure for this problem will be found.

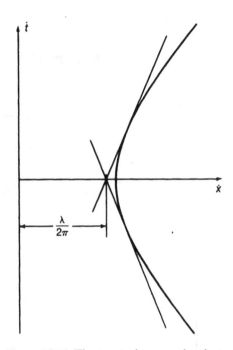

**Figure 29.16.** The two tachyon modes that couple to a source periodic in space with a wavelength λ indicated.

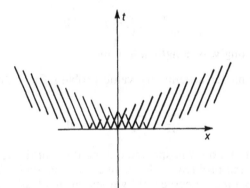

**Figure 29.17.** Two tachyon wave packets as represented in Figure 29.16 coming from a source of finite extent. One is going to the left and one to the right. (Reduced scale.)

PROBLEMS

29.1. (14) For the waves described by the wave diagram drawn in Figure 29.19, is it short waves or long waves that are unstable? What is the critical wavelength separating stable and unstable modes?

29.2. (12) Sketch the right-going wave packet whose wavelength and wave diagram are shown in Figure 29.20.

29.3. (20) *Wave crests* at time $t = 0$ are marked in Figure 29.21 for a right-going wave. Draw the world lines of these wave crests using the wave diagram sketched.

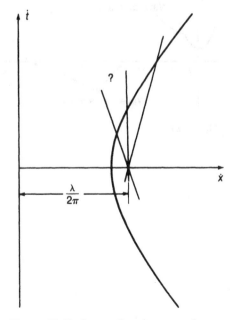

**Figure 29.18.** Loss of tachyon modes as the space wavelength increases.

**Figure 29.19.**

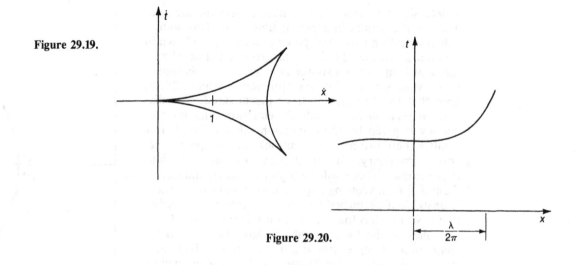

**Figure 29.20.**

29.4. (26) Show directly that the equation

$$\frac{\partial^2 \psi}{\partial t^2} - \frac{\partial^2 \psi}{\partial x^2} - \mu^2 \psi = 0$$

has unstable long-wavelength solutions.

29.5. (29) Consider a boundary-value problem for the ordinary wave equation

$$\frac{\partial^2 \psi}{\partial t^2} - \frac{\partial^2 \psi}{\partial x^2} + \mu^2 \psi = 0,$$

in which the function $\psi$ is specified along the world line $x = 0$. This is called a signaling problem. It looks like the tachyon initial-value problem rotated by ninety degrees. Why does this not lead to an instability?

29.6. (31) (For those who have had a course in electricity and magnetism.) Describe the idea of a "cutoff frequency" in a waveguide in terms of wave diagrams.

29.7. (26) In Example 2 here, for some speeds there may be no reflected wave. Show that the condition for this is

$$(u + v) < \sqrt{2}u.$$

29.8. (34) Suppose the moving obstacle in Example 2 here both allows waves to pass through it and reflects them. Show that under some conditions two additional refracted waves are possible.

29.9. (30) In the above problem we lose two wave modes at some stage. Unlike the situation in the tachyon problem, this loss does not mean that the problem is unstable. Why?

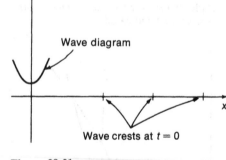

Wave diagram

Wave crests at $t = 0$

**Figure 29.21.**

# 30. Gravitation

*"One had to be a Newton
to notice that the Moon is falling,
when everyone sees that it doesn't fall."*

PAUL VALÉRY

Gravity has been ignored thus far. This puts us in a logically sloppy position — the special-relativity theory developed so far should not necessarily apply on the surface of the Earth. True, for a $\mu$ meson that lives for only a few microseconds, we can certainly ignore gravity; yet it is disturbing to have one's fundamental assumptions only approximately satisfied. The effect of gravity is to leave us with no free particles. Gravity, unlike any other force, cannot be excluded or canceled by shielding. Nothing that we know escapes the pull of gravity.

The resolution of this situation is both surprising and elegant. We do not *add* gravity to our theory. Instead we go back and build it in right in the foundations. The clue to this is contained in Problem 1.8. There you were asked to discover that the map

*New idea of free particle*

$$(x, t) \mapsto (X, T) = (x - \tfrac{1}{2}t^2, t) \tag{30.1}$$

(or any linear version of it) turns the family of parabolae

$$x = a + bt + \tfrac{1}{2}t^2 \tag{30.2}$$

into the straight lines

$$X = a + bT. \tag{30.3}$$

Objects falling in a uniform gravitational field follow such a family of parabolae; so we can redefine our notion of a free particle to refer to these particles without violating the free-particle postulates. This turns out to be a useful and fruitful idea, and henceforth we follow it.

---

**Free Particle: A particle moving under the influence of no forces except possibly gravity.**

---

[Dodson and Poston used a curved space model for light propagation to describe a mirage. See page 423 of their book for a nice figure.]

In the spirit of the epigraph, we can say that it took the genius of Einstein to see that the Moon is moving in a straight line, when everyone sees that it doesn't.

With this new idea of a free particle, we can use special relativity even in uniform gravitational fields. Since all smooth fields are uniform over small regions, we can now use special relativity near every point to find the local behavior of particles, clocks, and light signals. This situation is a vast improvement over having special relativity apply only when gravity is absent. It is in this sense that we can say that special relativity describes the local geometry of spacetime, just as a plane describes the local geometry of any smoothly curved surface.

Once we decide to set up inertial reference frames using the above notion of a free particle, then we must study the local behavior of these free particles in various gravitational fields. The dispersion relation for particles might no longer be the special-relativity hyperbola. This is a hard question to study experimentally. We are pretty much stuck at one value of the gravitational field. Still, the Earth's rotation moves us in and out of the Sun's field daily, and the eccentricity of the Earth's orbit moves us even further annually. Careful searches have been made for effects with these periodicities. So far they have turned up no evidence that any modification of the special-relativity dispersion relation is needed. Space experimenters hope to send a clock close to the Sun to look more closely for such effects.

*Local theory is still special relativity*

Some metaphysical reasoning leads us to believe that the special-relativity dispersion relation will not be modified by gravitational fields. Suppose that in some theory there is a dimensionless parameter $\theta$, which we measure in some situation to have the value $1.473 \pm .001$. It would be easy to conceive that in some nearby situation it could have some slightly different value. On the other hand, suppose we measure the value of $\theta$ to be $2.001 \pm .001$. Then we would suspect that we really have $\theta = 2$ exactly, and that $\theta$ would no longer be able to change slowly with a changing situation. The special-relativity dispersion relation is as special among dispersion relations as the integers are among real numbers. The special-relativity dispersion relation is a quadratic curve; so it can be expressed in terms of a second-rank tensor. This is a fantastic simplification. If it is really true here and now, then it is probably true everywhere.

*General relativity*

General relativity is the theory of gravitation in which the motion of waves and particles near any point is given by a tensor field $\mathcal{G}$,

$$\mathcal{G} = g_{\alpha\beta} dx^\alpha \otimes dx^\beta, \tag{30.4}$$

and free particles of mass $m$ are described by a wave diagram

$$g_{\alpha\beta}\dot{x}^{\alpha}\dot{x}^{\beta} = -m^2 \qquad (30.5)$$

or its equivalent dispersion relation. The effects of gravity show up in the variation of $\mathcal{G}$ from point to point. This variation of $\mathcal{G}$ is related to the matter content of spacetime by equations called the Einstein field equations. That is all there is to it. Of course, we have invested a lot of effort in gaining this viewpoint from which general relativity can be described so simply. The manner in which the matter content determines $\mathcal{G}$ is quite technical, and involves a further kit of mathematical tools that would take us weeks to develop. We will leave that problem to a special course on general relativity. Here we will take the spacetimes as given and study their consequences. This will allow us to study the standard cosmological models and their interpretations in all the detail needed.

*Example* | The spacetime in the outer regions of a nonrotating black hole immersed in an otherwise flat spacetime is given by the metric tensor

$$\mathcal{G} = -\left(1 - \frac{2m}{r}\right)dt^2 + \frac{dr^2}{\left(1 - \frac{2m}{r}\right)} \qquad (30.6)$$

$$+ r^2(d\theta^2 + \sin^2\theta \, d\phi^2).$$

The most difficult feature of these general-relativity spacetimes is that their coordinates have no *a priori* meaning. One interprets the coordinates by studying the spacetime itself: answer first, question later.

*Geometric units*

The mass unit used in Equation 30.6 is the geometric unit of mass which results from setting both $c = 1$ and $G = 1$. **The dimension of mass in geometric units is seconds.** A physical definition of a mass of one second is given by Kepler's law,

$$a^3\omega^2 = m, \qquad (30.7)$$

where the orbital semimajor axis $a$ is measured in light seconds and the orbital frequency in radians/second. The geometric mass of the Sun is $4.92 \times 10^{-6}$ seconds, the mass of the Earth is $1.48 \times 10^{-11}$ seconds, and the mass of one kilogram is $2.48 \times 10^{-36}$ seconds. [See Table 30.1.]

[See page 140 for a discussion of
this inverse.]

*Light signals*    What about light signals? In our special-relativity discussion of light signals, we did not distinguish between group and phase velocities. That holds in general relativity as well, as we now show. The dispersion relation for a light-signal wave packet is given by the tensor equation

$$g^{\alpha\beta} g_\alpha p_\beta = 0, \tag{30.8}$$

where here the $g^{\alpha\beta}$ are the components of the inverse of the metric tensor, and the $p_\alpha$ are the components of the phase gradient

$$p = d\theta. \tag{30.9}$$

Equation 30.8 can be taken as either a postulate, a statement of experimental observations, or a derivation from Maxwell's equations. For us it is a convenient starting point.

**TABLE 30.1.   GEOMETRIC UNITS**

*Lengths:*

| | |
|---|---|
| foot | $1.02 \times 10^{-9}$ sec. |
| meter | $3.34 \times 10^{-9}$ sec. |
| radius of the Earth | $2.12 \times 10^{-2}$ sec. |
| radius of the Sun | $2.32$ sec. |
| astronomical unit | $499$ sec. |
| light-year | $3.16 \times 10^{7}$ sec. |
| parsec | $1.03 \times 10^{8}$ sec. |

*Mass:*

| | |
|---|---|
| kilogram | $2.48 \times 10^{-36}$ |
| mass of the Earth | $1.48 \times 10^{-11}$ |
| mass of the Sun | $4.92 \times 10^{-6}$ |

*Miscellaneous:*

| | |
|---|---|
| density of water | $6.67 \times 10^{-8}$ sec$^{-2}$ |
| h | $1.82 \times 10^{-86}$ sec$^{2}$ |
| $(50 \text{ km/sec/mpc})^{-1}$ | $6.18 \times 10^{17}$ sec. |

**Example**

For the special-relativity metric

$$\mathscr{G} = -dt^2 + dx^2, \qquad (30.10)$$

the inverse $\mathscr{G}^{-1}$ is given by

$$\mathscr{G}^{-1} = -\frac{\partial}{\partial t} \otimes \frac{\partial}{\partial t} + \frac{\partial}{\partial x} \otimes \frac{\partial}{\partial x}. \qquad (30.11)$$

[See page 132 for this specific inverse.]

If we write

$$p = k\, dx + \omega\, dt, \qquad (30.12)$$

then our dispersion relation

$$\mathscr{G}^{-1} \cdot (p,\, p) = 0 \qquad (30.13)$$

can be written out in terms of components in index notation. By using

$$\mathscr{G}^{-1} = g^{\alpha\beta} \frac{\partial}{\partial x^\alpha} \otimes \frac{\partial}{\partial x^\beta} \qquad (30.14)$$

and

$$p = p_\alpha dx^\alpha, \qquad (30.15)$$

we find that

$$g^{\alpha\beta} p_\gamma p_\delta \frac{\partial}{\partial x^\alpha} \otimes \frac{\partial}{\partial x^\beta} \cdot (dx^\gamma,\, dx^\delta)$$

$$= g^{\alpha\beta} p_\alpha p_\beta = 0. \qquad (30.16)$$

Thus we have recovered Equation 30.8. In terms of $k$ and $\omega$, we have

$$\mathscr{G}^{-1} \cdot (p,\, p) = k^2 - \omega^2 = 0 \qquad (30.17)$$

and this is the familiar dispersion relation.

This dispersion relation has a peculiarity that makes it necessary for us to go back and repeat the group-velocity derivation. To find the group velocity $\sigma$, we take the space derivative of the dispersion relation

*Group velocity for light*

[This derivation is a nice exercise in index shuffling.]

$$\frac{\partial}{\partial x^\gamma} (g^{\alpha\beta} p_\alpha p_\beta) = \frac{\partial g^{\alpha\beta}}{\partial x^\gamma} p_\alpha p_\beta + 2 g^{\alpha\beta} p_\alpha \frac{\partial p^\beta}{\partial x^\gamma}. \qquad (30.18)$$

Because $p$ is already a gradient, we have

$$\frac{\partial p_\beta}{\partial x^\gamma} = \frac{\partial p_\gamma}{\partial x^\beta}, \qquad (30.19)$$

since partial derivatives commute. Thus we find an equation

$$\frac{\partial p_\gamma}{\partial x^\beta}(g^{\alpha\beta}p_\alpha) = -\frac{1}{2}\frac{\partial g^{\alpha\beta}}{\partial x^\gamma}p_\alpha p_\beta.$$ (30.20)

The lefthand side is the derivative of the $p_\gamma$ in the direction which we call the group-velocity direction,

$$\sigma = g^{\alpha\beta}p_\alpha\frac{\partial}{\partial x^\beta}.$$ (30.21)

The special property of these waves shows up in the fact that

*Cannot normalize*

$$p \cdot (\sigma) = 0.$$ (30.22)

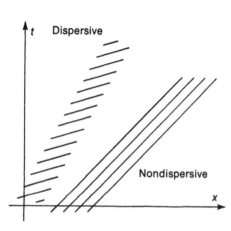

**Figure 30.1.** The difference between dispersive and nondispersive waves.

This foils our attempt to find the usual wave diagram for such waves. The wave diagram turns into a ray pointing straight out from the origin. Any attempt to normalize it would unfortunately turn it into a single point. This ray is our dashed light-signal world line. For these waves the energy flows at the same speed as the wave crests. In Figure 30.1 we compare such wave packets with ordinary dispersive wave packets. Because of the lack of dispersion in light, we can take either an entire wave packet or a particular wave crest as a light signal. They both travel at the same speed. This fact justifies the procedure that we used in Section 13 to compute the Doppler shift. There we computed the frequency shift from the spacing between two light-signal world lines. Doing so is valid only for waves with no dispersion. A similar procedure will be used in Section 32 to compute the frequency shift in the gravitational field of the Earth, and in Section 40 for the shift in an expanding universe.

PROBLEMS

30.1. (10) Verify Kepler's Law for the Earth-Sun system in geometric units.

30.2. (12) Show that the geometrical mass $m_g$ is related to the mass in kilograms, $m_k$, by

$$m_g = \frac{Gm_k}{c^3}.$$

30.3. (23) The Euclidean metric $\mathscr{E}$ written in polar coordinates is

$$\mathscr{E} = dr^2 + r^2\,d\theta^2.$$

This is a simple example of a metric tensor whose components are not independent of the coordinates. Derive this, sketch the metric, and discuss.

30.4. (28) Show that the two-dimensional spacetime $(u, v)$ with metric structure

$$\mathcal{G} = du^2 - u^2\, dv^2$$

is just Minkowski spacetime in strange coordinates. Use the transformation [See Problem 1.5.]

$$t = u \cosh v.$$

$$x = u \sinh v.$$

# 31. Spacetime Near the Earth

In the next few sections we will study the external gravitational field of a spherically symmetric body like the Earth. This will show us how curved spacetime can indeed describe gravity, tell us what spacetime curvature means physically, and give us one experimental test of general relativity.

The metric $\mathcal{G}$ outside the earth or any spherically symmetric body is given approximately by

*Spacetime metric*

$$\mathcal{G} = -dt^2 + dr^2 + r^2(d\theta^2 + \sin^2\theta\, d\phi^2) + \frac{2m}{r}(dt^2 + dr^2). \quad (31.1)$$

On the surface of the Earth, the components in Equation 31.1 differ from the special-relativity values by about 1 part in $10^9$. I have neglected many small corrections to this already small term, among them one due to the oblateness of the Earth (an error of about 1 in $10^3$), one due to the mass of the atmosphere (about 1 in $10^6$), one due to quadratic terms such as $(m/r)^2$ (about 1 in $10^9$), and others due to the influence of the Sun and the Moon (about 1 in $10^{17}$). The general-relativity calculation that gives Equation 31.1 is beyond our present skills, but it yields a unique spacetime outside any spherically symmetric body of a given mass, independent of its internal structure, be it a planet or a black hole. The above metric is the same as that given in Section 30 for a black hole, simplified by neglecting the $(m/r)^2$ terms.

[I am deliberately using an approximate expression here, even though the exact expression is hardly more complicated, in order to emphasize that there are many astrophysical approximations that are more important. The precision of the exact general-relativity solution is empty precision. These astrophysical approximations are only rarely mentioned, partly because so much general relativity is done by mathematicians.]

Any general-relativistic spacetime contains within itself its own interpretation. Given an unknown photograph, you could study it and use internal evidence to discover where, when, and how it was taken. Study a spacetime in the same spirit. Thus, what does the above spacetime represent?

*Interpretation of the coordinates*

If we put in $m = 0$, then we are left with the Minkowski spacetime of Section 23, written in polar coordinates. The absence of a body leads to Minkowski spacetime.

***Proof***

Let us show explicitly that the metric

$$\mathcal{N} = -dt^2 + dx^2 + dy^2 + dz^2, \qquad (31.2)$$

when written in polar coordinates,

$$z = r \cos \theta,$$
$$x = r \sin \theta \cos \phi, \qquad (31.3)$$
$$y = r \sin \theta \sin \phi,$$

takes the above form. In $(r,\theta,\phi)$ coordinates, the old coordinates $(x,y,z)$ are each functions. These functions have gradients, 1-forms given by

$$dz^* = \cos \theta \, dr - r \sin \theta \, d\theta,$$
$$dx^* = \sin \theta \cos \phi \, dr + r \cos \theta \cos \phi \, d\theta$$
$$\qquad - r \sin \theta \sin \phi \, d\phi, \qquad (31.4)$$
$$dy^* = \sin \theta \sin \phi \, dr + r \cos \theta \sin \phi \, d\theta$$
$$\qquad + r \sin \theta \cos \phi \, d\phi.$$

The metric $\mathcal{N}$ is written in terms of the 1-forms $dx$, $dy$, $dz$. To transform to the new coordinates, we just use the expression for the old 1-forms in terms of the new ones. Substituting, we have

$$\mathcal{N} = -dt^2 + (\cos \theta \, dr - r \sin \theta \, d\theta)^2$$
$$+ (\sin \theta \cos \phi \, dr + r \cos \theta \cos \phi \, d\theta$$
$$- r \sin \theta \sin \phi \, d\phi)^2 \qquad (31.5)$$
$$+ (\sin \theta \sin \phi \, dr + r \cos \theta \sin \phi \, d\theta$$
$$+ r \sin \theta \cos \phi \, d\phi)^2,$$

and after a flurry of cancellation, we find the desired result:

$$\mathcal{N} = -dt^2 + dr^2 + r^2 \, d\theta^2 + r^2 \sin^2 \theta \, d\phi^2. \quad (31.6)$$

***Symmetries***  The most important features of a spacetime are its symmetries. Here the time $t$ variable does not appear in $\mathcal{G}$, and so the transformation

$$(t,r,\theta,\phi) \mapsto (t + a,r,\theta,\phi) \qquad (31.7)$$

for any value of the parameter $a$ leaves $\mathscr{G}$ unchanged. The vector field generating this transformation is just $\partial/\partial t$. Any spacetime with such a time-shift symmetry is called stationary. The metric $\mathscr{G}$ is also unchanged under a reflection which takes $t \mapsto -t$ and $dt \mapsto -dt$. Stationary spacetimes with this additional symmetry are called static. Finally, the angular coordinates $\theta$ and $\phi$ appear in the form describing the geometry on the surface of a sphere, and so the spacetime is also spherically symmetric.

*Time-shift symmetry*

*Spherical symmetry*

Let us look at one of these surfaces of spherical symmetry, given by $r = r_0$ and $t = t_0$. The size of this spherical surface as measured by instruments that lie in the surface itself will be $r_0$, because $(r_0)^2$ is the coefficient of the part of the metric $(d\theta^2 + \sin^2\theta \, d\phi^2)$ that represents the surface of the sphere.

*Spherical shells*

*Example*

Let's calculate the circumference of this sphere, that is, the proper length around the equator or any other great circle. The equator is a curve given by

$$u \mapsto [T(u), R(u), \Theta(u), \Phi(u)] = (t_0, r_0, \frac{\pi}{2}, u),$$

$$(31.8)$$

where $0 \leq u \leq 2\pi$. A short piece of this curve, from $u$ to $u + \Delta u$, is represented by a vector $\Delta u \, (\partial/\partial\phi)$, and the proper length of this vector is found from the metric $\mathscr{G}$ to be

$$(\Delta l)^2 = \mathscr{G} \cdot \left( \Delta u \frac{\partial}{\partial\phi}, \Delta u \frac{\partial}{\partial\phi} \right) = (\Delta u)^2 r_0^2. \quad (31.9)$$

[Note that once you see what is going on, you can just read this right off from $\mathscr{G}$.]

Adding up proper lengths all along the curve gives us

$$l = \int_0^{2\pi} \Delta l = \int_0^{2\pi} r_0 \, du = 2\pi r_0. \quad (31.10)$$

Because the circumference is $2\pi r_0$, we say that the size of the sphere is $r_0$ as measured internally. The area would come out to be $4\pi r_0^2$ as well. Note that we cannot measure the diameter of this sphere. If we try to do so, we run into the Earth, where the above expression for $\mathscr{G}$ is no longer valid. That is why we used internal measurements to find its size. An analogous situation will come up for four-dimensional spacetimes. There too we will discuss only their internal geometry. We will say nothing about whether this spacetime is imbedded in some larger space.

[Remember that for us "internally" means staying on the surface of the sphere.]

*Spacetime curvature*   We can also measure the spacing between different $r$=constant shells. A vector connecting two shells at radii $r_0$ and $r_0 + \Delta r$ will be $\Delta r\,(\partial/\partial r)$, and the length of this vector will be the proper distance $\Delta l$ between the two shells in the limit $\Delta r \mapsto 0$. Using our expression for the metric gives us

$$(\Delta l)^2 = \mathscr{G} \cdot \left(\Delta r \frac{\partial}{\partial r}, \Delta r \frac{\partial}{\partial r}\right) = \left(1 + \frac{2m}{r_0}\right)(\Delta r)^2, \qquad (31.11)$$

and so, to our level of approximation,

$$\Delta l = \left(1 + \frac{m}{r_0}\right)\Delta r. \qquad (31.12)$$

Such a result could not happen in Euclidean geometry. There two spheres of proper sizes $r_0$ and $r_0 + \Delta r$ must be a distance $\Delta r$ apart. This departure from Euclidean geometry is what is called spacetime curvature, and this calculation shows that in this sense spacetime near the Earth is curved.

*Another view of curvature*   The large number of intermediate steps in the above discussion might make you feel that spacetime curvature is in some way mysterious. Not so. Let me repeat the discussion in a different and more concrete form. Imagine the following experiment. At an event $E$ an observer following a world line $r =$ constant, $\theta =$ constant, and $\phi =$ constant sends out light signals in all directions. At a short proper time $\Delta\tau$ later, where are these signals? Look first only in the radial direction, so that we can draw a two-dimensional spacetime diagram, Figure 31.1. For events close to event $E$, the coordinates $r$ and $t$ form an inertial reference frame but not a canonical reference frame. The departures from a canonical reference frame are terms that are linear in $(m/r)$. The events which the observer sees to be simultaneous at a time $\Delta\tau$ later than $E$ form the unit contour of a 1-form,

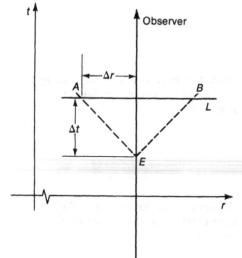

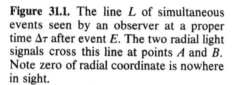

**Figure 31.1.** The line $L$ of simultaneous events seen by an observer at a proper time $\Delta\tau$ after event $E$. The two radial light signals cross this line at points $A$ and $B$. Note zero of radial coordinate is nowhere in sight.

$$\omega = \left(1 - \frac{m}{r_0}\right)\frac{dt}{\Delta\tau}, \qquad (31.13)$$

again using the binomial theorem to simplify a square root. A vector representing the world line of a light signal from the point of emission to the event $\Delta\tau$ later must satisfy two conditions,

$$\omega \cdot (\sigma) = 1, \qquad (31.14)$$

$$\mathscr{G} \cdot (\sigma, \sigma) = 0, \qquad (31.15)$$

and writing

$$\sigma = \Delta t \frac{\partial}{\partial t} + \Delta r \frac{\partial}{\partial r}, \tag{31.16}$$

we have

$$\left(1 - \frac{m}{r_0}\right)\Delta t = \Delta \tau, \tag{31.17}$$

$$-\left(1 - \frac{2m}{r_0}\right)(\Delta t)^2 + \left(1 + \frac{2m}{r_0}\right)(\Delta r)^2 = 0 \tag{31.18}$$

which says that

$$\Delta r = \pm\left(1 - \frac{m}{r}\right)\Delta \tau. \tag{31.19}$$

[Note here another instance of why we must distinguish $dr$, a 1-form, from $\Delta r$, a small coordinate increment.]

It is easy enough to repeat this argument using one more space dimension. We will work in the equatorial plane, $\theta = \pi/2$, and use $\phi$ as the additional coordinate. The events where the light signals are found a time $\Delta\tau$ after emission are now given by

*Light signals*

$$\sigma = \Delta t \frac{\partial}{\partial t} + \Delta r \frac{\partial}{\partial r} + \Delta\phi \frac{\partial}{\partial \phi}, \tag{31.20}$$

subject to the conditions

$$\left(1 - \frac{2m}{r_0}\right)(\Delta t)^2 = (\Delta\tau)^2, \tag{31.21}$$

$$\left(1 + \frac{2m}{r_0}\right)(\Delta r)^2 + r_0^2(\Delta\phi)^2 = (\Delta\tau)^2. \tag{31.22}$$

To visualize what is going on here, we will now draw a space-space diagram, drawing just those events that are a time $\Delta\tau$ later than $E$ in Figure 31.2. The events where the light signals are found lie on a ellipse in this diagram. Such a representation of light signals is quite useful.

The physical interpretation of Figure 31.2 is straightforward. Light propagation in this spacetime behaves as if the index of refraction were anistropic. Since spacetime distances are measured in terms of light signals, this anistropy appears in our discussion of the geometry, and leads to the non-Euclidean effects that we call curvature.

[Note how this argument slips back and forth between tangent vectors and nearby events in spacetime. This works because tangent vectors were defined by the local behavior of curves of events. It is valid in the limit as $\Delta\tau$ goes to zero.]

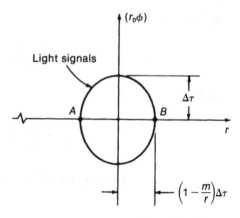

**Figure 31.2.** A spatial section containing events that are simultaneous and a time $\Delta\tau$ after event $E$. Line $L$ of Figure 31.1 appears here as the $r$-axis, and events $A$ and $B$ are shown as well. The azimuthal scale was chosen so that as $m \to 0$, the light-signal locus becomes a circle.

| | |
|---|---|
| *Example* | Look at Figure 31.2 and check the sign in Equation 31.22. From Equation 31.12 we see that nearby spheres of constant $r$ and $t$ are further apart in proper distance than Euclidean geometry would lead us to expect. In Figure 31.2 we see that the metric-figure ellipses are flattened in the radial direction. Since our length standard is the distance that light travels in some standard time interval, and a shorter length standard means a longer measured distance, this is all consistent. |

This discussion of the geometry of the spacetime near the Earth was simplified considerably by the time-shift symmetry. Without it we would have not been able to discuss the spatial geometry separately. This symmetry will also be used in the next section to simplify the discussion of the gravitational redshift.

### PROBLEMS

31.1. (14) Show that, according to our definitions, a spinning disc is stationary but not static. Invent a spacetime that is stationary but not static.

31.2. (22) Use the arguments of this section to show that a plane is flat and that a sphere is curved.

31.3. (18) For the example on page 201, calculate the circumference along a meridian rather than along the equator.

31.4. (25) Discuss the experimental precision needed to measure the space curvature near the surface of the Earth.

31.5. (20) Derive the first three estimates given on page 199.

31.6. (27) Show that the space described by the metric tensor

$$\mathcal{G} = \frac{dr^2}{1 - r^2} + r^2\, d\theta^2$$

is really the surface of a sphere in unusual coordinates. Discuss the range of the $r$ coordinate in particular.

# 32. Gravitational Redshift

The conceptually simplest experiment in a curved spacetime will involve only clocks. Clocks are simpler than rigid rods in our formalism, although not in practice. The rates of clocks at different locations can be compared, and the effects of gravity on observed clock rates is called the gravitational redshift. The shift can be to either higher or lower frequencies. The first observations were on light coming to us from the photospheres of white dwarf stars, which was shifted to the red; hence the name. The effect has also been observed with gamma rays in the laboratory and by using a precision clock in a rocket.

[This will be your very first general-relativity calculation!]

If you are learning to be cautious even with old ideas, you now expect a discussion about how to compare the rates of clocks at different speeds. If so, good. One approach would be to agree that the rates are to be compared by exchanging light signals between the clocks. This will work in most circumstances. Here an even broader class of comparisons can be considered and will all give the same result. The clock rates can be compared against the time-shift symmetry of the spacetime.

*Symmetry used to compare rates*

As long as the clocks are at rest—and "at rest" has meaning here only in terms of the time-shift symmetry—the vector field $\partial/\partial t$ which generates the symmetry provides a natural time unit along every clock's world line.

[Although I could have avoided this "heavy" statement, it is important for you to really come to grips with the idea of symmetry. I have gone out of my way to bring it in, hoping that practice will make its abstract ideas tangible.]

*Example* | Suppose you want to use some process, such as the postal service, to compare clock rates. As long as this process shares the time-shift symmetry, that is, the letters go as slowly now as later, then the clock rates can be compared by sending two letters. The first says, "Start counting ticks." The second says, "Stop counting them. I found $N$ ticks myself." The recipient then need only compare this number $N$ to his observed number to find the relative rate of the two clocks. More accuracy can be obtained by waiting longer between letters.

We will compute the gravitational redshift using light signals. Bear in mind, however, that the time-shift symmetry not only simplifies the calculation, but also allows it to be extended to any other process of comparison that shares the time symmetry. You should learn from this section not only about the gravitational redshift, but also more about symmetry and its uses.

*Rotation ignored*

To be honest with you, the calculation that we will do here is not directly applicable to the terrestrial experiments that we will cite. They do not use clocks at rest, but (with good reason) allow them to rotate with the Earth. To correct the rates for this motion we need to use only the special-relativity Doppler-shift expression. This is not an instructive complication; so it is left to the problems.

[I suggest you read over the Doppler-shift calculation of Section 13. The calculation here will be done in a parallel fashion.]

The straightforward way to compare two clock rates by using light signals is first to find the world lines of the light signals, and then to find how local observers at rest would describe these light signals by using their clocks. For simplicity, I will discuss only the case where the two clocks are in a vertical line. The light signals will then move in the $(r, t)$ plane and keep constant values of $\theta$ and $\phi$. Any such light signal is described by the parametrized curve

*Spacetime geometry*

$$u \mapsto [u, R(u), \theta_0, \phi_0], \tag{32.1}$$

whose tangent vector $\sigma$ is

$$\sigma = \frac{\partial}{\partial t} + \frac{dR}{du}\frac{\partial}{\partial r}. \tag{32.2}$$

A light signal must have a tangent vector which satisfies

$$\mathscr{G} \cdot (\sigma, \sigma) = 0, \tag{32.3}$$

since special relativity is still true locally. This condition gives us an ordinary differential equation whose solutions specify the light-signal world lines:

[All through here I am ruthlessly tossing out terms quadratic in $(m/r)$.]

$$\frac{dR}{du} = 1 - \frac{2m}{r}. \tag{32.4}$$

Note that our time-shift symmetry is still with us. If $R(u)$ is any solution to Equation 32.4, then so is $R(u - u_0)$ for any fixed value of $u_0$. This is sketched in Figure 32.1. The vector field $\partial/\partial t$ pushes one light-signal world line into another one. Because the $t$-coordinate used here is directly related to the time-shift symmetry, the coordinate interval $\Delta t$ between the two light-signal world lines is constant.

In fact, once we realize that the coordinate interval between two light-signal world lines is constant, we need no further details about the world lines. An observer at rest at a radius r will measure a proper time interval $\Delta\tau$ corresponding to a coordinate interval $\Delta t$ given by

$$(\Delta\tau)^2 = -\mathcal{G} \cdot \left(\Delta t \frac{\partial}{\partial t}, \Delta t \frac{\partial}{\partial t}\right) = \left(1 - \frac{2m}{r}\right)(\Delta t)^2. \tag{32.5}$$

Different observers, at radii $r$ and $r'$, will find a ratio between their clock rates of

$$\frac{\Delta\tau'}{\Delta\tau} = 1 + \frac{m}{r} - \frac{m}{r'}. \tag{32.6}$$

For a laboratory experiment these radii will be nearly equal. If we write

$$r' = r + H, \tag{32.7}$$

where $H$ is the height of the drop, satisfying

$$H \ll r, \tag{32.8}$$

then we can approximate the above expression by

$$\frac{\Delta\tau'}{\Delta\tau} = 1 - \frac{mH}{r^2}. \tag{32.9}$$

We will verify in the next section that the parameter $m$ here is really the mass of the gravitating body, measured in geometric units. The acceleration of gravity now has units of $(\text{second})^{-1}$, the reciprocal of the time required at such an acceleration to reach speeds comparable to the speed of light, and is given by

$$g = \frac{m}{r^2}. \tag{32.10}$$

The acceleration of the Earth's gravity is the reciprocal of a year (approximately). Our frequency shift can be written

*Symmetry of light-signal world lines*

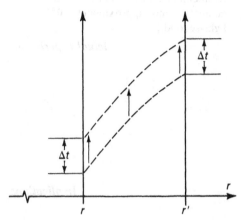

**Figure 32.1.**

*Laboratory experiment*

[See Table 30.1 for these units. The mass of the Earth is $1.48 \times 10^{-11}$ sec.]

$$\frac{\Delta\tau'}{\Delta\tau} = 1 - gH. \qquad (32.11)$$

[Remember that the foot is a good geometric unit, approximately $10^{-9}$ light-second.]

*Actual experiments*

For a 30-foot drop, as used in a typical laboratory experiment, we will have a $gH$ of about $10^{-15}$. This is not an easy experiment.

The experiment has actually been done. The light signals were the various wavecrests of gamma rays emitted by decaying nuclei, and the Mossbauer effect was used to compare the frequencies. The effect observed agreed with the predictions of general relativity to about 1%, which is as accurate as the experiment could be made. (Actually, the initial results did not agree well with the theory. Then it was noticed that the temperature of the nuclei gives them a motion which slows their clocks down. This must be carefully monitored and corrected for. Thus not only is the twin paradox observed for nuclear clocks, but it is a side effect that must be corrected for in other experiments.)

*Implications*

It is comforting that the experiment gives results that agree with predictions. Does this in itself mean that general relativity is the correct theory of gravitation? No, of course not. One must ask if any of the competing theories also predict the same gravitational redshift. Since we do not have in hand all possible competing theories, we must look around and see how much this derivation depends on the details of general relativity. The answer to that is, unfortunately, not very much. We will present two more derivations of the gravitational redshift, slowly reducing the number of assumptions. The final conclusion will be that nearly any theory of gravity will lead to the same frequency shift.

*Energy conservation*

The gravitational frequency shift can be "derived" from energy-conservation arguments alone. Of course, we have not demonstrated that general relativity is compatible with energy conservation. In fact, energy conservation is a difficult question in an expanding universe. In the static spacetime being discussed here, there is no problem.

*A gravitational engine*

Let us look at two small systems, each one capable of storing energy and emitting light signals. Think of the systems as nuclei if you wish, of masses $M$ and $m$; the light signals can then be gamma rays. Each system must be capable of absorbing a gamma ray, and the masses of the excited systems will be larger; call them $M^*$ and $m^*$.

[At this point we have used the mass-energy relation of special relativity, which applies to inertial mass, also to the gravitational mass. Some theories of gravity escape our argument at this point.]

We consider the following closed cycle of operations, diagrammed in Figure 32.2. Start with $m$ excited and $M$ not. M is above $m$. This is state 1 in the figure. In state 2 the lower nucleus has sent out its energy as a gamma ray of energy $h\nu$, given by

[Here $h$ is Planck's constant, numerically, $1.83 \times 10^{-86}$ sec².]

$$h\nu = m^* - m. \qquad (32.12)$$

In state 3 this gamma ray has traveled up to $M$, arriving there with a new frequency $\nu'$. Assume that $M$ is such that this new energy is able to exactly excite the upper nucleus. Thus we must have

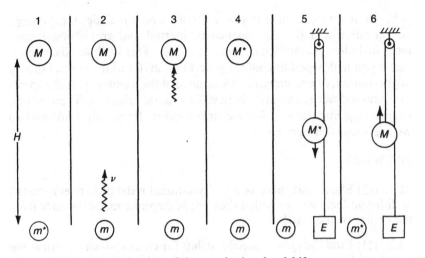

**Figure 32.2.** Quick derivation of the gravitational redshift.

$$hv' = M^* - M, \qquad (32.13)$$

which is state 4. In state 5 we slowly lower down the excited nucleus, $M^*$, collecting its gravitational energy in some collector, an energy

$$E_{\text{down}} = M^* gH. \qquad (32.14)$$

(We call the height $H$ to avoid confusing it with Planck's constant.) At the bottom we collect the excitation energy from $M^*$, and raise $M$ back up to the top. It now weighs less, and so we end up with some net energy left over at the bottom,

$$hv' + E_{\text{down}} - E_{\text{up}} = (M^* - M)gH + hv'. \qquad (32.15)$$

This must be enough to exactly reexcite $m$. If not, then one could construct a source of energy by running the cycle one way or the other. Thus we must have

$$hv = hv' \, (1 + gH); \qquad (32.16)$$

that is,

$$\frac{v}{v'} = 1 + gH, \qquad (32.17)$$

the same result found earlier.

*Redshift from energy conservation*    From this we conclude that any theory which includes local energy conservation and the equivalence of inertial and gravitational mass must include a gravitational frequency shift. This is either disappointing, if you had hoped to confirm general relativity as the correct theory of gravitation, or encouraging, if you needed the frequency-shift expression and did not necessarily believe in general relativity. There are no engineering applications for the gravitational frequency shift, and so we are mainly disappointed.

PROBLEMS

32.1. (15) Show that there is no gravitational redshift between clocks at different locations provided they are at the same radial distance from the center of the Earth.

32.2. (11) Find the gravitational redshift for clocks which are not along a vertical line.

32.3. (33) Calculate the effects of rotation on a laboratory-size redshift experiment.

32.4. (31) Calculate the gravitational redshift in the exact metric given in Section 30. Do not assume $(m/r)$ to be small.

32.5. (27) Explain how to use the wave diagram to calculate the Doppler shift for ordinary dispersive waves, such as water waves or elastic waves. Use regular clocks, not the intrinsic clocks introduced in Section 28.

## 33. Huygens' Construction and the Falling Apple

We are now ready to show that the curved spacetime

$$\mathscr{G} = \mathscr{N} + \frac{2m}{r}\,(dr^2 + dt^2) \tag{33.1}$$

*Gravity attracts*    (where $\mathscr{N}$ is the Minkowski metric) describes gravity. We must show that objects accelerate downward in such a spacetime. We do this here using a geometric argument. Besides showing that objects do fall, this geometric argument will verify that the parameter $m$ in Equation 33.1 above is the gravitational mass of the attracting body. It will also show that any theory of gravity that can be discussed in terms of waves and wave packets must have the same gravitational frequency shift that we have just described.

I am calling the geometric construction of the motion of wave packets Huygens' construction, although our construction is a bit different from his. Also, we are restricting our attention only to the high-frequency limit, whereas he intended to discuss the general case. For electrons the high-frequency limit demands that the gravitational field not change rapidly over a distance of a Compton wavelength $(3 \times 10^{-11}$ cm), which is certainly true. Heavier particles have even shorter Compton wavelengths.

The motion of a wave packet is determined once either the wave diagram or the dispersion relation is known at every point. The geometric construction is given in terms of the wave diagram. Suppose one knows the phase of the wave at every point. To be a solution of the original partial differential equation, the phase gradient 1-form must be tangent to the wave diagram at every point. Such a phase function is drawn in Figure 33.1, and in Figures 33.2 and 33.3 we have drawn wave diagrams and phase gradients for events $A$ and $B$. Because the tangent spaces drawn in Figures 33.2 and 33.3 are local approximations to the manifold, they can be drawn right on top of the manifold, which we have done in Figure 33.4.

*Semiclassical treatment*

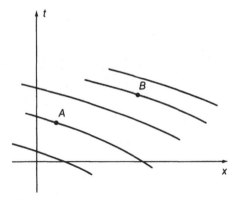

**Figure 33.1.** Wave crests for a randomly chosen wave train. This is a spacetime diagram.

[Look back at Figure 26.7 for this construction.]

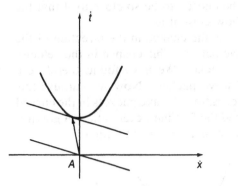

**Figure 33.2.** The tangent space at $A$.

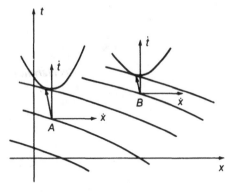

**Figure 33.3.** The tangent space at $B$.   **Figure 33.4.**

Huygens' construction is the reverse of the above. We start with a single given wave front. The next wave front is constructed by drawing wave diagrams centered on this wave front, and then drawing their envelope, as in Figure 33.5. This envelope is the next wave front. We will use this construction most often in a situation where we have only minimal "sideways" information. After all, if our wave packets are to

*Huygens' construction*

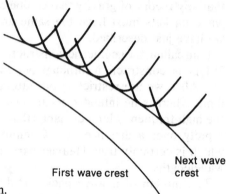

**Figure 33.5.** Huygens' construction.

represent particles, they should not be too spread out. In Figure 33.6 we use just two nearby events. Start from event $A$ and a given wave crest there, labeled number 1 in the figure. Pick a nearby event $A'$ on the wave crest, and draw the wave diagrams for both of the events. Find wave crest number 2 as the mutual tangent to these wave diagrams. The group velocity vector $v$ can now be found, and the next event, $B$, on the trajectory of the wave packet can be established. We now have the same information at $B$ that we had at $A$, and so the construction can be continued. Huygens' construction is really the limit of this process. Event $A'$ must be chosen to be so close to $A$ that the result does not depend on just how close it is.

*Acceleration*    The result of this construction is the change in the direction of the group-velocity vector of the wave packet. This change in the velocity is the acceleration of the wave packet. We have found therefore a description of the dynamics of wave packets. Now we can use this description to show that the acceleration of wave packets in the curved spacetime near the Earth is indeed the familiar acceleration of gravity; but first we need to practice with a more familiar example.

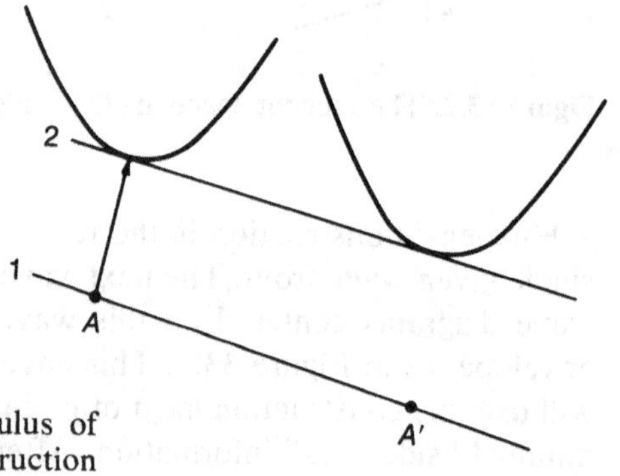

**Figure 33.6.** The differential calculus of wave diagrams: Huygens' construction over small distances.

**Example** | Let us try to explain mirages. The dispersion relation for light propagating in a medium with a spatially variable index of refraction is

**Light bending**

$$k_x{}^2 + k_y{}^2 + k_z{}^2 = n^2\omega^2, \qquad (33.2)$$

where the phase gradient is given by

$$d\theta = k_x\,dx + k_y\,d_y + k_z\,dz + \omega\,dt, \qquad (33.3)$$

and this covector equation defines $k_x$, $k_y$, $k_z$, and $\omega$. The index of refraction $n$ can depend on $x$, $y$, $z$, and $\omega$. If you are familiar with optics, then you have probably seen a derivation of this. (If not, please accept it as a reasonable starting point.) Furthermore, let us discuss only monochromatic light. The mirage that you actually see is a composite of many colors. It may be colored, as in the green flash, or not. For monochromatic light, the frequency $\omega$ will be the same everywhere, even if the index of refraction varies with position. The time-dependence of a linear equation whose coefficients do not depend on time can be broken up into exponentials.

Since nothing of interest is happening in time, we should forget about it and look at the problem only in space. This leads to a reduced problem, with a structure similar to that of our wave-propagation problems, but now only in space rather than in spacetime. This reduced problem is described by a dispersion relation

$$k_x{}^2 + k_y{}^2 + k_z{}^2 = \frac{n^2}{\lambda^2}, \qquad (33.4)$$

where $\lambda$ is a constant, the vacuum wavelength. It is this dispersion relation that we want to discuss, in two spatial dimensions. The wave diagram for it will be a circle, and the radius of the circle will be proportional to $(n/\lambda)$. A mirage refers to a situation where the air near the ground is heated and expands, and thus has a lowered index of refraction. The situation in terms of wave diagrams is sketched in Figure 33.7. Huygens' construction can now be used to show, as in Figure 33.8, that light rays, the trajectories of wave packets, are curved upward in such a situation. People looking toward the ground far in front of them will actually see light from the sky because of this bending; it looks like the reflection from a puddle of water. This is the mirage familiar to automobile drivers.

[Reader please verify; see Problem 15.1.]

Let us now use this geometric construction to verify that the spacetime given by Equation 33.1 does predict the fall of Newton's apple, and check that the parameter $m$ there is the mass of the attracting

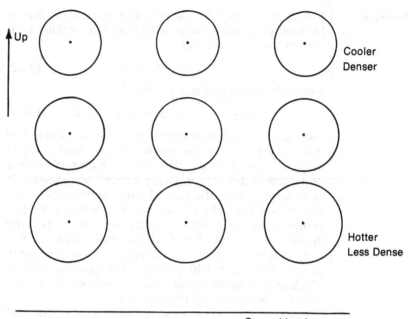

**Figure 33.7.** Wave diagrams for light propagation over a heated surface, grossly exaggerated.

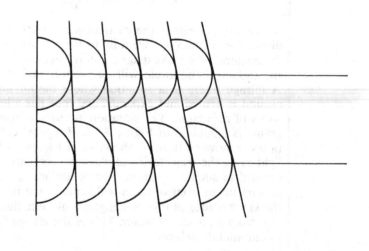

**Figure 33.8.** Using Huygens' construction to find the tilt in the wave crests.

body. Really, we are only going to show that single particles fall. To deduce that a composite body falls requires further discussion. The falling bodies considered here are assumed to be of such small mass that they do not affect the gravitational field. Two such small masses, called "test bodies," will not influence each other. Each will fall with the same acceleration, according to the calculation that will follow, and so will a composite body composed of such particles. We now proceed to translate the above geometric arguments into algebraic relations.

We start by asking about the local behavior of wave diagrams for the above spacetime. We shall ignore the angular coordinates and consider only radial motion. We need only the $dt^2$ and $dr^2$ parts of the metric

*Local behavior of spacetime*

$$\mathcal{G} = -\left(1 - \frac{2m}{r}\right) dt^2 + \left(1 + \frac{2m}{r}\right) dr^2. \tag{33.5}$$

This metric determines the wave diagram for wave packets. The wave diagrams change as we go out in radius, and we can easily sketch these changes if we recognize that the 1-forms

$$\left(1 - \frac{m}{r}\right) dt, \quad \left(1 + \frac{m}{r}\right) dr \tag{33.6}$$

are unit 1-forms, duals to the unit vectors

$$\left(1 + \frac{m}{r}\right) \frac{\partial}{\partial t}, \quad \left(1 - \frac{m}{r}\right) \frac{\partial}{\partial r}, \tag{33.7}$$

(ignoring here quadratic terms in $m/r$). The radial vector expands as one goes out in radius, and the timelike vector contracts. The 1-form *representations* do likewise. See Figure 33.9. The metric can be written in terms of these unit 1-forms,

$$\mathcal{G} = (-)\left(1 - \frac{m}{r}\right) dt \otimes \left(1 - \frac{m}{r}\right) dt$$

$$+ \left(1 + \frac{m}{r}\right) dr \otimes \left(1 + \frac{m}{r}\right) dr, \tag{33.8}$$

as sketched in Figure 33.10.

To discuss a laboratory or apple-tree-scale experiment, we can further simplify things by noting that $r$ changes only a little within the

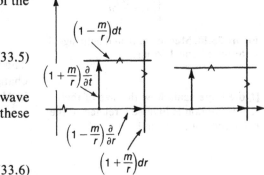

**Figure 33.9.** The 1-forms $[1 + (m/r)]dr$ and $[1 - (m/r)]dt$ and the vectors $[1 - (m/r)](\partial/\partial r)$ and $[1 + (m/r)](\partial/\partial t)$. Axis scales are chosen to make the left figure square. Rectangularity of righthand figure is exaggerated.

[Remember that a numerically larger 1-form has its contour lines closer together, and conversely.]

*Laboratory-scale approximation*

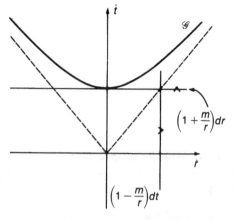

**Figure 33.10.** Metric figure and light signals in terms of unit 1-forms.

[Note there again how the size of the representation is inversely related to the magnitude of a 1-form.]

region of discussion. At the event $A$ of Figure 33.6, we have, say, unit 1-forms

$$\left(1 - \frac{m}{R}\right)dt, \quad \left(1 + \frac{m}{R}\right)dr, \tag{33.9}$$

taking $R$ as some reference radius. At the nearby event $B$, at radius

$$r = R + H, \tag{33.10}$$

with $H \ll R$, we have, using a binomial expansion, unit 1-forms

$$\left(1 - \frac{m}{R} + \frac{mH}{R^2}\right)dt, \quad \left(1 + \frac{m}{R} - \frac{mH}{R^2}\right)dr. \tag{33.11}$$

The effect we are looking for is an acceleration, which is the linear change of velocity with time. We need calculate only up to linear terms in $H$ to find this acceleration. The situation is sketched in Figure 33.11.

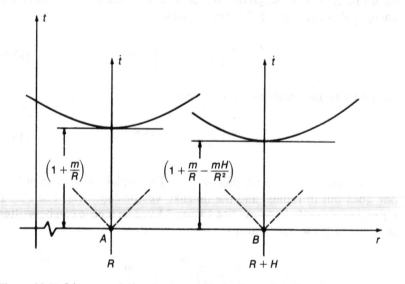

**Figure 33.11.** Linear variation in the metric in the radial direction.

***Wave-packet motion***

We now use our geometric construction to see what happens to a wave packet in this spacetime. We discuss only the simple case of a wave packet released from rest. Such a wave packet has one wave crest purely horizontal in the diagram. The next wave crest is found as the tangent to the wave diagrams for events $A$ and $B$ centered on the previous wave crest, as shown in Figure 33.12.

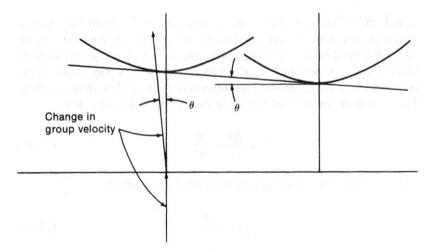

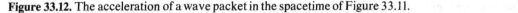

**Figure 33.12.** The acceleration of a wave packet in the spacetime of Figure 33.11.

The position of this new wave crest in spacetime is easily calculated. Remember that

$$\frac{m}{R} \ll 1. \tag{33.12}$$

The tilt angle $\theta$ will be just the difference in the vertical size between the wave diagrams, which is here $mH/R^2$, divided by the spatial separation $H$:

$$\theta = \frac{m}{R^2}. \tag{33.13}$$

This is the tilt angle at a unit time later. Actually, our wavelength must be much shorter than one light-second, and the tilt at one wave period $\Delta t$ later will be proportionally smaller:

$$\theta = \frac{m\,\Delta t}{R^2}. \tag{33.14}$$

Because $m/R \ll 1$, we can ignore the small differences between coordinate time and proper time in all of this.

So far so good, but the wave packet appears to be deflected upward, to larger radii, rather than downward. But be careful. One must use the group velocity to tell where the wave packet is going, and it comes from the wave diagram. Just as in special relativity, the angles behave

*Tilt of wave crests*

[Note how the separation $H$ between $A$ and $B$ has dropped out. This is what was meant on page 212 by "close enough not to matter how close."]

*Acceleration*

[This was discussed in the example on page 168.]

in a peculiar manner. Since these diagrams differ from the special-relativity diagrams by only one part in $10^9$, we can use here our knowledge of special-relativity wave diagrams, for which the group velocity tilts by the same angle as the wave front, but in the opposite direction. A wave front tilting upward corresponds to a particle that is falling. The change in velocity of the wave packet in a time $\Delta t$ is just

$$\frac{\Delta v}{\Delta t} = \frac{m}{R^2},$$

(33.15)

and this, of course, is called the acceleration of gravity;

$$g = \frac{m}{R^2}.$$

(33.16)

*Eötvös experiment*    Note that it is independent of the actual wavelength of the wave packet. Particles of different masses have different wavelengths, but they all have the same acceleration in a gravitational field. This deep result is called the Principle of Equivalence. We have extremely good experimental evidence for this. For laboratory-sized objects, the differences in their gravitational accelerations is less than one part in $10^{11}$. The experiment is called the Eötvös experiment. (His original result, obtained in 1905, was that there was no difference larger than one part in $10^8$.)

*Overview*    Step back for a moment and look at the big picture. In order for special relativity to be locally correct, any theory of gravity must have wave diagrams that are nearly like those of special relativity. To make the particles fall, we must tilt the wave fronts, and by just the right amount. This requires that the wave diagrams be compressed in the time direction as we go out in radius. But this compression is exactly what is needed to give us the gravitational redshift. Thus we have an even more general derivation of the redshift. Any theory of gravity which allows wave equations to be written for the particles will have a gravitational redshift of the correct amount.

### PROBLEMS

33.1. (19) For a mirage formed over a hot roadway, one might reasonably have $n \approx 1$ and its gradient $dn/dz \approx .00005 \ m^{-1}$. Find the deflection angle for light passing over a kilometer of road.

33.2. (27) Note that only the *tt*-component of the metric entered into the calculation of the gravitational redshift. Show that the *rr*-component can be changed by coordinate transformations which preserve both the explicit spherical symmetry and time-shift symmetry of the spacetime, whereas the *tt*-component cannot be thus changed.

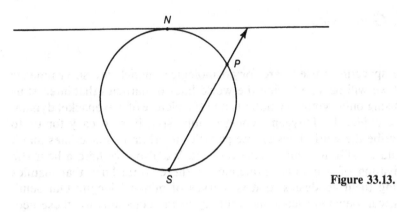

**Figure 33.13.**

33.3. (19) A convenient system of coordinates on the surface of a sphere is stereographic coordinates, found by projecting the points of the sphere onto a plane tangent at the North pole, from a projecting point at the South pole, as shown in Figure 33.13. In such coordinates the wave diagrams are circles with radii proportional to $(1+r^2)$. Study geodesics in this space graphically.

33.4. (28) Take the free-particle dispersion relation

$$g^{\alpha\beta}p_\alpha p_\beta + b^2 = 0,$$

[Here $b$ is proportional to the mass of the particle.]

and expand it for small distances above the surface of the Earth. Rescale the vertical coordinate to put it into the simpler form

$$\left(1 - \frac{2mx}{R^2}\right)\omega^2 = k^2 + b^2,$$

where here $x$ is a vertical coordinate.

33.5. (37) Using the simplified dispersion relation of Problem 33.4, find the group velocity and the wave diagram. Supposing that a particle has zero velocity at $x=0$, find $k$ and $\omega$ for $x < 0$. Sketch carefully both the wave crests and the rays.

33.6. (38) Extend the falling particle calculation in the text to particles moving with velocities comparable to the speed of light.

33.7. (18) What acceleration is shown in Figure 33.12 if the metric diagram $\mathcal{G}$ corresponds to a time interval of $10^{-9}$ seconds?

33.8. (04) Where is the energy going in Figure 33.8?

33.9. (26) Study geodesics in the upper half-plane, the part of the $x, y$-plane where $y > 0$, with metric figures that are circles of radius $y$.

## 34. Geodesics

The spacetimes used here for cosmological models are so symmetric that we will be able to find the world lines of particles that interest us by using only symmetry notions and the picture of wavepacket dynamics provided by Huygens' construction. Still, it is so easy for us to describe the world lines of free particles in arbitrary spacetimes that it would be silly to omit discussion of it. We also give here a heuristic derivation of a minimal principle for these world lines that justifies calling them geodesics, that is, curves of minimal length. Our semiclassical viewpoint leads us naturally to the equations for these geodesics that correspond to Hamilton's Equations in classical mechanics.

A particle in a nonuniform gravitational field is described by a dispersion relation

*Dispersion relation*

$$g^{\alpha\beta} p_\alpha p_\beta = -b^2. \tag{34.1}$$

Here we are using the index notation and the summation convention, both for practice and convenience. The $p_\alpha$ are the components of the phase gradient

$$d\theta = p_\alpha dx^\alpha. \tag{34.2}$$

The $g^{\alpha\beta}$ are the components of the inverse of the metric tensor, defined by

$$g^{\mu\alpha} g_{\alpha\nu} = \delta^\mu_\nu, \tag{34.3}$$

where the metric is given by

$$\mathcal{G} = g_{\mu\nu} dx^\mu \otimes dx^\nu. \tag{34.4}$$

Both the $p_\alpha$ and the $g^{\alpha\beta}$ can depend on position in spacetime. The constant $b$ depends on the mass of the particle, and drops out of the calculation when we differentiate, since it is constant. For simplicity we set $b$ equal to one for the rest of this section. We said before that all particles follow the same trajectories regardless of their masses.

*Spatial geodesics*   *Example 1*   | Let us set up the problem of finding the geodesics on the surface of the sphere. This is just like the problem of finding the world lines of wave packets through spacetime except for a few sign changes. The metric on the surface of a sphere is

$$\mathcal{G} = R^2(d\lambda^2 + \cos^2\lambda\, d\phi^2), \tag{34.5}$$

using here a latitude coordinate measured from the equator, rather than the usual $\theta$-coordinate, measured from the North pole. For this metric we have

$$g_{\lambda\lambda} = R^2,$$

$$g_{\phi\phi} = R^2\cos^2\lambda, \tag{34.6}$$

$$g_{\lambda\phi} = g_{\phi\lambda} = 0,$$

and

$$g^{\lambda\lambda} = \frac{1}{R^2},$$

$$g^{\phi\phi} = \frac{1}{R^2\cos^2\lambda}, \tag{34.7}$$

$$g^{\lambda\phi} = g^{\phi\lambda} = 0.$$

Let us call the components of the phase gradient $l$ and $n$:

$$p = l\,d\lambda + n\,d\phi. \tag{34.8}$$

Thus we have

$$p_\lambda = l, \quad p_\phi = n. \tag{34.9}$$

The dispersion relation is

$$\frac{1}{R^2}\left(l^2 + \frac{n^2}{\cos^2\lambda}\right) = 1. \tag{34.10}$$

[We will continue this example after we have developed more of the general theory.]

Because the phase gradient $p$ is already a gradient,

$$p_\alpha = \frac{\partial\theta}{\partial x^\alpha}, \tag{34.11}$$

**Group velocity**

its second derivatives are symmetric:

$$\frac{\partial p_\alpha}{\partial x^\beta} = \frac{\partial p_\beta}{\partial x^\alpha}. \tag{34.12}$$

Now both the $p_\alpha$ and the $g^{\alpha\beta}$ are functions of position, and when we take the gradient of the dispersion relation, we find that

$$\frac{\partial g^{\alpha\beta}}{\partial x^\sigma}p_\alpha p_\beta + 2g^{\alpha\beta}\frac{\partial p_\alpha}{\partial x^\sigma}p_\beta = 0, \tag{34.13}$$

and interchanging partial derivatives gives us

$$\frac{\partial p_\sigma}{\partial x^\alpha}(-g^{\alpha\beta}p_\beta) = \frac{1}{2}\frac{\partial g^{\alpha\beta}}{\partial x^\sigma}p_\alpha p_\beta. \tag{34.14}$$

[Directional derivatives were discussed on page 94.]

The lefthand side is the directional derivative $p_\sigma$ in the direction

$$v = -g^{\alpha\beta}p_\beta\frac{\partial}{\partial x^\alpha}, \tag{34.15}$$

and this direction is what we call the group-velocity vector. We have, using the dispersion relation,

[Remember that the constant $b$ was set equal to one.]

$$p_\alpha v^\alpha = 1; \tag{34.16}$$

so our group velocity is already properly normalized. The vector $v$ defines curves through spacetime that are the world lines of our wave packets. If we represent these world lines by the map

$$s \mapsto X^\alpha(s), \tag{34.17}$$

then we must have

$$\frac{dX^\alpha}{ds} = v^\alpha = -g^{\alpha\beta}p_\beta. \tag{34.18}$$

Our equation for the $p_\alpha$ gives us the derivative of the $p_\alpha$ along these rays:

**Equation for the change in group velocity**

$$\frac{dp_\alpha}{ds} = \frac{1}{2}\frac{\partial g^{\beta\gamma}}{\partial x^\alpha}p_\beta p_\gamma. \tag{34.19}$$

Together 34.18 and 34.19 are eight simultaneous ordinary differential equations for the world lines of the wave packets.

[Example 1 continues.]

With our Euclidean-like metric, we use the other sign for $v$, and so use equations

$$\frac{dX^\alpha}{ds} = g^{\alpha\beta}p_\beta, \tag{34.20}$$

$$\frac{dp_\alpha}{ds} = -\frac{1}{2}\frac{\partial g^{\beta\gamma}}{\partial x^\alpha}p_\beta p_\gamma. \tag{34.21}$$

For our example these are the equations

$$\frac{d\lambda}{ds} = \frac{l}{R^2}, \tag{34.22}$$

$$\frac{d\phi}{ds} = \frac{n}{R^2 \cos^2 \lambda}, \tag{34.23}$$

$$\frac{dl}{ds} = -\frac{n^2 \sin \lambda}{R^2 \cos^3 \lambda}, \tag{34.24}$$

$$\frac{dn}{ds} = 0. \tag{34.25}$$

We have here four ordinary differential equations for four unknowns.

Suppose we want to find the geodesic that starts from the point $(\lambda_0, \phi_0)$ in a direction such that

$$\frac{d\lambda}{ds}(0) = 0. \tag{34.26}$$

To use the above system of equations, we need starting values for $l$ and $n$ also. From the first equation, we find that

$$l_0 = 0. \tag{34.27}$$

From the dispersion relation, we find that

$$n_0 = R \cos \lambda_0. \tag{34.28}$$

With these initial values, it is now possible to find solutions to the above equations and so to find the geodesics on the surface of the sphere.

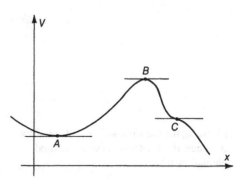

**Figure 34.1.** Points $A$, $B$, and $C$ are all extrema of the function $V(x)$. Only $A$ corresponds to stable equilibrium.

[These differential equations will be studied further in the next section when we discuss the curvature of the sphere.]

By setting up the equations as shown above, we can find the trajectories of free particles in any spacetime, provided that we can solve the resulting differential equations. These eight first-order ordinary differential equations can be viewed geometrically as a vector field in the eight-dimensional space whose coordinates are all of the $q$'s and all of the $p$'s. This is a fruitful approach, but one that would lead us too far afield to pursue.

It is also possible to describe the dynamics of wave packets in terms of a minimal principle. This idea will be familiar to those who have studied classical mechanics. (If you have not come across this before, then you should only skim this discussion.) Minimal principles do play a large role in physics, and they will repay study.

Minimal principles are common and efficient statements of physical laws. A simple example is the statement that the static equilibrium of a particle occurs at an extremum (i.e., a point having no first derivative) of the potential energy. The key idea is that a small displacement of to the system in any direction leads to only a second-order change in the potential energy, as sketched in Figure 34.1.

### Geodesic equations

[The reader familiar with classical mechanics should recognize here Hamilton's equations. Also you find here an interpretation for the canonical momenta. They are usually dealt with abstractly. With our geometric background, we see that they are the components of the wave-packet phase gradient.]

### Minimal principle

[Actually, we should call them extremal principles, since they refer to maximal and stationary values as well as minima. But for historical reasons they are often called minimal principles.]

The dynamics of dispersive wave packets can be described by a minimal principle. In Figure 34.2 we sketch the world line of a wave packet. At any point on the path the phase gradient 1-form can be found. For any given path we can compute the total phase change along the path merely by adding up the phase changes along the path. This *Action* total phase change is called the *action*, and it can be computed as follows. Break the path up into many small segments, and treat each small piece as a straight line. We will finish by letting the number of these pieces go to infinity. How do we find the phase change along one of these pieces? These short pieces can be represented in the limit by vectors, say, $u$. The geometry is sketched in Figure 34.3. The phase along a short segment of the path will be

[Note again the distinction that we draw between $d\theta$, a 1-form, and $\Delta\theta$, a small change in $\theta$.]

$$\Delta\theta = d\theta \cdot u. \tag{34.29}$$

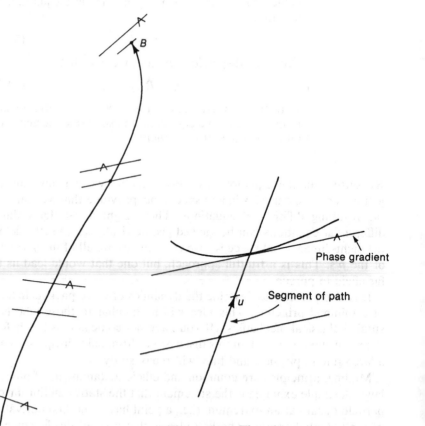

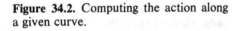

Figure 34.2. Computing the action along a given curve.

Figure 34.3. The phase change along a short segment $u$ of the path. The phase change shown here is $\frac{1}{2}$.

*Example 2*

For a theory with a wave diagram

$$\dot{t}^2 - \dot{x}^2 = 1, \qquad (34.30)$$

let us compute the phase change over a straight line path from $(0, 0)$ to $(t,x) = (2,1)$. See Figure 34.4. The phase gradient at any point on this path will be given by the Minkowski metric $\mathcal{N}$ according to

$$d\theta = -\mathcal{N} \cdot v, \qquad (34.31) \qquad \text{[See page 169 for this.]}$$

where $v$ is the group velocity. This vector $v$ can be determined by the conditions that it (1) be parallel to the path, and (2) satisfy the wave-diagram condition, Equation 34.31. This gives us

$$v = \frac{1}{\sqrt{3}}\left(\frac{\partial}{\partial x} + 2\frac{\partial}{\partial t}\right). \qquad (34.32)$$

The phase gradient is thus

$$d\theta = (-dt^2 + dx^2) \cdot \frac{(-1)}{\sqrt{3}}\left(\frac{\partial}{\partial x} + 2\frac{\partial}{\partial t}\right) \qquad (34.33)$$

$$= \frac{1}{\sqrt{3}}(2\,dt - dx).$$

Because the path is a straight line and the phase gradient is constant, there is no change as we take the limit. The total phase change along the path, the action $S$, is thus

$$S = d\theta \cdot \left(\frac{\partial}{\partial x} + 2\frac{\partial}{\partial t}\right) \qquad (34.34)$$

$$= \frac{1}{\sqrt{3}}(4 - 1) = \sqrt{3}.$$

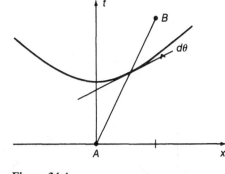

**Figure 34.4.**

In fact, the action can be calculated for any path between $A$ and $B$ whatsoever. *The true path followed by the wave packet is such that, for small changes in the path, the action is stationary.* We now give a graphical argument for this. A careful calculation requires more technical machinery that we are justified in developing here.

*Least action*

Let us consider two events and any world line whatsoever connecting them. At every event on the path we can find the phase gradient that a wave packet moving in that direction would have, and so we can sketch in wave crests in the immediate neighborhood of the path. If this path is the true one followed by a wavepacket, then these local wave crests will follow Huygens' construction. To discuss how this relates to the action, we first simplify our representation of this general path by making suitable coordinate transformations, just as we used

*Graphical argument*

linear transformations to simplify problems in Euclidean and Minkowski geometry. We sketch this simplification process in Figure 34.5. Suitable shearing will let you straighten out the world line. These will not be linear shears, of course. Further shearing transformations can then be used to make the wave crests all appear perpendicular to the path. Finally, we can make the wave crests equidistant by differential stretching along the path. These simplifications are not essential, and with a little practice you could follow this argument without using these simplifying transformations, but it is usually a good idea to make one's representa-

***Simple representation***      tions as simple as possible.

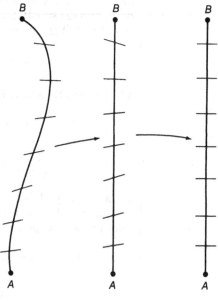

**Figure 34.5.**

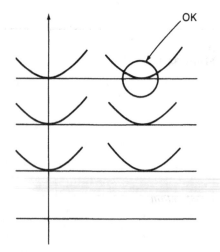

**Figure 34.6.** Correct behavior of nearby wave diagrams and constant phase surfaces.

In Figure 34.6 we draw a part of a path, its phase gradients, and some relevant wave diagrams. For all of this we assume that the space-time has been specified so that we can construct the wave diagram at any point. We also draw nearby wave diagrams and from these we can tell that this path does indeed satisfy Huygens' construction. In Figure 34.7 we draw a similar path that does not. In both figures the lefthand wave diagrams are tangent to the wave crests, because that is how we constructed the wave crests. Huygens' construction demands that the nearby wave diagrams also be tangent to the wave crests. Those in Figure 34.6 are, those in Figure 34.7 are not. One example of this tangency or lack of tangency has been circled in each figure.

***Variation of the path***

Now we consider a small variation in the path, pushing it a little bit to one side, and look for changes in the phase that depend linearly on

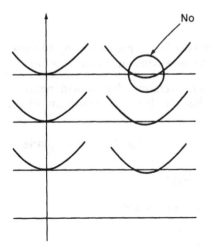

**Figure 34.7.** Incorrect behavior.

this variation of the path. Take a small piece of the path, say, between points $P$ and $Q$ in Figure 34.8, and push it sideways. To compute the phase change between the new points $P'$ and $Q'$, we need to construct the wave diagram at $P'$. If the path satisfies Huygens' construction, then this new wave diagram is tangent to the old wave crest, and there will be the same phase change between $P'$ and $Q'$ as there was between $P$ and $Q$, up to second-order terms. If the path does not satisfy Huygens' construction, as in Figure 34.9, then there will be a phase difference for the variation that will be linear in the magnitude of the variation. A doubled variation with attendant doubled phase difference is also shown in the figures. Of course, it is only strictly linear for infinitesimal displacements. Thus we see that the action is indeed stationary on the true paths, as we set out to show.

Such a principle of least action is very useful. It allows someone familiar with the calculus of variations to write down differential equations for the world lines of wave packets almost by inspection. Furthermore, it gives you an intuitive picture of these world lines as curves of stationary length. Such curves are called geodesics. The geodesics in spacetime are usually lines of longest time interval (remember the twin paradox). This extremal property of geodesics gives them many other applications. Figure 34.10 shows an idealized version of an elastic rod. Compare this with Figures 34.8 and 34.9. The minimum-energy state of an elastic rod is along a geodesic. Figure 34.11 shows a miniature ox cart. The two wheels are firmly fixed to the axle. Each wheel must travel the same distance. The ox cart follows a geodesic along a surface if it does not slip.

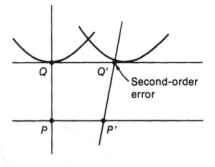

**Figure 34.8.** The phase change of $PQ$ and $P'Q'$ agree up to terms linear in the distance $PP'$.

[Small variations in wave diagrams were used in the discussion of the falling apple in Section 33.]

*Shortest lines*

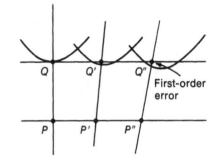

**Figure 34.9.** Here this is a linear disagreement.

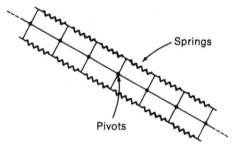

**Figure 34.10.** Idealized elastic rod.

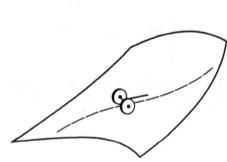

**Figure 34.11.** Miniature ox cart.

PROBLEMS

34.1. (14) Show that the dispersion relation is preserved by the geodesic equations. This always gives us one constant of the motion.

34.2. (18) Suppose that the dispersion relation does not depend on some particular coordinate. Show that this leads to a constant of the motion.

34.3. (30) Solve the geodesic equations for the Euclidean plane in polar coordinates

$$\mathcal{G} = dr^2 + r^2 \, d\phi^2.$$

34.4. (15) Find the geodesic equations for the metric

$$\mathcal{G} = \frac{dx^2 + dy^2}{y^2}.$$

34.5. (33) Show that the geodesics of the above metric are circles centered on the $x$-axis.

34.6. (15) Set up the geodesic equations as in the example on page 222 for the metric

$$\mathcal{G} = R^2(d\theta^2 + \sin^2 \theta \, d\phi^2).$$

34.7. (22) Do the same as in 34.6 for the metric

$$\mathcal{G} = R^2(d\chi^2 + \sinh^2 \chi \, d\phi^2).$$

34.8. (37) Extend Problem 31.5 to the effects of the Sun and the Moon. Do not forget that a coordinate system on the Earth is falling freely toward the Sun and the Moon.

## 35. Spacetime Curvature

The natural step to take next in the development of spacetime geometry would be to discuss how nearby geodesics behave. The non-Euclidean behavior of such geodesics is called spacetime curvature. For our purposes here, we need no more than an intuitive picture of curvature. Its efficient manipulation requires an extensive investment in mathematical tools, an investment well-justified in a general-relativity course, but not for us. Already we are in danger of paying too much attention to pretty mathematics. The brief discussion of curvature here should give you some intuitive feel for Einstein's equations. That they can be manipulated to produce the Friedmann equations which describe the dynamics of the standard cosmological models, you will have to take on faith.

One can discuss spacelike geodesics as well as timelike geodesics. In a stationary spacetime these spatial geodesics describe the shapes of straight rods. The best picture that I can give you for spacetime curvature involves a construction of straight rods that I call a *geodesic square*.

It is important to realize that one can give an operational construction for the idea of a straight rod. In addition to the rod itself, you need a matching straight hole into which it just fits (see Figure 35.1). If the rod fits snugly into the hole in all eight possible orientations, then we will call it a straight rod. The cross section will also have to be square, and there can be no twist along the rod.

**Figure 35.1.** Jig for proving a straight rod.

Now let us take four such straight rods, all of the same length, and attempt to build a square, as in Figure 35.2. As I indicate there, you will find that, despite the hopes raised by Euclidean geometry, the

### Geodesic square

[In a curved spacetime such a construction cannot be really square — that's the whole point — but I refuse to call it a geodesic pseudosquare. Nor do I have much patience with those who will point out that I am really discussing a rectangle.]

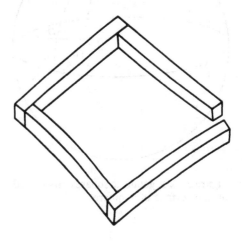

**Figure 35.2.** A geodesic square.

[To prove that there is no twist you will have to lay two rods side by side.]

### Twisted square

[Since I am about to show you that the idea of a square does not exist, you might want to come back and think about the internal consistency of this step.]

[I am going to calculate this in a very straightforward fashion. More efficient ways would be so tricky they might confuse you.]

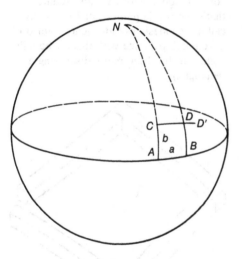

**Figure 35.3.** A geodesic square constructed on the surface of a sphere.

square usually fails to close. Not only does it fail to close, but the ends will be relatively twisted as well. Such a device is a machine for measuring spacetime curvature.

*Example 1*     Let us construct a geodesic square on the surface of a sphere. The geodesics are just the familiar great circles. For simplicity, we orient our square as in Figure 35.3, placing one side on the equator and two other sides on lines of longitude. To find the fourth side takes a little computation.

Take the same coordinates $(\lambda, \phi)$ on the sphere as used in Section 34. In such coordinates, three sides of the square are given by the parametrized curves:

$$\overline{AB}: s \mapsto (\lambda, \phi) = \left(0, \frac{s}{R}\right), \qquad (35.1)$$

$$\overline{AC}: s \mapsto \left(\frac{s}{R}, 0\right), \qquad (35.2)$$

$$\overline{BD}: s \mapsto \left(\frac{s}{R}, \frac{a}{R}\right), \qquad (35.3)$$

where the size of the square (really a rectangle) is given by the lengths $a$ and $b$, the size of the sphere is $R$, and we consider only small figures for which $a, b \ll R$. The parameter $s$ in all these curves is arc length.

To complete the square, we need to find the side $\overline{CD'}$. The differential equations for this curve were given in the example in Section 34:

$$\frac{d\lambda}{ds} = \frac{l}{R^2}, \qquad (35.4)$$

$$\frac{d\phi}{ds} = \frac{\cos \lambda_0}{R \cos^2 \lambda}, \qquad (35.5)$$

$$\frac{dl}{ds} = -\frac{\cos^2 \lambda_0 \sin \lambda}{\cos^3 \lambda}, \qquad (35.6)$$

with initial values

$$\lambda_0 = \frac{b}{R}$$

$$l_0 = 0, \qquad (35.7)$$

$$\phi_0 = 0,$$

Since we need only a short bit of the curve, we look for a power-series solution,

$$\lambda = \frac{b}{R} + C_1 s^2, \tag{35.8}$$

$$\phi = C_2 s + C_3 s^2, \tag{35.9}$$

$$l = C_4 s, \tag{35.10}$$

where the $C_i$ are coefficients to be determined from the above equations. The differences we seek involve terms like $ab^2$ and $a^2 b$. The pure cubic terms $a^3$ and $b^3$ must cancel out, since they remain as the rectangle shrinks to zero area.

Routine substitution and approximation gives us the values of the coefficients. The curve $\overline{CD'}$ is given by

[This is why we have been discussing a rectangle rather than a square. The rectangle requires less work.]

$$\overline{CD'}: s \mapsto \left( \frac{b}{R} - \frac{bs^2}{2R^3}, \frac{s}{R} + \frac{sb^2}{2R^2} \right) \tag{35.11}$$

and thus the point $D'$ by

$$(\lambda, \phi)_D \quad \left( \frac{b}{R} - \frac{ba^2}{2R^2}, \frac{a}{R} + \frac{ab^2}{2R^2} \right). \tag{35.12}$$

This geodesic square is represented in Figure 35.4. The ends miss by a distance proportional to the cube of its size, and the angular twist between the ends is proportional to the area of the square.

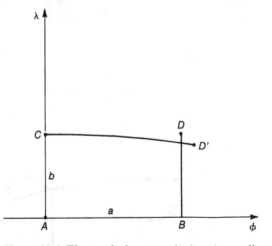

**Figure 35.4.** The geodesic square in $(\lambda, \phi)$ coordinates.

How do we represent this curvature of spacetime mathematically? The twist in the square will be a linear function of the lengths of the sides of the square. If we take another vector to give the direction

*Representation of the twist*

whose twist we want to measure, then the change in this vector will also be linear in its length. This suggests a linear operator, a tensor, mapping three vectors into a fourth. It is beyond us here to prove that such an operator exists and that it is a tensor, but such is the case. If we call the vectors along the sides of the square $A^\mu$ and $B^\mu$, and call the vector whose twist we measure $\eta^\mu$ then the change in $\eta^\mu$ going around the square, $\delta^\mu$, is given by

**Curvature tensor**

$$\delta^\mu = R^{\mu}_{\;\nu\sigma\tau}\eta^\nu A^\sigma B^\tau. \tag{35.13}$$

The coefficients $R^{\mu}_{\;\nu\sigma\tau}$ here are the components of the Riemann curvature tensor.

**Example 2**

For our geodesic square on the surface of the sphere, let us look at the change in a vector $\partial/\partial\phi$. This vector is directed along the sides of the top and bottom of the square, and is at right angles to the side (see Figure 35.5). Its twist going around the square appears as the difference vector $\delta$ shown there. The angle between the vectors at $D$ and at $D'$ is $ab/R^2$; so the difference vector $\delta$ will be

[Now you should see why we made the sides square rather than round, in order to preserve this orientation information.]

$$\delta = -\frac{ab}{R^2}\frac{\partial}{\partial\lambda}. \tag{35.14}$$

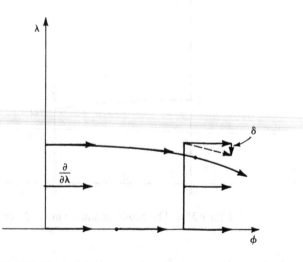

**Figure 35.5.** Transport of a tangent vector around a geodesic square.

For Equation 35.13 to be true, we must have

$$-\frac{ab}{R^2}\frac{\partial}{\partial\lambda} = R^{\mu}_{\cdot\phi\phi\lambda}\left(\frac{\partial}{\partial\phi}\right)\left(\frac{a}{R}\frac{\partial}{\partial\phi}\right)\left(\frac{b}{R}\frac{\partial}{\partial\lambda}\right)\frac{\partial}{\partial x^{\mu}}, \qquad (35.15)$$

and from this we can find two of the components of the Riemann tensor on the sphere:

$$R^{\phi}_{\cdot\phi\phi\lambda} = 0, \qquad (35.16)$$
$$R^{\lambda}_{\cdot\phi\phi\lambda} = -1.$$

If we now carry around a vector $\partial/\partial\lambda$, we can find two more components:

$$R^{\lambda}_{\cdot\lambda\phi\lambda} = 0, \qquad (35.17)$$
$$R^{\phi}_{\cdot\lambda\phi\lambda} = +1.$$

Going around the loop the other way just changes the sign; so we must have

$$R^{\lambda}_{\cdot\phi\lambda\phi} = +1, \quad R^{\phi}_{\cdot\phi\lambda\phi} = 0. \qquad (25.18)$$
$$R^{\phi}_{\cdot\lambda\lambda\phi} = -1, \quad R^{\lambda}_{\cdot\lambda\lambda\phi} = 0,$$

Finally, going around a loop of zero area must give us zero; so we have

$$R^{*}_{\cdot *\lambda\lambda} = 0, \qquad (35.19)$$
$$R^{*}_{\cdot *\phi\phi} = 0$$

for all possible indices *. Thus we have found all six-teen components of the Riemann tensor. We have explicitly

**Riemann** =

$$\left(\frac{\partial}{\partial\lambda}\otimes d\phi - \frac{\partial}{\partial\phi}\otimes d\lambda\right)\otimes(d\lambda\otimes d\phi - d\phi\otimes d\lambda).$$

$$(35.20)$$

[This method of calculating the curvature tensor will work only for symmetric spaces if the dimension is more than two. We must leave the general case for more advanced books.]

[Note how much more compact the explicit statement is compared to the list of all components.]

*Riemann tensor is local curvature*

The local curvature of a manifold is described by the Riemann tensor, a linear operator that takes in three vectors and outputs a fourth. Its natural dimensions must be $1/(\text{length})^2$, since, when multiplied by $(\text{length})^3$, it gives us length. Outside a body of mass $M$, we expect on dimensional grounds to have a Riemann tensor going like $M/R^3$. Another interpretation of the Riemann tensor is that it is the tidal gravitational field. Indeed, the tides do go like $M/R^3$.

*Light bending   Example 3*

We can make a crude estimate of the light deflection by a massive body such as the Sun by using the above ideas. The crudity of the argument comes from using a light ray as a rigid rod. A precise calculation shows that causes an error of a factor or two. Also, to use the above expression for a small rectangle for the large rectangle appearing in this problem (see Figure 35.6), we must consider it to be broken up into many small ones. The net effect of this is to sum the Riemann tensor over the rectangle. This involves adding tensors at different points and really needs to be done carefully.

If the Riemann tensor drops off like $1/r^3$, then most of the contribution to this sum will come from a region right around the body of size $R$, where $R$ is the radius at closest approach to the body. Thus we estimate the deflection angle $\theta$ to be

$$\theta \approx R^2 \times \frac{M}{R^3} \approx \frac{M}{R}. \tag{35.21}$$

By not doing the integral properly, we are missing some numerical factor here. For the Sun, this back-of-the-envelope calculation gives us a deflection angle,

$$\theta \approx \frac{M}{R} \approx \frac{5 \times 10^{-6}}{2} \approx 2 \times 10^{-6} \text{ radians} \approx \frac{1}{2} \text{ arc sec.} \tag{35.22}$$

A careful calculation gives the precise result

$$\theta = \frac{4M}{R}. \tag{35.23}$$

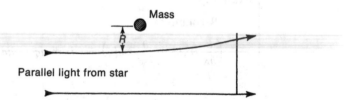

**Figure 35.6.** An enormous geodesic square in the sky used to calculate the gravitational light bending. Both light rays, in fact, curve.

*Einstein equations*

The Riemann tensor is important because it is the dynamical quantity appearing in Einstein's equations. These equations describe the dynamics of spacetime much as Maxwell's equations describe the dynamics of the electromagnetic field. Einstein's equations are

$$R^{\mu}_{.\alpha\nu\beta}g^{\alpha\beta} - \frac{1}{2}R^{\gamma}_{.\alpha\delta\beta}g^{\alpha\beta}\,\delta^{\delta}_{\gamma}\,\delta^{\mu}_{\nu} = 8\pi T^{\mu}_{\nu}. \qquad (35.24)$$

Here $R^{\mu}_{.\alpha\nu\beta}$ is the Riemann tensor, $g^{\alpha\beta}$ the inverse of the metric tensor, $\delta^{\alpha}_{\beta}$ the Kronecker delta tensor, and $T^{\mu}_{\nu}$ is a tensor composed of energy density, momentum density, and stress density. It is called the stress-energy tensor for short, and it acts as a source for gravity, just as charge and current densities act as sources for electromagnetism. The manipulation and solution of these equations are far beyond us, but you should at least get a chance to see them in terms of quantities that you know.

PROBLEMS

35.1. (14) By using a geodesic square, show that the surface of a cylinder has no curvature as we have defined it.

35.2. (16) Do the same as in 35.1 for a cone.

35.3. (25) In Problem 35.2, what about the tip of the cone?

35.4. (12) Is the claim true that the largest Riemann tensor in the solar system is found here on Earth?

35.5. (33) Construct a geodesic square on the sphere with one vertex at the North pole.

35.6. (36) Construct a geodesic square on the pseudosphere

$$\mathcal{G} = R^2(d\chi^2 + \sinh^2\chi\,d\phi^2).$$

Show that its curvature is the opposite of that of a sphere.

35.7. (28) On the space with metric

$$\mathcal{G} = \frac{dx^2 + dy^2}{y^2}$$

(see Problems 34.4. and 34.5), find a graphical construction for a right pentagon, a regular five-sided figure with a right angle at every vertex.

35.8. (29) Find the components of the Riemann tensor everywhere on a sphere. Use symmetry, not brute force.

35.9. (36) Extend the idea of a geodesic square to one having two time-like sides. Consider one in the radius-time directions on the surface of the Earth. The timelike legs will be falling particle trajectories. Is there spacetime curvature here? What Riemann-tensor components can be calculated from this, and what are their values?

35.10. (13) In Example 1 on page 231, why do we keep only linear terms in the power series for $l$?

# CHAPTER FOUR

# Cosmology

The theory of gravitation developed by Einstein in the years 1906–1915 resolved the conflict between Newtonian gravity and special relativity. In his theory the effects of gravity are modeled in a spacetime with a nonconstant metric, a curved spacetime. In addition to its compatibility with special relativity—which is not surprising, since this compatibility was one goal of its invention—it has two additional features. The first of these is that we can give a well-defined and natural description of an infinite universe. The second feature, even more surprising, is the possibility that the universe may be finite in size and yet without an edge or boundary. These are some of the boldest, most exciting, and most courageous ideas about physical reality to arise in a precise scientific theory. They are subtle ideas, and nearly all the effort that we have invested in the first three chapters is needed to honestly apprehend these ideas. Now for the fun!

[A fine history of that exciting time can be found in the book by Lanczos (1974).]

## 36. The Whole Universe Catalog

We now turn our attention to the problem of constructing a model of the universe using general relativity. This will involve us in a tremendous amount of coarse-graining. "The stars like dust" is detailed by comparison. For us an entire galaxy will be one particle of the model, to be represented by a single world line, and these particles themselves will be approximated by a continuous fluid. Before starting this chapter, you should read one or another of the excellent descriptive

[The collection of *Scientific American* reprints edited by O. Gingerich is also useful.]

books on astronomy and cosmology. It would be wasteful to repeat here what is done so well elsewhere. As a guide to what to look for in such reading, I survey the significant contents of the universe. These are to be incorporated into our model. Ideally one would hope to use a small number of the known facts as input to a model to establish values for the few unspecified parameters that appear. The model would then explain the rest, and so gain credence. A good model has a high ratio of output to input. Alas, our knowledge of the universe is so scanty and imperfect that even using it all we can answer very few questions. We can take some comfort that the universe is not an outrage: the observed properties are at least compatible with our models. The best we can do is to show you the framework in which such questions as "What is the ultimate fate of the universe?" can be asked and answered as a matter of physics rather than of philosophy or religion.

*Universe models*

*Dark night sky*

One of the most significant cosmological observations is absolutely certain. The sky at night is dark. The precise background brightness is not known. We must look out at the universe not only through the haze of the Earth's atmosphere, but also through the haze of our own galaxy. The spaces between the stars are certainly much darker than the surface of the Sun. Even this poses a difficult problem for anyone making a model of the universe.

Before the Copernican revolution, a dark night sky caused no trouble. It could be dark at night just because nothing is out there. Early astronomical thinkers had no reason to suspect that the universe was not finite and centered around us. Now such a view is unpopular. If we are not to be in any special place in the universe, then a model of the universe based on Euclidean geometry would require an infinite universe. Such an infinite universe poses several immediate difficulties. For one, Newtonian gravity is unable to describe an infinite amount of matter. Worse yet, the infinite number of stars in such a universe would rain light down on us from every direction. The night sky would be as bright as the surface of the sun. An infinite, static, Euclidean universe has no way out of this dilemma, which is called *Olber's Paradox.*

*Stars*

Among the smaller objects in the universe that are of interest to cosmologists there are stars, planetary systems (or at least one), and globular clusters of stars. These latter are clusters of about $10^5$ stars, dynamically bound together by their mutual gravity, and presumably sharing a common origin. For each of these objects one can determine an age, contingent on our knowledge of physics and of the initial conditions. Barring unusual initial conditions, the universe must have existed for a proper time greater than any of these ages. In the solar system we can date the Earth, the Moon, and meteorites, and we get ages like $1.6 \times 10^{17}$ sec (5 billion years). From detailed computer simulations of the evolution of the Sun, we think it has a similar age. There are uncertainties here because we do not know what the composition is at the center of the Sun. If instead we observe all the stars in a globular cluster, we are seeing a collection of stars with different masses but, if they have a common origin, similar compositions. The shape of the H-R diagram for these clusters tells us about the unknown chemical composition. This allows us to date the stars in such a cluster more accurately. These clusters also contain very old stars, perhaps the oldest stars in the universe. Ages for these globular clusters range from 2 to $4 \times 10^{17}$ sec.

*Star clusters*

[If you do not know what an H-R diagram is, then you have not done enough outside reading.]

These nearby objects carry information to us from long ago. Many cosmological models make some predictions about the abundances of the various chemical elements, and these abundances can be studied for many of these local objects. In particular, a big-bang origin for the universe inevitably turns a considerable amount of hydrogen into helium and a small amount of deuterium. The proportions of these elements will depend on the details of the early universe, when it was changing its properties on the same time scale as the neutron lifetime. Deuterium is especially interesting, because it is very hard to produce in any other place but in the early universe. If you make deuterium in the center of a star, it will be burned up by nuclear processes before it can get to the surface of the star.

*Elements*

*Helium*

Other objects of cosmological interest are galaxies and clusters of galaxies. The galaxies are the elementary particles of cosmology. Looking at a galaxy, one can often measure its gross redshift, its apparent luminosity, its apparent angular size, and the sizes of its HII regions. One can study the statistics of the distribution of galaxies, their clustering, uniformity, and possible anisotropy. The dominant feature of galaxies is that smaller and dimmer galaxies, presumably farther away, have systematic spectral displacements of their light to the red, as if they were moving away from us. As we will see, to say that they are or are not moving away from us is meaningless. It is a harmless statement, but operationally meaningless. People adopt it for the convenient discussion of galaxies, and we too will often talk as if the distant galaxies were moving away from us. This systematic "motion" of the galaxies is called the Hubble flow. As far as one can tell, the Hubble flow is smooth and isotropic, without any large systematic fluctuations.

*Galaxies*

*Redshift*

The universe is much more transparent to radio waves than it is to light. By using radio waves, we can observe to very large distances. There are in the sky sources of radio noise that we think are indeed at great distances. Were these objects better understood, they would be useful cosmological indicators. Were they all of uniform and con-

*Radio sources*

stant brightness, then counting their numbers as a function of brightness would give us important information about the geometry of the universe. Unfortunately, they are not a uniform population of objects, nor does each radio source remain at constant brightness throughout its lifetime. We also see pairs of radio sources. If the properties of these pairs could somehow be deduced, then they could be used as standard lengths, and from their apparent angular sizes we would again learn about the global geometry of the universe.

*Quasars*      Finally, among the strong radio sources there are the quasars, about which so much has been written. These objects are visible in the optical as well, and have high redshifts which, if they are due to the Hubble flow, put them at enormous distances. If we knew what they were, they would be very useful in cosmology.

*Microwave background radiation*      A final "object" in the universe important for cosmology is a universal bath of blackbody microwave radiation. It comes from every direction, has always a temperature of 3K to better than 0.25 per cent, and is very smooth in its angular distribution, down to scales of a milliradian. The only viable explanation for this radiation is that it is left over from a compressed and therefore hotter phase of the universe. Among other things, the microwave background radiation defines for us an absolute rest frame. Only for one velocity will the radiation appear to us to have an isotropic temperature distribution. For other states of motion, the radiation will be redshifted in one direction and so appear there colder, or be blueshifted in the opposite direction, and so appear warmer there. With our present technology we can just measure the Sun's peculiar velocity relative to the microwave background radiation. These microwave photons vastly outnumber the other particles in the universe. Most of the entropy of the universe resides in our ignorance of their phases and positions.

*Missing mass*      One can attempt to account directly for the matter in the universe by taking an inventory. On whatever scale this is done, one never finds enough matter. Both the galaxy itself and also clusters of galaxies behave dynamically as if they contain more matter than their visible components can account for. Very little room remains in conventional physics for this "missing mass." Perhaps a spirit of humility, so often a hindrance in astrophysics, is called for. We just may not know all the physical possibilities yet.

We can now proceed to construct a mathematical model which incorporates the above observational material, which is compatible with general relativity, and which can answer such sweeping questions as what was the origin and what will be the ultimate fate of the universe. Vital clues are provided by the Hubble flow and the dark night sky. We expect to find a dynamic universe rather than a static one. This gives us an enormous advantage over the ancients. The future, no doubt, has us at a similar advantage.

## 37. Roberston-Walker Spacetimes

The spirit of our times is to assume that the universe is not constructed for us or centered around us. Following the idea of the Copernican revolution, we feel that we should be in a typical location rather than in a special, unique location. To find today an astronomy book with the title *In the Center of Immensities* causes us to smile. Of course, on a small-enough scale we are in a special position. We are, for example, right on the surface of the Earth. Also, our planet is not in a typical location in the galaxy. It is much closer to a star than would be a point chosen at random. But where else would you find the energy flows necessary for life? Our location in the galaxy is more typical, but even there star formation happens only in certain places. And the stars must be second-generation stars in order to have heavy elements available for planets. This book could not have been written during just any old time in the evolution of the universe. It must now be late enough for stars to have planets, but early enough for the stars still to be shining. Within that broad window, however, our spacetime location is nothing special. Our galaxy is pretty ordinary. It is neither an isolated galaxy nor a member of one of the great clusters. These clusters of thousands of galaxies must be the most prominent landmarks in the universe. On the scale of these clusters of galaxies, the universe seems to be featureless. This is not surprising. In a dynamic universe, one of finite age, structures involving such large sizes may not yet have had time to form. Unfortunately, numerical estimates show that even objects as small as galaxies have not yet had time to form spontaneously. We are left with the puzzle of where they came from, and why no larger structures were also formed.

*Just a typical spot*

Our model of the universe deals only with this largest of scales. No individual galaxies appear in it, nor do even the great clusters. We consider only a cosmic fluid whose particles are the galaxies. At this level of coarse graining the universe is featureless. Its only prominent feature, to paraphrase the famous detective, is its lack of prominent features. This coarse-grained model should describe the average behavior of the universe. If such a simplification is made, then we can easily find models for the universe. Such models are called Robertson-Walker spacetimes. The assumption of spherical symmetry is an enormous simplification. Without it little progress can be made, and nearly all cosmological discussions today take place in the context of these Robertson-Walker spacetimes. As far as we can tell, the over-all structure of the universe is indeed symmetric and featureless. Some caution is needed here. The near symmetry observed for the present universe need not imply symmetry at earlier times. The symmetric fireball of a nuclear explosion does not imply a symmetric nuclear bomb.

*A blurred view*

*Symmetry*

[More detail on homogeneity and isotropy in the real universe will come in Section 50.]

*Homogeneity and isotropy*

[In Section 21 the example of electrical conductivity used a material homogeneous but not isotropic.]

What is the maximum amount of symmetry that a spacetime can have? First off, one should not be able to tell one point from another. Second, one should not be able to tell one direction from another. The first of these properties is called *homogeneity,* the second, *isotropy.* A plane is a set whose Euclidean geometry shows this high degree of symmetry. Another example is the surface of a sphere.

*Sphere*    ***Example 1***

To prove that the surface of a sphere is homogeneous, we find transformations which leave the metric unchanged, but which can move any point to any other point. Such transformations that preserve the metric are called *isometries.* One such transformation is a rotation about the North pole. Another one is a rotation about a pole on the equator. By combining such transformations, any point may be moved to any other. Thus the surface of a sphere is a homogeneous space.

Once homogeneity has been demonstrated, it is sufficient to show isotropy at a single point, since all points are equivalent. Let us look at the North pole. Rotations about the North pole are isometries, and can transform any given direction into any other. Thus the surface of a sphere is also isotropic.

*Half-plane*    ***Example 2***

What about the set of points in the upper half-plane, $y > 0$, with metric

$$\mathscr{G} = \frac{dx^2 + dy^2}{y^2}?\qquad(37.1)$$

One isometry of this manifold is the finite translation, $T_d$, through a distance $d$,

$$T_d: (x,y) \mapsto (x + d,y).\qquad(37.2)$$

Because of the explicit $y$-dependence, a $y$-translation is not an isometry. You might not expect any others, but there are some. Consider the "expansions" $E_\alpha$,

$$E_\alpha: (x,y) \mapsto (\alpha x,\alpha y).\qquad(37.3)$$

These also turn out to be isometries. We show this by finding that the representation of the metric $\mathscr{G}$ in the new coordinates is the same as its representation in the old coordinates. From the coordinate transformations

$$\begin{aligned} x' &= \alpha x, \\ y' &= \alpha y, \end{aligned}\qquad(37.4)$$

we have

$$dx' \mapsto \alpha \, dx,$$ (37.5)

$$dy' \mapsto \alpha \, dy,$$

and so

$$\mathscr{G} \mapsto \frac{dx'^2 + dy'^2}{(y')^2}.$$ (37.6)

The metric in the new, primed coordinates has the same functional form as the metric did in the old coordinates, and so we have an isometry.

The action of this transformation is shown in Figure 37.1. The curve $AB$, which is a geodesic (Problem 34.5), is moved into the curve $A'B'$. Since the metric is unchanged by this transformation, the lengths of these two curves must be the same.

There are exactly three different homogeneous and isotropic $n$-manifolds. One is Euclidean $n$-space, the $n$-dimensional version of a plane, one is an $n$-dimensional version of a sphere, and the third is a space of opposite curvature called a pseudosphere. These three types of symmetric spaces correspond to Euclidean geometry and to the two non-Euclidean geometries invented by Gauss, Lobachevsky, and Bolyai. One representation of the two-dimensional version of a pseudosphere is the space described in Example 2 above. In that representation it is not obviously isotropic, but other coordinates can be found for which the isotropy is manifest. We will devote the next several sections to exploring the geometry of the 3-sphere and the 3-pseudosphere. For the moment, let me just give you the metric that describes each of them, in coordinates that are as much like spherical polar coordinates as can be arranged. The spaces $\mathbb{R}^3$ with Euclidean metric can be represented by the metric tensor

$$\mathscr{G} = R^2[d\chi^2 + \chi^2(d\theta^2 + \sin^2\theta \, d\phi^2)].$$ (37.7)

We use $\chi$ for the radial coordinate in order to keep clear the distinction between this radial coordinate, which is dimensionless, and the usual one. The tensor

$$d\Omega^2 \equiv d\theta^2 + \sin^2\theta \, d\phi^2$$ (37.8)

describes the metric on the 2-sphere, that is, on the surface of an ordinary sphere. Because of isotropy, all our symmetric spaces have sections that look like 2-spheres.

[Recall the discussion in Section 20 about this if need be.]

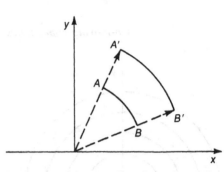

**Figure 37.1.** An expansion about the origin. This is an isometry of the Poincaré half-plane.

*Symmetric spaces*

[We are not going to prove here that there are only three such symmetric spaces. Ohanian gives an informal proof in his Chapter 10.]

*Euclidean spaces*

*Spheres*

*Example 3* | Both the Euclidean plane and the 2-sphere itself have sections that are circles, 1-spheres. These are shown in Figures 37.2 and 37.3.

*Co-moving coordinates*

The galaxies will move along lines $(\chi, \theta, \phi) = $ constant. Such coordinates that move along with the particles are called co-moving coordinates. In all our symmetric spaces, the sections of constant $\chi$ will be 2-spheres whose size depends on $\chi$, but whose metric is proportional to $d\Omega^2$.

The space called the 3-sphere has a metric

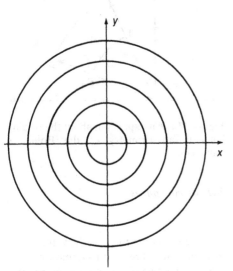

$$\mathscr{G} = R^2(d\chi^2 + \sin^2 \chi \, d\Omega^2), \qquad (37.9)$$

where now the $\chi$ coordinate ranges only from 0, one pole, to $\pi$, the other pole. $R$ is a constant with the dimensions of length, called the radius of the 3-sphere. The radii of the 2-sphere sections now goes like $R \sin \chi$. The largest 2-sphere in this space has a radius $R$.

*Example 4* | When the 2-sphere is sectioned into 1-spheres, their radii are given by $R \sin \theta$, where $R$ is the radius of the 2-sphere, and $\theta$ is the usual spherical coordinate. Again, these 1-spheres have a maximum radius of $R$, on the equator. The 3-sphere thus has an equatorial 2-sphere given in our coordinates by $\chi = \pi/2$.

**Figure 37.2.** The Euclidean plane as the union of circles.

The 3-pseudosphere has a metric

*Pseudospheres*

$$\mathscr{G} = R^2(d\chi^2 + \sinh^2 \chi \, d\Omega^2). \qquad (37.10)$$

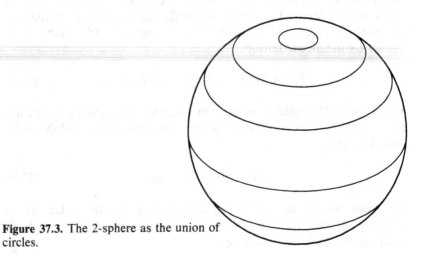

**Figure 37.3.** The 2-sphere as the union of circles.

Here $\chi$ ranges from zero to infinity. To avoid continually having to write three nearly identical equations, we define a function $S(\chi)$, which is to be $\chi$, $\sin \chi$, or $\sinh \chi$, depending on whether we are talking about Euclidean 3-space, the 3-sphere, or the 3-pseudosphere. In both the 3-sphere and the 3-pseudosphere, the coordinate $\chi$ has intrinsic meaning. For example, the distance "around" the 3-sphere in terms of the $\chi$-coordinate is exactly $2\pi$. For the Euclidean case, there is no intrinsic definition of $\chi$. Only the product $(R\chi)$ enters into that case. The Euclidean case is more symmetric. This shows up in its geometry in the existence of similar figures, which do not exist on either of the other two symmetric spaces.

| | | |
|---|---|---|
| ***Example 5*** | One can see this nonexistence of similar figures easily on the 2-sphere. In Figure 37.4 we draw a spherical triangle whose sides are great circles with right angles at all three vertices. One cannot construct a triangle similar to this one but half the size. The angles will then be less than right angles. | *Similar figures* |

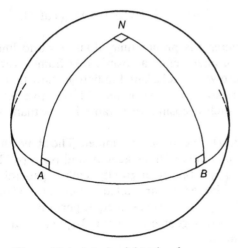

**Figure 37.4.** A truly right triangle.

We will be using models of the universe that all have space sections *Dynamic models* which are symmetric 3-spaces. The sizes of these space sections can depend on time, and since we expect to construct an evolutionary universe, they indeed will depend on time. If we try to extend the symmetry of space to a symmetry of spacetime, then we would seek models in which all times are equivalent as well. Such models would have no evolution. No such models compatible with general relativity exist.

Models with this symmetry, using other theories of gravity, have been made. They are called steady-state models. That they conflict with general relativity is unfortunate, but not a fatal flaw. We really have no solid evidence that general relativity is the correct theory of gravitation for objects as large as the universe. However, these steady-state models are not able to explain the evolutionary appearance of the universe, and they are not much considered today.

*Cosmic time*    Once one admits a dynamic universe, then a special time coordinate is singled out. Use any changing property of the universe, and you have a "clock." For example, the temperature of the microwave background radiation can be used as a universal time coordinate. This time is not measured by a clock in the sense of special relativity. These cosmic clocks are not local. The microwave radiation measured now came from long ago and far away. There is no conflict with special relativity in using such a "clock" to define a universal time. It is very much like the Newtonian absolute time.

*Robertson-Walker spacetimes*    The most general spacetime which is dynamic and which has symmetric space sections has a metric

$$\mathcal{G} = -dt^2 + R^2(t)[d\chi^2 + S^2(\chi)\,d\Omega^2]. \tag{37.11}$$

The time parameter $t$ is proper time along a world line of constant $\chi$, $\theta$, and $\phi$. Such lines define a cosmic rest frame. Recent measurements of the microwave background radiation show that the world line of our galaxy is nearly such a world line. The relative velocity between our galaxy and such a cosmic rest frame is less than $10^{-3}$ times the speed of light.

The function $R(t)$ is yet to be determined. The above metric includes the kinematic constraints of homogeneity and isotropy. The dynamics must be provided by a theory of gravity such as general relativity. All spacetimes in the above form are called Robertson-Walker spacetimes. Before we add the dynamics as specified by general relativity, we pause to explore the curious geometry of the 3-sphere and the 3-pseudosphere in some detail.

PROBLEMS

37.1. (08) In every direction in our universe, we see a greater density of quasars far away from us. Does this put us in a special place in the universe?

37.2. (12) Invent a spacetime that is homogeneous but not isotropic.

37.3. (18) Is the space with metric

$$\mathcal{G} = y^2 (dx^2 + dy^2)$$

isotropic?

37.4. (15) Why was a spacetime with two unknown functions

$$\mathcal{G} = -A^2(t) \, dt^2 + R^2(t) [dx^2 + S^2(x) \, d\Omega^2]$$

not considered in this section?

37.5. (24) Find the metric of the Poincaré half-plane

$$\mathcal{G} = \frac{dx^2 + dy^2}{y^2}$$

in new coordinates $(u, v)$ defined by

$$z \equiv x + iy,$$

$$w \equiv u + iv,$$

$$w = \frac{2}{z + i} + i.$$

All maps represented by complex variables like this are conformal. See any book on complex variable theory for a discussion. The proof is easy. Use this conformality to simplify the calculation.

37.6. (19) Use the $(u, v)$ representation of Problem 37.5 to show that the Poincaré half-plane is isotropic.

37.7. (31) Show that Minkowski spacetime can be written as a Robertson-Walker spacetime either with $S(\chi) = \chi$ or with $S(\chi) = \sinh \chi$.

## 38. The Global Structure of the 3-Sphere

To appreciate the structure of all but trivial 3-spaces is difficult. Although one can learn to visualize 2-spaces in quite a few dimensions, four or five at least, to visualize objects even in Euclidean 3-space is very difficult. Architects spend years trying to develop such skills. The visualization problems posed by the 3-sphere will be a challenge to your imagination. To reliably manipulate what cannot be easily visualized, we must invent faithful representations. Here we will give

*Visualization problems*

[See the August 1976 *Scientific American* for a short article on curved space. It is reprinted in *Cosmology + 1*, edited by Gingerich.]

several representations, first for the 2-sphere, for practice, and then for the 3-sphere.

*Imbedding*

[Read this "*S* three."]

The most natural representation of the 3-sphere, which we will call $S^3$ for short, uses a fictitious Euclidean 4-space. Call its rectangular coordinates $(x,y,z,w)$. The set of points that are a unit distance from the origin

$$x^2 + y^2 + z^2 + w^2 = 1 \qquad (38.1)$$

[What I mean by global structure will slowly emerge in this section.]

has the global structure of the 3-manifold $S^3$. This is the natural generalization of the 2-sphere, the set of points at a unit distance from the origin in Euclidean 3-space. We should be careful not to be too concrete here. $S^3$ is not this set of points. $S^3$ is any set with the same global structure as this set of points.

*Spherical polar coordinates*

We can introduce coordinates on $S^3$—and, in fact, on all of the $S^n$— which are analogs of familiar spherical polar coordinates. Before we do so, for practice let us carefully follow the procedure for the case of $S^2$. (One usually pursues the lower-dimensional analogs first.)

On $S^2$ ordinary spherical polar coordinates are defined by

*The 2-sphere*

$$x = \sin\theta\cos\phi,$$
$$y = \sin\theta\sin\phi, \qquad (38.2)$$
$$z = \cos\theta.$$

Since

$$x^2 + y^2 = \sin^2\theta, \qquad (38.3)$$

we have

$$x^2 + y^2 + z^2 = 1 \qquad (38.4)$$

for all $\theta$ and $\phi$. Every $(\theta,\phi)$ gives us a point on $S^2$, but different values of $(\theta,\phi)$ can give the same point on $S^2$. Here $z$ and $\theta$ are directly related. Each value of $z$ occurs once and only once as $\theta$ sweeps over the range

$$0 \le \theta \le \pi. \qquad (38.5)$$

For fixed $z$, the points sweep out a circle, given by

$$x^2 + y^2 = 1 - z^2, \qquad (38.6)$$

and so the proper range for $\phi$ must be

$$0 \le \phi \le 2\pi. \qquad (38.7)$$

These are the proper ranges for our coordinates. What happens at the edges? For any value of $\theta_0$, the points $(\theta_0, 0)$ and $(\theta_0, 2\pi)$ are the same point on $S^2$. On the Earth this seam is called the international dateline. At the poles $(0, \phi_1)$ and $(0, \phi_2)$ represent the same point for any values of $\phi_1$ and $\phi_2$. A $(\theta, \phi)$ chart with the correct rules for the edges is shown in Figure 38.1. It is straightforward to verify that no other points of $S^2$ are multiply represented in this chart. I will use here the convention that seams where identifications are to be made will be shown dashed. Lines to be collapsed into points will be shown as double lines. Such a chart is one useful representation for any set with the global structure of $S^2$.

### Global topology

[This convenient geographical language is about the only remaining "geo" in geometry.]

[Remember that technically a chart should be an open set, and the question of the edges does not arise. We find it better to include the edges and identify repeated points rather than to use several charts for sets known to be manifolds.]

[Now you should begin to see what I mean by "global structure." It is this pattern of edge identifications.]

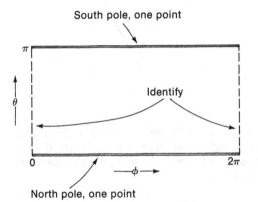

South pole, one point

Identify

North pole, one point

**Figure 38.1.** A coordinate chart for the 2-sphere. The double lines are really single points. The dashed lines are really the same line.

*Example* | The totally right-angled triangle shown in Figure 37.4 appears in our chart in Figure 38.2.

The above chart and its rules for identifying common points on its boundaries is a convenient representation of $S^2$ because it lies in the plane and so is amenable to graphical constructions and illustrations. Every geography book is full of variants of such a representation. We can find other representations which have less of this strange behavior at the edges while still remaining in the plane. They are intermediate steps in the reassembly of $S^2$ from the above chart.

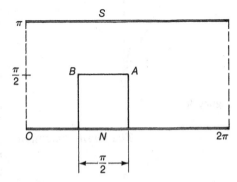

**Figure 38.2.** Our truly right triangle from Figure 37.4.

***Solid-disc model***

[I am not going to give these
transformations explicitly since we
do not need them. You should be able
to write them down out of your head.]

One possible first step in modifying the representation above is to
collapse the line $\theta = 0$, which is really only a single point of $S^2$, into an
actual point. This results in the pie-shaped sector shown in Figure 38.3.

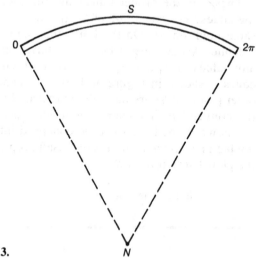

**Figure 38.3.**

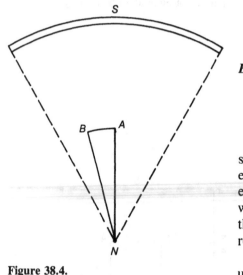

**Figure 38.4.**

***The fourth dimension***

***Example***

Figure 38.4 shows our truly right triangle in this representation. The triangle now looks like a triangle should.
The neighborhood of point $N$ is still not correctly
represented, however.

A reasonable second step is to fan this wedge out and then glue the
seam together, as in Figure 38.5. We now have a representation faithful
everywhere but around the point $S$, which appears spread out over the
entire rim. Such a chart is the familiar polar projection so popular
with military strategists. This solid-disc representation is useful and
the best that we can do if we stay in two dimensions. The analogous
representation of $S^3$ will be its most useful one.

Figure 38.6 shows a late stage in the reassembly of $S^2$. By pushing it
up into the third dimension, we can pinch off the South pole, making it
into a point and putting its correct neighborhood around it.

Although we can easily draw this final stage for $S^2$, since we have a
rudimentary ability to visualize objects in three dimensions, for $S^3$ this
stage will involve a fourth dimension. For visualizing this, we need
a trick. Let us use color to represent an additional dimension. For
example, a 2-surface which lies in three dimensions can be represented

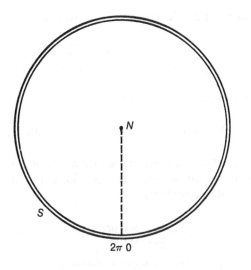

Figure 38.6.

*The 3-sphere*

by a colored subset of the plane. For each point, its locations in the plane gives two coordinates and its color gives the third. A different extra variable is texture. Less easy to describe verbally, this is better suited to sketching.

Using color for the third dimension, we can describe the final stage of the reassembly of $S^2$ without leaving the plane. Start with the solid-disc model of $S^2$. "Lift" it into the third dimension by coloring it blue at the rim and reddening the color as one goes in to the center. In three dimensions this turns the disc into a bowl. Now we can shrink the South pole "circle" into smaller and smaller circles until it becomes a true point lying right on top of the North pole if you wish. We now have a solid disc where every point but the rim has two colors, one from each hemisphere.

We now take these ideas and use them to represent $S^3$. A coordinate parametrization analogous to spherical polars will be

$$w = \cos \chi,$$
$$z = \sin \chi \cos \theta,$$
$$x = \sin \chi \sin \theta \cos \phi,$$
$$y = \sin \chi \sin \theta \sin \phi.$$

(38.8)

Again this parametrization ensures that our definition is satisfied:

$$x^2 + y^2 + z^2 + w^2 = 1.$$

(38.9)

What are the coordinate ranges? Each value of $w$ corresponds to a unique $\chi$ over the range

$$0 \leq \chi \leq \pi. \tag{38.10}$$

Knowing $\chi$ gives us a set satisfying

$$x^2 + y^2 + z^2 = 1 - w^2, \tag{38.11}$$

and for $w \neq 1$ this is a 2-sphere with spherical polar coordinates $(\theta, \phi)$. Thus $\theta$ and $\phi$ must have the usual ranges and identifications. Figure 38.7 shows this chart and gives its identification rules.

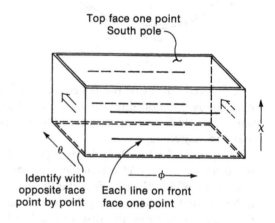

**Figure 38.7.** The 3-sphere in $(\chi, \theta, \phi)$ coordinates.

*Solid-ball model*

[Note that only the true, $w = 1$ North pole will be capitalized.]

Again we can find a more convenient representation by starting to reassemble $S^3$. Look at the face $\theta = 0$. For each $\chi$, this face represents not a line but a point. Collapse these lines so that they really are points. These points are the north poles of the little $S^2$s into which $S^3$ can be sliced. We now have a wedge, like a piece of cheese, shown in Figure 38.8. Following the two-dimensional example, we next fan it out into the full cheese wheel of Figure 38.9. The top and bottom faces of this wheel are single points. The curved side is a one-dimensional family of the south poles, $\theta = \pi$.

We can repeat this process. Let us shrink the bottom face, the North pole, to a point. This gives us the solid cone of Figure 38.10. Finally we expand this cone out and around and pinch off the line of south poles to a true line of points. This finally correctly represents the neighborhood of the North pole. I call this the solid-ball model of $S^3$.

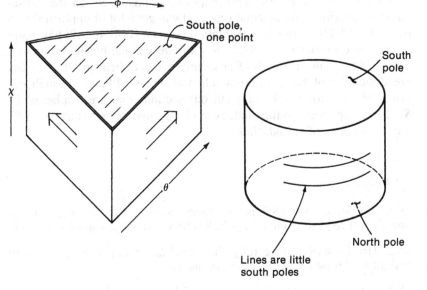

Figure 38.8 labels: South pole, one point; χ; φ; θ

Figure (cylinder) labels: South pole; North pole; Lines are little south poles

**Figure 38.8.** Collapse the little north poles to true points.

**Figure 38.9.** Join together all the international date lines of the 2-spheres.

See Figure 38.11. The radial coordinate inside this ball is χ, and the angular coordinates $(\theta, \phi)$ are the usual angular coordinates on spheres. We will make extensive use of this representation of $S^3$.

We can finish off the reassembly of $S^3$ by pushing this up into the fourth dimension. Color it blue on the surface and redden it in to the center. Then pull the South pole back in to lie on top of the North pole.

*Into four dimensions*

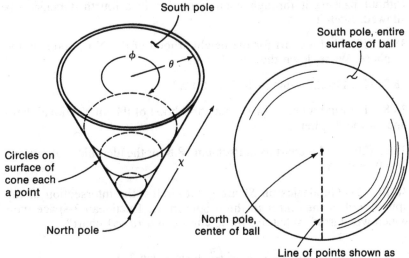

Figure (cone) labels: South pole; φ; θ; χ; Circles on surface of cone each a point; North pole

Figure (ball) labels: South pole, entire surface of ball; North pole, center of ball; Line of points shown as circles in preceding figure

**Figure 38.10.** Collapse the North pole to a point.

**Figure 38.11.** Collapse all the little south poles to points.

Every point in space now represents two points of $S^3$. In the subsequent discussions and calculations, you will get a lot of opportunity to play with $S^3$. Using this representation one can verify that $S^3$ has many of the same properties as $S^2$. It has a finite volume but no boundary. Similar figures do not exist. For example, a tetrahedron whose sides are doubled will have different dihedral angles. Planes through the origin of the fictitious 4-space cut out sections that are 2-spheres. In $S^3$ these 2-spheres are the analogs of planes, just as great circles on $S^2$ are the analogs of straight lines.

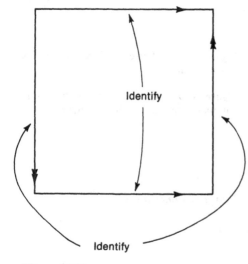

Figure 38.12.

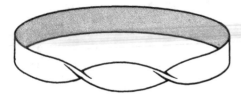

Figure 38.13.

## PROBLEMS

38.1. (06) In Figure 38.8 is the sharp edge of the pie, the vertical line closest to you, a line or a point? Where is it in the solid-ball model?

38.2. (16) Describe more fully the transition between Figures 38.10 and 38.11. Draw some intermediate figures.

38.3. (28) Reassemble $S^3$ starting from Figure 38.7 by first collapsing $N$ to a point, then gluing the identification closed, and then collapsing circles. What do you end up with? Sketch.

38.4. (20) Assemble the Klein bottle, a square region of the plane identified along the edges as shown in Figure 38.12. Use color for the fourth dimension. What about using time for the fourth dimension?

38.5. (22) In Figure 38.13 we show a doubly twisted strip. A singly twisted one would be a Moebius strip. Show how to untwist this strip without passing it through itself if motion in a fourth dimension is allowed. Sketch.

38.6. (10) Find a chart for the neighborhood of the North pole on the 2-sphere. Write down the actual maps.

38.7. (12) Do the same as in 38.6 for $S^3$.

38.8. (15) Find a chart for the neighborhood of the international date-line on the 2-sphere.

38.9. (20) Find a chart as in Problem 38.8 for the identification surface $\phi = 0, 2\pi$ on $S^3$.

38.10. (23) Geodesics on $S^2$ are great circles, the intersection of the sphere with planes through the origin on the Euclidean 3-space with which we started. What are these curves in a $(\theta, \phi)$ chart?

$$u \to (\theta, \phi) = \left(\frac{\pi}{2} + \sin s, \sin 3s\right).$$

38.11. (30) Same as 38.10, but for $S^3$.

38.12. (13) The two lines, *A* and *B*, shown in Figure 38.14, both pass through the North pole. What is the angle between them? Give both a graphical argument and a careful argument using a chart valid in the neighborhood of the pole. Where would the continuation of line *B* extend?

38.13. (33) Consider the set of all geodesics on $S^2$. This set is a manifold. Show that it is a part of $S^2$, with dimension two, but with some identifications. This manifold is called $RP^2$, real projective 2-space.

38.14. (30) Like the Moebius strip, $RP^2$ has only one side. An intrinsic (within the surface) description of this property can be made by trying to fill $RP^2$ with handed letters (*B*s but not *A*s, say) consistently. Show that this can be done on $S^2$ but not on $RP^2$.

38.15. (28) Discuss the curve in $S^2$,

$$u \mapsto (\theta, \phi) = \left(\frac{\pi}{2} + \sin s, \sin 3s\right).$$

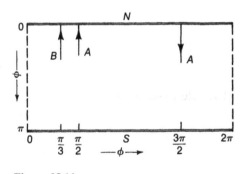

**Figure 38.14.**

## 39. *The Metric Structure of the 3-Sphere*

We turn now from the global structure of $S^3$ to its local metric structure. The homogeneity and isotropy of $S^3$ refer to its metric structure. We can find this metric structure, which I gave you out of the blue in Section 37, by using the representation of $S^3$ as a subspace of Euclidean 4-space. In the next section we can begin to study cosmological models based on $S^3$.

We have a representation of $S^3$ as a three-dimensional subspace of Euclidean 4-space. We can use the methods of Section 20 to pull back the basis 1-forms of this 4-space into $S^3$. Since the metric on the 4-space can be written in terms of these basis 1-forms, it too can be pulled back. In this way we can find a natural metric on $S^3$.

*Metric from imbedding*

*Example 1* | Let us look at the sphere as a subspace of Euclidean 3-space, and calculate the pullback of the 1-form $dz$. The $z$ coordinate is a function on this subspace,

*The 2-sphere*

$$z^*: S^2 \to \mathbb{R}; \ (\theta, \phi) \mapsto \cos \theta. \qquad (39.1)$$

As a function on $S^2$, it has a gradient, call it $dz^*$. Thus

$$dz^* = -\sin \theta \, d\theta. \qquad (39.2)$$

This 1-form on $S^2$ is the pullback of $dz$ as a 1-form on the 3-space. Once the pullbacks of the basis 1-forms are defined, any 1-form can be pulled back.

Note that our formulae are set up in just the right form for this computation. The subspaces are defined just like parametrized curves, as maps into the larger space; 1-forms can always be pulled back in the direction opposite to the maps.

[Example 1 continued]

We can also pull back $dx$ and $dy$, finding

$$dx^* = \cos\theta\sin\phi\,d\theta + \sin\theta\cos\phi\,d\phi, \quad (39.3)$$

$$dy^* = \cos\theta\cos\phi\,d\theta - \sin\theta\sin\phi\,d\phi, \quad (39.4)$$

and from these we pull back the metric

$$\mathscr{E} = dx^2 + dy^2 + dz^2 \qquad (39.5)$$

[We continue to abuse precise notation by using the same symbol for $\mathscr{E}$ and its pullback. We could call these $^{(3)}\mathscr{E}$ and $^{(2)}\mathscr{E}$ if we were pedantic.]

to find the metric on $S^2$:

$$\mathscr{E} = d\theta^2 + \sin^2\theta\,d\phi^2. \qquad (39.6)$$

We can pull back the 4-space metric

$$\mathscr{E} = dx^2 + dy^2 + dz^2 + dw^2 \qquad (39.7)$$

in the same manner as in the above example to find the metric on $S^3$,

$$\mathscr{E} = d\chi^2 + \sin^2\chi\,d\Omega^2, \qquad (39.8)$$

which we gave earlier.

*Other coordinates*    Because we are using a fully covariant notation, we are free to pick the most convenient coordinates for a particular problem. Sometimes it is handy to use the sizes of the 2-spheres as a radial coordinate instead of $\chi$. This is related to the doubly covered ball representation, and many authors use it. Define a new radial coordinate $\rho$ by

$$\rho \equiv \sin\chi; \qquad (39.9)$$

then

$$d\rho = \cos\chi\,d\chi \qquad (39.10)$$

and

$$d\chi^2 = \frac{d\rho^2}{1 - \rho^2}. \qquad (39.11)$$

The metric $\mathscr{E}$ in these coordinates is

$$\mathscr{E} = \frac{d\rho^2}{1 - \rho^2} + \rho^2\,d\Omega^2. \qquad (39.12)$$

Equation 39.9 does not have a unique solution for $\chi$ in terms of $\rho$. As in the doubly covered disc, there are two points in $S^3$ for every $(\rho,\theta,\phi)$ coordinate, except for $\rho = 1$.

The $(\chi,\theta,\phi)$ coordinates preserve the proper spacing between the $(\theta,\phi)$ 2-spheres. The $(\rho,\theta,\phi)$ coordinates preserve the proper sizes of the spheres. Because $S^3$ is a curved space, no coordinates can be found which make the 2-spheres both the proper size and also the proper distance apart.

We can represent this metric structure by constructing the metric-tensor representation in the tangent spaces. Such a three-dimensional picture is a bit hard to draw. A slice through the center of $S^3$ gives us an $S^2$. We show the metric structure of such an $S^2$ in Figure 39.1, using the solid-disc representation. Note how the metric figures are expanding as we approach the rim, which is a single point. This is a clue that

**Space curvature**

[There is an excellent discussion of the 3-sphere in Reichenbach. He gives you a seven-page guided tour of the 3-sphere, complete with diagrams of what you would see if you visited one.]

**Metric figures**

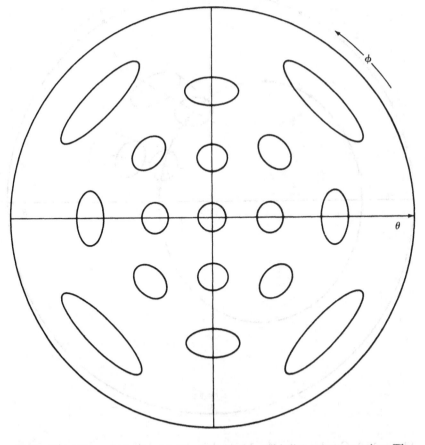

**Figure 39.1.** The metric on the 2-sphere in the solid-disc representation. The scale of the tangent space is reduced by a factor of five.

something is funny there. The metric figures of $S^3$ are ellipsoids which come from revolving the ellipses of Figure 39.1 about the radial direction.

*Geodesics*    If we think of the metric figures as wave diagrams, Huygen's construction tells us how the geodesics behave. Figure 39.2 shows a geodesic in the solid-disc representation of $S^2$, along with some constant-phase surfaces. These geodesics are great circles.

We will discover more about the geometry of $S^3$ by actually using it. We turn now to cosmological models based on $S^3$. We postpone the corresponding discussion of the pseudosphere to Section 44. Its geometry is even less familiar. Both more practice with $S^3$ and experience with the uses of these symmetric spaces will help you with it.

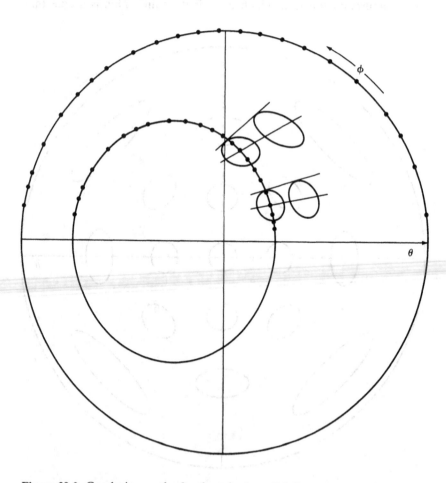

**Figure 39.2.** Geodesics on the 2-sphere in the solid-disc representation.

PROBLEMS

39.1. (17) What is the surface area of a great sphere on $S^3$?

39.2. (16) Calculate the pullback of the 1-form

$$x\, dx + y^2\, dy + z^3\, dz$$

onto the submanifold of $\mathbb{R}^3$ given by

$$(u,v) \mapsto (x,y,z) = (u,v,v^2).$$

39.3. (16) Show how to calculate the sizes of the ellipses in Figure 39.1.

39.4. (30) Find a conformal system of coordinates for $S^2$ where the metric is in the form

$$\mathcal{G} = f(u,v)\,(du^2 + dv^2).$$

39.5. (22) Redraw Figure 39.2 in the coordinates of 39.4.

39.6. (28) Define and calculate the volume of $S^3$.

# 40. Light Propagation

*"It is better to light a candle than to curse the darkness."*

CONFUCIUS

Much of our information about the universe comes from optical observations. To describe such observations, we need a description of how light propagates in our Robertson-Walker spacetimes. For our purposes we only need to discuss motion in the $(t,\chi)$ plane, keeping $\theta$ and $\phi$ constant. These light signals move out radially from the North pole. Such a light signal in an expanding universe suffers a frequency shift to the red, as we in fact observe in the universe today. Because of the homogeneity and isotropy of the Robertson-Walker universes, all light signals are equivalent, but the radial ones are nicely aligned with our coordinate system.

*Radial light signals*

The discussion of light propagation will be simplified if we introduce a new time coordinate specially adapted to the light signals. This new

*Arc-time*

coordinate will be called arc-time, $\eta$, defined by the ordinary differential equation

$$\frac{d\eta}{dt} = \frac{1}{R(t)}, \tag{40.1}$$

where $R(t)$ is the function in the Robertson-Walker spacetime which describes the evolution of the universe. In terms of arc-time, the metric (Equation 37.11) can be written

$$\mathscr{G} = R^2(t)[-d\eta^2 + d\chi^2 + S^2(\chi)\,d\Omega^2]. \tag{40.2}$$

In $(\eta,\chi)$ coordinates, radial light-signal world lines are lines at 45°, just as in special relativity, because their tangent vectors $\sigma$ must satisfy

$$\mathscr{G}(\sigma,\sigma) = 0; \tag{40.3}$$

writing

$$\sigma = \dot{\eta}\frac{\partial}{\partial\eta} + \dot{\chi}\frac{\partial}{\partial\chi}, \tag{40.4}$$

this becomes

$$\dot{\eta}^2 = \dot{\chi}^2. \tag{40.5}$$

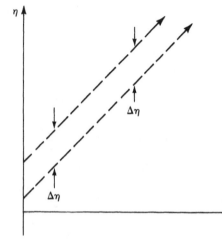

**Figure 40.1.**

*Arc-distance*    There is a natural measure of distance on $S^3$, the arc-distance, corresponding to a great circle on the Earth. The arc-distance around the equator of $S^n$ is always defined to be $2\pi$. In the radial direction, arc-distance in $S^3$ is given directly by $\chi$. The parameter $\eta$ is called arc-time because it is closely related to this arc-distance. In a given amount of arc-time, a light signal travels exactly that amount of arc-distance. This is an important relation that we will use repeatedly.

Look at two radial light signals sent out in the same $(\theta,\phi)$ direction
*Frequency shifts*    at different times. We can use these to compute the frequency shift between a source and an observer at rest with respect to the symmetry rest frame of the universe. Figure 40.1 shows that the arc-time interval between the two light signals must be constant. Just as in the gravitational redshift calculation, this allows us to write down the relationship between $\Delta\tau_1$, the period between the two light signals when they are sent out, and $\Delta\tau_2$, the period between them when they are received, for a source and observer moving in the direction $\partial/\partial t$. For others one

[I am using $\tau$ here instead of $t$ to emphasize that this is proper time, not coordinate time.]

must also include the effects of the special-relativity Doppler shift. Arc-time converts to proper time according to

$$\Delta\tau = R\,\Delta\eta, \tag{40.6}$$

[This formula is easy to remember. Both $\tau$ and $R$ have dimensions of seconds, while $\eta$ is dimensionless.]

provided that the universe does not change its size appreciably during one cycle of the wave. The wavelength of the light is proportional to this period, and so we have

$$\frac{\lambda_{\text{REC}}}{\lambda_{\text{EMIT}}} = \frac{R(t_{\text{REC}})}{R(t_{\text{EMIT}})}. \tag{40.7}$$

If $R$ is larger when the light is received, then the received wavelength is longer, and we say that the light appears redshifted. A widely used redshift parameter is $z$, defined by

$$(1 + z) = \frac{\lambda_{\text{REC}}}{\lambda_{\text{EMIT}}}. \tag{40.8}$$

$z$

For small $z$, $z$ is merely the velocity that would cause such a redshift.

Such a proportional wavelength shift is interesting. It turns radiation with a blackbody spectrum into radiation with a blackbody spectrum for a different temperature, according to

*Blackbody radiation*

$$TR = \text{constant}. \tag{40.9}$$

As light propagates through the expanding universe, the number of photons does not change. Yet each photon gets shifted to longer wavelengths and so has less energy. Where does the energy go? No one can really say. There is no way to define the total energy of a spacetime. To say that the energy of the light goes into the expansion of the universe is to mumble some pleasant and harmless words, but ones that do not have any meaning. They do not explain; they only reassure.

*Photons conserved*

# 41. Friedmann Universes

To proceed further, we must use a theory of the dynamics of gravitation to specify the function $R(t)$ in the Robertson-Walker spacetimes. Here will be a gap in our treatment. It is beyond our present skills to

*Spacetime dynamics*

[A derivation can be found in Chapter 10 of Ohanian.]

apply the Einstein equations, given on page 235, to the Robertson-Walker spacetimes. The computation of curvature tensors is very complicated unless one takes the time to learn quite a bit of geometric formalism. The Einstein equations relate the dynamics of spacetime to its energy, momentum, and stress content. The result will be ordinary differential equations for the scale factor $R(t)$ and for the matter and pressure content of the universe. From these ordinary differential equations, called the Friedmann equations, we will be able to proceed with our discussion of cosmology without further gaps.

*Cosmic rest frame*

The relationship between spacetime and its matter content, as given by Einstein's equations, is one in which a symmetric spacetime must have a similarly symmetric matter distribution. Let us consider things in a cosmic rest frame, with its time axis in the $\partial/\partial t$ direction. If there were any net energy flow in this frame, it would single out some particular direction. This would conflict with the assumed isotropy of spacetime. Similarly, the only stress allowed is an isotropic pressure. Other stresses, such as a shear stress, would conflict with isotropy. The only possible material variables compatible with our assumed symmetry are an energy density $\rho(t)$ and a pressure (momentum-flux density) $p(t)$. Both must be independent of $\chi$, $\theta$, and $\phi$, although they will be time-dependent.

*Matter*

Different types of matter are characterized by different relationships between pressure and density. Such a relationship is called an equation of state. We will study in detail only two general types of matter. The

*Dust*

first we call dust, and it has the equation of state

$$p = 0. \tag{41.1}$$

*Radiation*

The second we will call radiation, and it has

$$p = \tfrac{1}{3}\rho. \tag{41.2}$$

*Galaxies*

We use the dust equation of state to model the smoothed fluid whose particles are the ordinary galaxies. To verify that their pressure is negligible will be a useful way to clarify the idea of pressure for you. Start with momentum. Momentum is the flow of mass (energy). Similarly, stress (pressure) is the flow of momentum. The flow of momentum is what you usually call force per unit area. In a fluid of particles of mass $m$ moving with speed $v$ in all directions, the energy per particle will go like $\gamma m$. The momentum is the flow of this energy. For a single particle this will be the speed times the energy density, $v\gamma m$. This is a vector quantity, and in a gas with particles moving in all possible directions, the net momentum flux is zero. Similarly, the pressure is the flow of this momentum, and for a single particle it goes like $v^2\gamma m$. Even for particles moving in all directions it does not cancel because it goes like $v^2$. For particles like galaxies, we observe their random speeds to be about 300 km/s, that is, $10^{-3}$ of the speed of light. The pressure will be smaller than the energy density by a factor of $v^2$, that is, $10^{-6}$. Thus we see that the pressure of ordinary galaxies today does not play a significant role in the evolution of the universe.

For radiation we have a different situation. Suppose the particles are moving with a speed $v$ which is nearly the speed of light. Then the average of the pressure term will be an average of $v^2$ over all directions. Since there are three independent directions, the average comes out to be 1/3; hence the equation of state for particles moving at the speed of light is

["Hence" is too strong; but at least this argument reduces the surprise.]

$$p = \tfrac{1}{3}\rho. \qquad (41.3)$$

This is the equation of state for blackbody radiation, which is a random gas of photons.

Our cosmological model involves three functions of time, the scale-factor $R$, the matter density $\rho$, and the pressure $p$. Einstein's equations provide two ordinary differential equations for these, and the equation of state provides the third equation. The ordinary differential equations resulting from a proper general-relativity calculation are

*Friedmann equations*

$$2\frac{R''}{R} + \left(\frac{R'}{R}\right)^2 + \frac{\kappa}{R^2} + 8\pi p = 0, \qquad (41.4)$$

$$\left(\frac{R'}{R}\right)^2 + \frac{\kappa}{R^2} = \frac{8\pi\rho}{3}, \qquad (41.5)$$

where

$$R' \equiv dR/dt \qquad (41.6)$$

and $\kappa$ is defined by

$$S'' = -\kappa S; \qquad (41.7)$$

that is, $\kappa = +1$ for the 3-sphere, $\kappa = 0$ for 3-space, and $\kappa = -1$ for the 3-pseudosphere. A Robertson-Walker spacetime obeying the above equations and a suitable equation of state is called a Friedmann universe.

Besides dust, radiation, and any mixture of the two, other equations of state are sometimes discussed. Particle physicists have speculated

*Exotic matter*

that at enormous densities nuclear matter might obey an equation of state

$$p = \rho. \tag{41.8}$$

Such matters resist compression even more than a gas of radiation. Such matter is called ultrastiff. Were it any stiffer, the speed of sound would exceed the speed of light. A second fantastic form of matter comes from asking for a Lorentz-invariant equation of state. Such matter would have an equation of state

$$p = -\rho = -\frac{\Lambda}{8\pi}; \tag{41.9}$$

that is, would have the same constant negative pressure everywhere. This type of matter would appear in Einstein's equations in a term called the cosmological term, and $\Lambda$ is called the cosmological constant. Einstein himself made up this type of matter just to keep his models of the universe from expanding. He was making these models before Hubble observed the actual expansion of the universe. Today there is no need to postulate these exotic types of matter. Still, both of these are fun to play with.

## 42. The Closed Dust Universe

[This quotation, in one form or another, has been ascribed to a great many authors. For some of its history, see the essay "The Fearful Sphere of Pascal" in Jorge Luis Borges' *Labyrinths*.]

*"Nature is an infinite sphere,*
*whose center is everywhere*
*and whose circumference is nowhere."*

PASCAL

The first cosmological model we study will be one having 3-spheres for its spatial sections and dust for its matter content. This is the easiest and most familiar case. You have had a little experience with spherical trigonometry, but none with pseudospherical trigonometry.

For this dust universe we have only two functions to find, $R$ and $\rho$.

The Friedmann equations are

$$2\frac{R''}{R} + \left(\frac{R'}{R}\right)^2 + \frac{1}{R^2} = 0,$$ 

(42.1) *Friedmann equations*

$$\left(\frac{R'}{R}\right)^2 + \frac{1}{R^2} = \frac{8\pi\rho}{3}.$$ 

(42.2)

You might expect to go at these by solving the first one for $R(t)$, and then computing $\rho$ from the second. There is more to these equations than meets the eye, however.

For dust, the energy density $\rho$ is just the local mass density. For a *Matter content* fluid of particles, this will just be their mass times the local number density. The particles (for us, galaxies) will persist, and their total number should be a constant. That will be true only if the number density increases as $1/R^3$ as the universe expands. What about this? Is this compatible with the Friedmann equations? Let us compute $(R^3\rho)'$ to see if it is zero. Write Equation 42.2 in the form

$$R(R')^2 + R = \frac{8\pi\rho R^3}{3},$$ 

(42.3) *Mass conservation*

and differentiate both sides:

$$\frac{d}{dt}\left(\frac{8\pi\rho R^3}{3}\right) = (R')^3 + 2RR'R'' + R'.$$ 

(42.4)

Now the righthand side is just Equation 42.1 multiplied by $R^2R'$, and so we do have

$$\frac{d}{dt}(\rho R^3) = 0.$$ 

(42.5)

Mass conservation is contained in the Friedmann equations. To use this, we define a constant $a$ by

$$a \equiv \frac{8\pi\rho R^3}{3}$$ 

(42.6)

This $a$ is a constant during the evolution of the universe. Different values of $a$ will give us different models of the universe, models of different sizes. Knowing $a$, we can relate $\rho$ directly to $R$. We can find $R$ most easily from Equation 42.3:       *Reduced equation*

$$R(R')^2 + R = a.$$ 

(42.7)

[If you start inventing cosmological theories off the wall, you will find these internal consistencies rather difficult to arrange.]

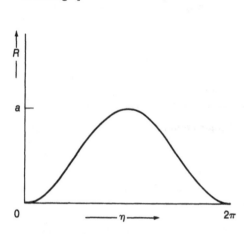

**Figure 42.1.** The evolution of a dust universe in terms of arc-time.

*Arc-time*

*Proper time*

[I hope my colleagues will not be offended by what seems to me to be a more logical division of these fields. Astronomy looks back on the light cone, whereas geology looks back on timelike world lines.]

Finding a constant of the motion reduced the differential equation from second order to first order.

The interpretation of the constant $a$ is contained in Equation 42.7. If we have at some epoch

$$R' = 0, \tag{42.8}$$

then we must have

$$R = a. \tag{42.9}$$

From Equation 42.1, we must then have also

$$R'' < 0; \tag{42.10}$$

this will be a point of maximum radius for the universe, and the constant $a$ will be the maximum radius of the universe. There can be no other stationary point for $R(t)$ and hence no minimum size for the universe.

Equation 42.7 can also be written in terms of arc-time, as

$$\left(\frac{dR}{d\eta}\right)^2 = R(a - R), \tag{42.11}$$

and this can be integrated to give

$$R(\eta) = \sin^2\left(\frac{\eta}{2}\right) = \frac{a}{2}(1 - \cos \eta). \tag{42.12}$$

The constant of integration was picked so that $t = 0$ when $R = 0$. This is graphed in Figure 42.1.

For astronomical purposes, arc-time is the most convenient time variable. For geological purposes, such as dating rocks and stars, proper time is the more convenient. The conversion can be found from

$$\frac{dt}{d\eta} = R(\eta) = \frac{a}{2}(1 - \cos \eta), \tag{42.13}$$

which upon integration leads to

$$t(\eta) = \frac{a}{2}(\eta - \sin \eta). \tag{42.14}$$

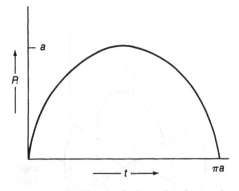

**Figure 42.2.** The evolution of a dust universe in terms of proper time.

We have also picked the constant of integration here so that time zero was when the universe had zero radius. The evolution of the universe in terms of proper time is shown in Figure 42.2.

To use such a model of the universe, we need to know two quantities. One is the parameter $a$, the maximum size of the universe, a quantity with the dimensions of length. The other is our present location in the universe, the arc-time or proper time of the present. Both of these are difficult to measure, the first because it involves a length scale, requiring us to relate our laboratory lengths to astronomical lengths. The astronomical distance scale involves a long and involved extrapolation. The second is difficult because it involves effects that are second-order in distance. Locally, every spacetime looks flat. Observational cosmology has been characterized as the search for these two quantities. The search is difficult, and rather little progress has been made to date. In the next section, we discuss observations that can in principle lead us to these parameters.

*Model parameters*

The evolution of this universe is such that, after a finite amount of arc-time and proper time, it collapses again to $R = 0$. It lives for a finite amount of proper time, $\pi a$, and for a finite amount of arc-time, $2\pi$. From this latter we see that a light signal can travel exactly once around the universe during the age of the universe. This prevents you from looking out and seeing the back of your head. For practice, let us make a picture of such a circumnavigating light signal in our various representations.

*Finite lifetime*

It will be difficult to find a faithful representation of this world line because all four coordinates change. Let us start by ignoring the angular coordinates $\theta$ and $\phi$. Figure 42.3 shows the $(\eta, \chi)$-history of such a world line, one that starts from the North pole at event $(\eta, \chi) = (0, 0)$, and proceeds at the speed of light to the South pole, an arc-distance $\pi$ away. It passes through the South pole, $(\eta, \chi) = (\pi, \pi)$, and

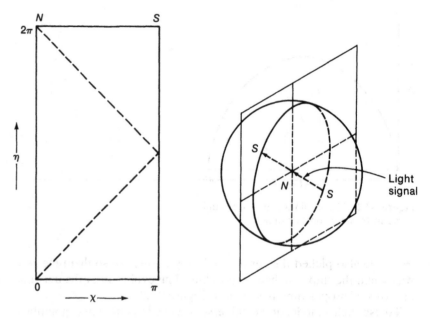

**Figure 42.3.** The around-the-universe light signal in $(\eta, \chi)$ coordinates.

**Figure 42.4.** The projection of the light signal in the solid-ball representation of the 3-sphere.

returns "around the backside of the universe" to the North pole. To see what I mean by the "backside," look at Figure 42.4, where the $\eta$ coordinate has been left out, and $\chi$, $\theta$, and $\phi$ are represented in the solid-ball model. The ray there leaves the North pole in the direction $(\theta, \phi) = (\pi/4, 0)$, picked for visual appearance, and returns to the North pole from the opposite direction, $(\theta, \phi) = (3\pi/4, \pi)$.

To smoothly represent the world line near the South pole, note that the ray in Figure 42.4 lies completely in what appears to be a plane. But remember that the rim is a single point. In Figure 42.5 we assemble

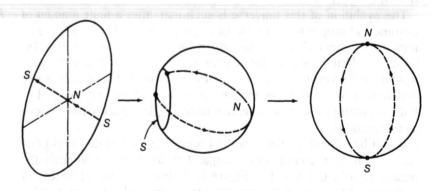

**Figure 42.5.** The path of the light signal in the great sphere, which is being reassembled from a solid-disc representation.

that apparent plane into its true shape, which is a sphere. By analogy with a great circle on $S^2$, I will call this a great sphere. We will make frequent use of it in what follows. On this great sphere we finally get a smooth representation of the circumnavigating light signal.

We will use the above three partial representations repeatedly. Used together, they will give us some ability to visualize situations in curved four-dimensional spacetime.

### PROBLEMS

42.1. (11) Suppose one observes a Hubble flow with a quadratic dependence of redshift on distance. Does this rule out a Friedmann universe?

42.2. (30) Discuss the problem of isometrically embedding these dust universes. Try both Euclidean and Lorentz spaces.

42.3. (19) Consider these two statements:
 (i) the 3° radiation comes from a fireball that is an arc-distance $\chi$ away;
 (ii) the fireball is everywhere.
Explain why they are not contradictory.

# 43. Observations in a Closed Dust Universe

*"Perhaps to the student there is no part of
 elementary mathematics so repulsive as is
 spherical trigonometry."*

P. G. TAIT
in the *Encyclopedia Britannica*, 11th ed.

To decide whether or not the closed dust model of the universe is an acceptable model, we must see if its predictions are in accord with actual observations. We will give a critical discussion of cosmological models in Section 50. Here we discuss some typical observations, and show how they are calculated in this model.

One observational number that we can hope to find is the present    *Age*
age of the universe, dating it by means of either rocks or stars. How much proper time has elapsed since the start of the universe? We want proper time, since that is what governs the rates of nuclear clocks. This age $t_0$ is given in terms of our model parameters $\eta_0$ and $a$ by

$$t_0 = \frac{a}{2}(\eta_0 - \sin \eta_0).  \tag{43.1}$$

We will follow the usual practice in cosmology, and use a subscript zero (not time = zero) to denote the present time.

*Expansion*   Another observation is the systematic redshift with distance, called the Hubble flow. This brings up the tricky question of distance. For nearby objects, distance can be measured by conventional surveying techniques. As long as the scale factor $R(t)$ changes only a little during a measurement, this proper distance is related to the arc-distance $\Delta\chi$ by

[I will use $\Delta\chi$ for arc-distance even when it is not measured along the $\chi$ direction.]

$$d = R(t)\,\Delta\chi,  \tag{43.2}$$

as can be seen from the metric $\mathscr{G}$. This gives us the best physical interpretation of the scale factor $R$. It is the constant with dimensions of length which converts arc-distance to proper distance. For galaxies at rest in the cosmic rest frame, their separation in terms of arc-distance is unchanging. This is why arc-distance is so useful. Their proper distance is changing because the scale factor $R(t)$ is changing.

Suppose that we observe the light from a nearby galaxy ($\Delta\chi \ll 1$), and compare its spectral lines with the spectral lines of the same elements produced in the laboratory (see Figure 43.1). The spectral lines will be redshifted by an amount

[The redshift in an expanding universe was discussed in Section 40.]

$$(1+z) = \frac{\lambda_{\text{REC}}}{\lambda_{\text{EMIT}}} = \frac{R(t_{\text{REC}})}{R(t_{\text{EMIT}})}.  \tag{43.3}$$

Since the galaxy is supposed to be nearby, we can approximate this with

$$\begin{aligned} R(t_{\text{EMIT}}) &\approx R(t_0 - d) \\ &\approx R_0 - dR_0', \end{aligned}  \tag{43.4}$$

[Here the prime indicates a time derivative.]

using $R_0$ for the present value of $R$, and so

$$(1+z) \approx 1 + d\frac{R_0'}{R_0}.  \tag{43.5}$$

For small distances, we have a linear redshift law of

$$z \approx \left(\frac{R_0'}{R_0}\right)d.  \tag{43.6}$$

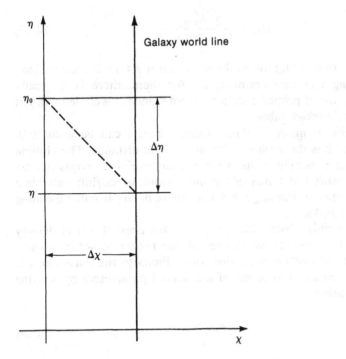

**Figure 43.1.** Observing the light from a nearby galaxy.

The constant of proportionality is called the Hubble constant, $H_0$, **_Hubble constant_**

$$H_0 \equiv \frac{R_0'}{R_0}, \tag{43.7}$$

and we can measure it in practice by observing the redshifts of galaxies that are near enough for us to find their distances, yet far enough away that they are not part of the local cluster of galaxies. The galaxies in a cluster are not free particles, and a cluster does not expand with the universe. The Hubble constant has dimensions of seconds$^{-1}$. Because of the manner in which the Hubble constant is measured, astronomers use a hybrid unit for the Hubble constant of kilometers/second/mega-parsec. The conversion factor for this is **_Hubble constant units_**

$$50 \text{ km/sec/mpc} = \frac{1}{6.18 \times 10^{17} \text{ sec.}} = \frac{1}{(19.6 \text{ Byr})}. \tag{43.8}$$

The parsec, remember, is almost exactly $10^8$ light-seconds. In terms of our model parameters $\eta_0$ and $a$, we have

$$aH_0 = \frac{2 \sin \eta_0}{(1 - \cos \eta_0)^2} \tag{43.9}$$

For objects that are far away, the scale factor $R(t)$ will change significantly during any measurement; so, for them, there is no really sensible measure of proper distance. Arc-distance is well defined, but it is not directly measurable.

*Redshift*     The redshift of many of these distant objects can be accurately measured, and it is the most suitable measure of distance. The Hubble constant can be thought of as the parameter which converts nearby distances measured in terms of the dimensionless redshift variable $z$ into proper distance. For large distances there is no particular meaning for $H_0 z$ or for $R_0 \Delta \chi$.

*Mass density*     Another possibly observable quantity is the present matter density in the universe. Whether we can see all the matter or not is an open question, but we can certainly find lower limits to the mass density. We find the density $\rho$ in terms of our model parameters by starting with the definition

$$a \equiv \frac{8 \rho \pi R^3}{3}, \tag{43.10}$$

and using Equation 42.12 for $R(t)$ to find

$$a^2 \rho_0 = \frac{3}{\pi (1 - \cos \eta_0)^3}. \tag{43.11}$$

Recall that $\rho$ has units of seconds$^{-2}$; so $a^2 \rho_0$ is dimensionless. A useful conversion factor is that the density of water is $6.67 \times 10^{-8}$ sec$^{-2}$. The matter density is often quoted in terms of the Hubble constant, using *Density parameter*     a dimensionless density parameter $\Omega_0$,

$$\Omega_0 \equiv \frac{8 \pi \rho_0}{3 H_0^2}, \tag{43.12}$$

which is related to our present arc-time by

$$\Omega_0 = \frac{2}{1 + \cos \eta_0}. \tag{43.13}$$

Being dimensionless, $\Omega_0$ can depend only on $\eta_0$ and not on $a$. For these closed dust universes, $\Omega_0$ has the range

$$\Omega_0 > 1. \tag{43.14}$$

Another observational quantity whose usefulness will be more apparent when we discuss the other models is the deceleration parameter $q_0$, defined by

$$q_0 \equiv -\frac{R_0 R_0''}{R_0' R_0'}. \qquad (43.15)$$

Remember that

$$R' \equiv \frac{dR}{dt}. \qquad (43.16)$$

Since $q_0$ is dimensionless, it must be related to $\Omega_0$ and $\eta_0$, and we have for dust universes

$$\Omega_0 = 2q_0. \qquad (43.17)$$

It is fairly straightforward to use these formulae. I give here a number of examples of gradually increasing difficulty.

*Example 1* | Suppose that light is received at arc-time $\eta_0$ from an object with a redshift $z$. How far away from us is it in terms of arc-distance?

Let $\eta$ be the arc-time when the object sent out the light, as shown in Figure 43.1. We have

$$(1 + z) = \frac{R(\eta_0)}{R(\eta)} = \frac{1 - \cos \eta_0}{1 - \cos \eta}, \qquad (43.18)$$

and this determines $\eta$. The arc-distance $\Delta\chi$ is then

$$\Delta\chi = \eta_0 - \eta, \qquad (43.19)$$

since the speed of light is unity in $(\eta,\chi)$ coordinates.

*Example 2* | You observe a distant quasar with a redshift of $z = 2$. Suppose we take a model of the universe with $q_0 = 1.9$. How old was the universe, assuming a closed dust model, when that light was sent out?

From $q_0$ one can find the present arc-time by using

$$q_0 = \frac{1}{1 + \cos \eta_0}. \qquad (43.20)$$

and we find

$$\eta_0 = 2.06 \text{ rad } (118°). \qquad (43.21)$$

In such a universe, a light signal could have made it about a third of the way around the universe since the beginning. From $\eta_0$ we can find the combination

$$(aH_0) = 0.81 \qquad (43.22)$$

from Equation 43.9. Without knowing some dimensioned quantity, we cannot hope to calculate a time. We will leave our answer in terms of the unknown Hubble constant.

The quasar sent out the light at an arc-time given by

$$(1 + z) = \frac{1 - \cos \eta_0}{1 - \cos \eta}; \qquad (43.23)$$

so we have

$$\eta = 1.03 \text{ rad } (59°). \qquad (43.24)$$

The proper age of the universe at the quasar when the light was sent out is given by

$$TH_0 = \frac{(aH_0)}{2} (\eta - \sin \eta), \qquad (43.25)$$

and we find for this

$$H_0 T = 0.0714. \qquad (43.26)$$

This can be written in terms of commonly used units as

$$(TH_0) = (1.3 \text{ Byr}) \times (50 \text{ km/sec/mpc}). \quad (43.27)$$

The present age of the universe in this model is

$$T_0 H_0 = (16 \text{ Byr}) \times (50 \text{ km/sec/mpc}). \quad (43.28)$$

If you want a Hubble constant of 100 km/sec · mpc, then the present age would be only 8 Byr, an uncomfortably short time since we think that we see stars as old as 15 Byr.

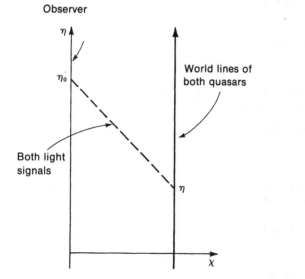

Observer

$\eta_0$

World lines of both quasars

Both light signals

$\eta$

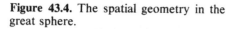

$\chi$

**Figure 43.4.** The spatial geometry in the great sphere.

*Example 3*

[I do not want to appear prejudiced on the question of quasars and whether all of their redshift is cosmological. Here I say quasar when I mean only a far-off, visible object with measurable cosmological redshift.]

[A less artificial, related problem would have one far behind the other, and ask how close the light from one passes to the other.]

Let us now go to a geometrically more complicated situation. Suppose an observer sees two quasars separated on the sky by a small angle, $\epsilon \ll 1$. Suppose further that they are at exactly the same distance, corresponding to a redshift $z$, from us. How far apart were they when the light was sent out?

The same steps that gave us a reasonable picture of the circumnavigating light signal in the last section also work here. Let us place ourselves at the North pole, $\chi = 0$. Since all points are equivalent, this is allowed. In the $(\eta, \chi)$ coordinates (Figure 43.2), the world lines of the two quasars and the two light signals coincide. In

Figure 43.3, we ignore the $\eta$-coordinate, and show things in the solid-ball picture of $S^3$. The observer is now at the center of the ball. Both quasars are at a radius $\chi = \eta_0 - \eta$, where $\eta$ is the arc-time when the quasars sent out the light. I have picked coordinates so that both quasars lie in the $\phi = 0$ great sphere. This great sphere appears as a plane through the center of the ball in Figure 43.3, and is drawn separately in Figure 43.4. The angle $\epsilon$ between the light rays shows up well in either Figure 43.3 or 43.4. The coordinates of the two quasars are thus

[The sequence in Figures 43.2, 43.3, and 43.4 corresponds to that in Figures 42.3, 42.4, and 42.5.]

$$(\chi, \theta, \phi) = (\eta_0 - \eta, 0, 0), \qquad (43.29)$$

and

$$(\chi, \theta, \phi) = (\eta_0 - \eta, \epsilon, 0). \qquad (43.30)$$

The coordinate separation between them is just

$$\Delta\theta = \epsilon; \qquad (43.31)$$

and using the metric

$$\mathcal{G} = R^2(\eta)\,[-d\eta^2 + d\chi^2 + \sin^2\chi\,(d\theta^2 + \sin^2\theta\,d\phi^2)] \qquad (43.32)$$

we can convert this coordinate separation to a proper distance

$$\delta = R(\eta)\,(\sin\chi)\epsilon. \qquad (43.33)$$

This can be written

$$(H_0\delta) = \epsilon\,\frac{(H_0 a)}{2}\,(1 - \cos\eta)\,\sin(\eta_0 - \eta). \qquad (43.34)$$

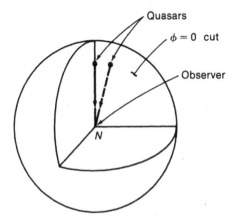

**Figure 43.2.** The $(\eta, \chi)$ plane does not show the spatial geometry.

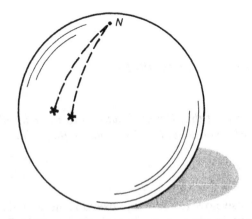

**Figure 43.3.** The spatial geometry shows up in the solid-ball representation.

[One second of arc corresponds to the smallest resolvable optical detail visible through the Earth's atmosphere, and so is a convenient angular unit. The conversion is 1 arc sec = $4.85 \times 10^{-6}$ radian.]

[The kiloparsec is a convenient length, $10^{11}$ light-seconds, since it is the size of typical features in a galaxy. The radius of our galaxy is about 10 kiloparsecs.]

For the universe with $q_0 = 1.9$ and a separation of $2''$ of arc, this comes out to be a length

$$(H_0\delta) = (1.05 \times 10^{12} \text{ sec}) \times (50 \text{ km/sec/mpc})$$

$$(43.35)$$

or

$$(H_0\delta) = (10.5 \text{ kpc}) \times (50 \text{ km/sec/mpc}).$$

$$(43.36)$$

PROBLEMS

43.1. (16) In a closed dust universe with $H_0 = 50$ km/sec/mpc and $\Omega = 2$, how far apart are galaxies in terms of arc-distance?

43.2. (26) In a closed dust universe with a density of galaxies given by $\mu$ galaxies/sec³, how many galaxies are there in total, in terms of $H_0$ and $\eta_0$?

43.3. (22) What is the interpretation of the number $(R_0 H_0)$?

43.4. (12) In a closed dust universe with $q_0 = 1.9$ and a Hubble constant of 50 km/sec · mpc, what is the present density, both in sec⁻² and in gm/cm³?

43.5. (27) Suppose one observes a cluster of objects all at the same distance, for example, a number of galaxies in a cluster. For each object a redshift is obtained, $z_i$. The $z_i$ are not all the same, because the objects in the cluster are moving relative to one another. How do you estimate the velocity dispersion of the objects from the observed redshift dispersion? Assume that $z_i - \langle z_i \rangle$ is small, but not the $z_i$ themselves. (The angled brackets indicate the average value.)

## 44. The Pseudosphere

*"I suspect that if you should go to the end of the world, you would find somebody there going further."*

H. D. THOREAU

We pause now to study the pseudosphere. This symmetric space will generate another set of cosmological models. The treatment here will parallel that of Sections 38 and 39 for the 3-sphere. The global structure of the pseudosphere is simpler than that of the 3-sphere. Its metric structure is no more complicated, only less familiar.

The pseudosphere of any dimension can be found as a subspace, this time of a space with a Lorentz metric rather than a Euclidean metric. For the 2-pseudosphere, we take a 3-space $(w,x,y)$ with a metric

$$\mathcal{G} = -dw^2 + dx^2 + dy^2, \tag{44.1}$$

and look at the subset of points that are at a unit timelike interval from the origin,

$$w^2 - x^2 - y^2 = 1, \tag{44.2}$$

in the positive $w$ direction. We expect such a subset to be homogeneous and isotropic, because it is invariant under the three isometries of $(w,x,y)$-space, these being the rotation

$$x\frac{\partial}{\partial y} - y\frac{\partial}{\partial x} \tag{44.3}$$

and the two Lorentz transformations

$$x\frac{\partial}{\partial w} + w\frac{\partial}{\partial x}, \tag{44.4}$$

$$y\frac{\partial}{\partial w} + w\frac{\partial}{\partial y}. \tag{44.5}$$

Each point on this 2-surface corresponds to a unit velocity vector, similar to our 4-velocity vectors in spacetime. We might expect that the natural measure on velocities, the rapidity, will be a useful coordinate here, and so it is. A natural parametrization for this surface is given by $(\chi,\phi)$ such that

$$w = \cosh \chi,$$

$$x = \sinh \chi \cos \phi, \tag{44.6}$$

$$y = \sinh \chi \sin \phi.$$

Here $\chi$ is the rapidity of a straight world line passing from the origin to the point given by the above equations.

The global structure here is straightforward. There is a seam where points with $\phi = 0$ and $\phi = 2\pi$ must be joined together. The rapidity $\chi$ can range over all positive values. When $\chi = 0$, we have a single point, the analog of the North pole on a sphere. The global structure here is

*Imbedding*

[Be careful! This Lorentz metric is as fictitious as the four-dimensional Euclidean metric of Section 38. It has nothing to do with spacetime.]

[I use $w$ here instead of $t$ to avoid any accidental connection with spacetime.]

*Symmetries*

[It would be consistent to call these pseudorotations, but they are already called Lorentz transformations.]

*Velocity space*

*Topology*

the same as for polar coordinates. Figure 44.1 shows the $(\chi,\phi)$ representation. It can also be glued together to form the analog of the solid disc representation.

*Metric*    This 2-pseudosphere inherits a metric by being a subspace of a larger manifold, and it is the metric

$$\mathscr{G} = d\chi^2 + \sinh^2 \chi \, d\phi^2. \tag{44.7}$$

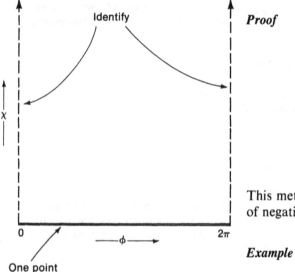

*Proof*    This metric is calculated from the parametric form given in Equations 44.6. The basis 1-forms on the 3-space pull back to 1-forms on the 2-surface given by

$$dw^* = \sinh \chi \, d\chi,$$
$$dx^* = \sinh \chi \cos \phi \, d\chi - \cosh \chi \sin \phi \, d\phi, \tag{44.8}$$
$$dy^* = \sinh \chi \sin \phi \, d\chi + \cosh \chi \cos \phi \, d\phi,$$

and the metric in 44.7 comes from substituting these into expression 44.1.

This metric is positive definite. There are no vectors of length zero or of negative length.

*Example*    Look at the metric at the point $(\chi,\phi) = (\alpha,0)$, where $\alpha$ satisfies

$$\sinh \alpha = 1,$$
$$\alpha \approx 0.89. \tag{44.9}$$

The metric there is

$$d\chi^2 + d\phi^2, \tag{44.10}$$

and the metric figure is a circle. It is clearly a Euclidean metric rather than a Lorentz metric. Since the space is homogeneous, this is true at every point.

**Figure 44.1.** The 2-pseudosphere.

*Stereographic representations*

[See Problem 33.3.]

There are stereographic representations of the pseudospheres just as there are for the spheres. One such projection is sketched in Figure 44.2, forming a chart of the 2-pseudosphere on the $x,y$-plane. This one chart covers the entire pseudosphere, and is enough to show that it is a manifold. These stereographic representations can have several interesting features. Points infinitely far away in the pseudosphere can appear at a finite distance in the projection. They can be conformal, that is, angle-preserving. Some can even take finite circles into finite circles.

*Example*

The projection onto the plane $w = 0$ from the point $(w,x,y) = (-1,0,0)$ has all the above features. It is the one shown in Figure 44.2. If we call the coordinates in the $x,y$-plane $(u,v)$ to avoid confusion, then we have

$$u = \frac{x}{w+1},\qquad(44.11)$$

$$v = \frac{y}{w+1},\qquad(44.12)$$

$$w^2 = 1 + x^2 + y^2,\qquad(44.13)$$

and these equations give the $(w,x,y)$ coordinates as functions of $u$ and $v$. We find, solving these,

$$x = \frac{2u}{\Delta},\qquad(44.14)$$

$$y = \frac{2v}{\Delta},\qquad(44.15)$$

$$w \equiv \frac{1 + u^2 + v^2}{\Delta},\qquad(44.16)$$

where

$$\Delta \equiv 1 - u^2 - v^2.\qquad(44.17)$$

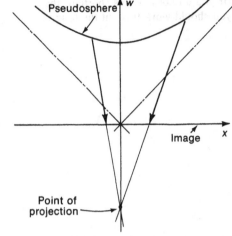

**Figure 44.2.** Mapping the pseudosphere on to the $x, y$-plane.

From these we can find out how the 1-forms $dw$, $dx$, and $dy$ pull back. For example, we have

[See Section 20 for pull-back.]

$$dx^* = \frac{2}{\Delta^2}[(1 + u^2 - v^2)\,du + 2uv\,dv].\quad(44.18)$$

Because of the axisymmetry, it is sufficient to calculate these along the line $v = 0$. There we have

$$dx^* = \frac{2(1 + u^2)\,du}{\Delta^2},\qquad(44.19)$$

$$dy^* = \frac{2\Delta\,dv}{\Delta^2},\qquad(44.20)$$

$$dw^* = \frac{4u\,du}{\Delta^2}.\qquad(44.21)$$

The metric can then be written

$$-dw^2 + dx^2 + dy^2 = \frac{4(du^2 + dv^2)}{\Delta^2}.\qquad(44.22)$$

[If you do not like this shortcut, then you should work this out the long way!]

[Another representation of the 2-pseudosphere is the Poincaré half-plane discussed in Problems 34.4 and 34.5. Both of these are discussed further in Robertson and Noonan (page 184), and in Arnold and Avez (Appendix 20).]

As written, this is already axisymmetric; so it is the correct expression for the metric everywhere.

The entire pseudosphere is mapped inside the unit circle in the $u,v$-plane. It differs from the solid-disc model $S^2$ in not including the rim. In fact, the distance from any point on the pseudosphere to the rim would be infinite. Just as with spheres, planes through the origin in $(w,x,y)$ space cut the pseudosphere in geodesics. Given this fact, we can easily find that in the $(u,v)$ representation given above, geodesics are circles which cut the unit circle at right angles. In this representation, it is easy to see the non-Euclidean features of the pseudosphere. Figure 44.3 shows two distinct lines, $A$ and $B$, neither of which intersect a third line $L$. In Euclidean geometry there is a unique line through a given point such that it does not intersect another given line. Figure 44.4 shows a triangle $ABC$ of non-zero area, all of whose vertex angles are zero. This representation of the pseudosphere is easy to manipulate with with Euclidean instruments. Play with it to develop your intuition. The Escher drawings titled "Circle Limit" are pseudospheres based on this representation.

**3-pseudosphere**

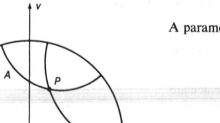

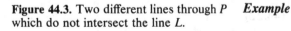

**Figure 44.3.** Two different lines through $P$ which do not intersect the line $L$.

The 3-pseudosphere is a straightforward generalization. In a 4-space with Lorentz metric, it can be represented as the set of points

$$w^2 - x^2 - y^2 - z^2 = 1. \qquad (44.23)$$

A parametrization of this set is given by

$$w = \cosh \chi,$$
$$z = \sinh \chi \cos \theta,$$
$$x = \sinh \chi \sin \theta \cos \phi, \qquad (44.24)$$
$$y = \sinh \chi \sin \theta \sin \phi,$$

and the metric that it inherits is given by

$$\mathcal{G} = d\chi^2 + \sinh^2 \chi \, d\Omega^2. \qquad (44.25)$$

The global structure in $(\chi, \theta, \phi)$ coordinates is the same as for ordinary $(r, \theta, \phi)$ spherical polar coordinates.

*Example*

The set of all 4-velocities in Minkowski spacetime is a 3-manifold. If distance is defined by relative velocity, then this velocity space has the geometry of a pseudo-sphere. The coordinate $\chi$ corresponds to the rapidity $\psi$ of Section 9.

## PROBLEMS

44.1. (19) Is the area of the triangle in Figure 44.4 finite or infinite?

44.2. (30) Give the equations for the projection shown in Figure 44.2. Describe the resulting image. Is it conformal?

44.3. (25) Find the three vector fields described in the text which generate the isotropy and homogeneity transformations on the 2-pseudosphere in $(\chi, \phi)$ coordinates.

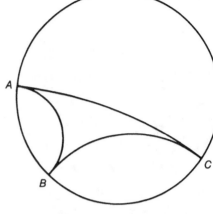

**Figure 44.4.**

44.4. (28) Show that the arc-distance $\chi$ between two points in the pseudosphere is given by

$$\cosh \chi = -A \cdot B,$$

where $A$ and $B$ are 4-vectors in the fictitious Lorentz 4-space. Use the symmetries of the pseudosphere to simplify your argument.

44.5. (35) Use arguments similar to those in Problem 12.8 to show that the angle $\theta$ which has a vertex $B$ and sides $BA$ and $BC$ is given by

$$\cos \theta = \frac{(A \cdot C) - (B \cdot A)(B \cdot C)}{[(B \cdot A)^2 - 1]^{1/2}[(B \cdot C)^2 - 1]^{1/2}}$$

where

$$A \cdot A = B \cdot B = C \cdot C = -1.$$

44.6. (26) Use the above result to find the analog of the Law of Cosines for pseudospherical triangles:

$$\cosh a = \cosh b \cosh c - \sinh b \sinh c \cos A,$$

where $a$, $b$, and $c$ are the arc-lengths of the sides, and $A$ is the angle opposite side $a$.

44.7. (28) Use the results of the preceding problems to find the analog of the Law of Sines:

$$\frac{\sin A}{\sinh a} = \frac{\sin B}{\sinh b} = \frac{\sin C}{\sinh c}.$$

44.8. (24) Show how to guess the Law of Cosines for Angles,

$$\cos A = -\cos B \cos C + \sin B \sin C \cosh a,$$

from the corresponding law for spherical triangles. What about the signs?

## 45. Dust and Radiation Friedmann Universes

We can also solve the dust Friedmann equations for the two other cases: the critical universe, whose spatial sections are Euclidean 3-spaces; and the open universe, whose spatial sections are 3-pseudospheres. Whereas at any time the volume of the 3-sphere universes is finite, and hence they are called closed, these other two cases always involve spatial sections that are infinite. The Friedmann equations can also be solved for a matter content that satisfies the equation of state of radiation. Here we list the properties of all these solutions, so that they too can be used as cosmological models.

For the other dust models the Friedmann equations can be solved as before. For all cases the function

$$a \equiv \frac{8\pi\rho R^3}{3} \tag{45.1}$$

is a constant of the motion. For the open case the solution involves hyperbolic functions; for the critical case it involves powers. The critical case is sometimes called the Euclidean model. This name is misleading, since it is a curved spacetime. It is also sometimes called the Einstein–de Sitter model. Table 45.1 lists these solutions. The scale factor in the open universe no longer relates to so obvious a property

**TABLE 45.1.  PROPERTIES OF DUST FRIEDMANN UNIVERSES.**

| Parameter | Closed | Critical | Open |
|:---:|:---:|:---:|:---:|
| $R$ | $\frac{a}{2}(1 - \cos \eta)$ | $\frac{a\eta^2}{4}$ | $\frac{a}{2}(\cosh \eta - 1)$ |
| $t$ | $\frac{a}{2}(\eta - \sin \eta)$ | $\frac{a\eta^3}{12}$ | $\frac{a}{2}(\sinh \eta - \eta)$ |
| $Ha$ | $\frac{2 \sin \eta}{(1 - \cos \eta)^2}$ | $\frac{8}{\eta^3}$ | $\frac{2 \sinh \eta}{(\cosh \eta - 1)^2}$ |
| $\frac{\pi\rho a^2}{3}$ | $\frac{1}{(1 - \cos \eta)^3}$ | $\frac{8}{\eta^6}$ | $\frac{1}{(\cosh \eta - 1)^3}$ |
| $\Omega$ | $\frac{2}{1 + \cos \eta}$ | $1$ | $\frac{2}{1 + \cosh \eta}$ |
| $q_0$ | $\frac{1}{1 + \cos \eta}$ | $\frac{1}{2}$ | $\frac{1}{1 + \cosh \eta}$ |

as the circumference, but it still defines the distance over which curvature effects become important.

Both the Euclidean and the open universes expand forever. Their pasts include only a finite amount of proper time before they have a state of vanishing radius as in the closed universes, but now the future is unbounded. If we can decide between open and closed models, then we can decide the future of the universe.

The density parameter $\Omega$ provides a natural measure of evolution in the open and closed models. For open models, $\Omega$ starts from one and decreases to zero. For the closed models, it starts from one and increases to infinity. A knowledge of $\Omega$ tells you not only what "time" it is, but also what kind of a universe you are in. When formulae can be written in terms of $\Omega$, they usually are valid in all three types of models, as we will see in the examples to come.

*Density parameter*

The critical models have $\Omega = 1$ at all times. There is no way to distinguish observationally between the Euclidean universe and a very young open or closed model. The critical models are self-similar in time and contain no intrinsic measure of evolution. They are even more symmetric than the open and closed models. Table 45.1 lists the important properties of dust Friedmann universes.

*Critical models*

*Example*

In a universe whose present age is given by a density parameter $\Omega_0$, what was the density parameter back at a redshift $z$?

First we consider the closed models, in which $\Omega > 1$. There we have

$$(1 + z) = \frac{R_0}{R} = \frac{1 - \cos \eta_0}{1 - \cos \eta}, \qquad (45.2)$$

and also

$$\cos \eta = \frac{2}{\Omega} - 1. \qquad (45.3)$$

This leads to

$$\Omega = \frac{\Omega_0 (1 + z)}{1 + \Omega_0 z}. \qquad (45.4)$$

The same expression holds also for the models in which $\Omega \leq 1$, as so often happens.

Unlike the situation for the closed models, the scale factor $R(t)$ here does not remain more or less about the same size as the constant $a$ that characterizes the model. Instead, the scale factor $R$ grows to infinity. This fact provides a solution to a philosophical dilemma. The universe is described by the constant $a$, which has the dimensions of

*The Planck universe*

length. An explanation of the universe should explain this constant. If you favor a quantum-mechanical explanation for the origin of the universe, then you expect that the length formed from Planck's constant, $G$, and $c$ would be important. Sadly, this length is quite inappropriate as a dimension for the universe. We have in our units (where $G = c = 1$)

$$(\hbar)^{1/2} = 1.35 \times 10^{-43} \text{ sec.} \qquad (45.5)$$

[The discussion in this paragraph is properly called philosophical, since it is an attempt to learn something about the universe by means of pure thought.]

The closed universe with these dimensions would live for only a time of that order. The open universe provides an escape from this dilemma, since even a universe with parameter

$$a \approx \sqrt{\hbar} \qquad (45.6)$$

has a final state not too different from the state of our present universe.

*Radiation universes*

The other simple and reasonable equation of state to consider is that of pure radiation,

$$p = \tfrac{1}{3}\rho. \qquad (45.7)$$

If this model is to be useful for the present universe, we must assume that it contains a small amount of matter, to account for the galaxies. The radiation itself would have to be in some form as yet undetected if it is to dominate the matter content. Radiation enclosed in a box exerts pressure on the walls and does work on the walls if the box expands. In this way the radiation in an expanding box loses energy. A slow expansion preserves not energy but rather the number of photons. Since their wavelength increases like $R(t)$, their energies ($E = h\nu$) must decrease like $1/R$. For radiation in a box, then, we have the conservation law

[I don't know of an elementary proof of this important result.]

*Conservation law for radiation*

$$\rho R^4 = \text{constant.} \qquad (45.8)$$

Such a conservation law is also true in an expanding universe. It is automatically contained in the Friedmann equations, just as the matter conservation law was in a dust universe.

[In Section 48 we will see that there is always a conservation law in the Friedmann equations.]

***Proof*** The Friedmann equations for a radiation-dominated universe are

$$2\frac{R''}{R} + \left(\frac{R'}{R}\right)^2 + \frac{\kappa}{R^2} + 8\pi p = 0 \qquad (45.9)$$

and

$$\left(\frac{R'}{R}\right)^2 + \frac{\kappa}{R^2} = \frac{8\pi\rho}{3}. \qquad (45.10)$$

The first and second can be combined to eliminate the matter term, to give a simple second-order equation:

$$\frac{R''}{R} + \left(\frac{R'}{R}\right)^2 + \frac{\kappa}{R^2} = 0 \qquad (45.11)$$

We differentiate our supposed constant of motion

$$\frac{d}{dt}\left(\frac{8\pi\rho R^4}{3}\right) = 2R(R')^3 + 2R^2 R' R'' + 2\kappa R R', \quad (45.12)$$

and this vanishes because it is $2R' R^4$ times Equation 45.11.

Let us call the constant of the motion $a^2$, so that $a$ again has the dimensions of length:

$$a^2 \equiv \frac{8\pi\rho R^4}{3}. \qquad (45.13)$$

To find $R(t)$ we need only the equation

$$(R')^2 = \frac{a^2}{R^2} - \kappa. \qquad (45.14)$$

This can be integrated directly, to give

$$R(t) = \begin{Bmatrix} \sqrt{t(2a-t)} \\ \sqrt{2at} \\ \sqrt{t(2a+t)} \end{Bmatrix} \quad \text{for} \quad \kappa = \begin{Bmatrix} +1 \\ 0 \\ -1 \end{Bmatrix}. \qquad (45.15)$$

From these, all the usual properties can be derived. They are listed in Table 45.2. The differences in the behavior of $\Omega$ and the differences between models as the universe evolves are just as before. The closed radiation universe lives for a proper time $2a$, and $a$ is the maximum radius of the universe. It lives for an arc-time of $\pi$. During the lifetime of a radiation universe, a light signal has time to make it only halfway around the universe.

**TABLE 45.2.** PROPERTIES OF RADIATION
FRIEDMANN UNIVERSES.

| Parameter | Closed | Critical | Open |
|:---:|:---:|:---:|:---:|
| $R$ | $a \sin \eta$ | $a\eta$ | $a \sinh \eta$ |
| $t$ | $a(1 - \cos \eta)$ | $\dfrac{a\eta^2}{2}$ | $a(\cosh \eta - 1)$ |
| $Ha$ | $\dfrac{\cos \eta}{\sin^2 \eta}$ | $\dfrac{1}{\eta^2}$ | $\dfrac{\cosh \eta}{\sinh^2 \eta}$ |
| $\dfrac{8\pi\rho a^2}{3}$ | $\dfrac{1}{\sin^4 \eta}$ | $\dfrac{1}{\eta^4}$ | $\dfrac{1}{\sinh^4 \eta}$ |
| $\Omega$ | $\dfrac{1}{\cos^2 \eta}$ | $1$ | $\dfrac{1}{\cosh^2 \eta}$ |
| $q_0$ | $\dfrac{1}{\cos^2 \eta}$ | $1$ | $\dfrac{1}{\cosh^2 \eta}$ |

***Radiation in the early universe***

Even if we do not make the unsurprising assumption that the universe is filled with undiscovered radiation, these radiation-dominated models will still be needed. If for no other reason than the microwave background radiation, the early universe must have been a radiation-dominated universe. The observed matter density in galaxies today is about five hundred times the energy density in the microwave background radiation. The energy density of radiation was more important in the past in proportion to $(1 + z)$. Thus, even with only the presently known contents, we would have had a radiation-dominated universe for times before $z = 500$. In addition, the galaxies today have random velocities of about $10^{-3}$. As the universe expands, such random velocities decrease like $1/R$. So, back before $z = 1,000$, the random velocities of galaxies, if galaxies existed then, would themselves be relativistic. Again, we would be dealing with $p = \frac{1}{3}\rho$ matter.

[See Ohanian, Chapter 10, for a more extensive discussion of the present energy content of the universe.]

[See Problem 45.12.]

***"Universal" formulae***

The separate columns in Tables 45.1 and 45.2 are not really independent. They can be collapsed into a single set of formulae by going to a notation that uses complex numbers. This is more useful in general calculations; whereas the formulae in Tables 41.1 and 41.2 are more useful for specific cases. We introduce another indicator, $\delta$, similar to $\kappa$, such that

[The only reason I introduced $\kappa$ was that it is in very wide use. Clearly $\delta$ will serve as well, since $\delta^2 = -\kappa$.]

$$\delta = \left\{ \begin{matrix} +1 \\ 0 \\ i \end{matrix} \right\} \text{ if } \left\{ \begin{matrix} \text{open} \\ \text{critical} \\ \text{closed} \end{matrix} \right\}. \qquad (45.16)$$

**TABLE 45.3.** UNIVERSAL FRIEDMANN-UNIVERSE PROPERTIES.

| Parameter | Dust | Radiation |
|---|---|---|
| $\dfrac{R}{a}$ | $\dfrac{1}{\delta^2}\sinh^2\left(\dfrac{\delta\eta}{2}\right)$ | $\dfrac{\sinh\delta\eta}{\delta}$ |
| $\dfrac{t}{a}$ | $\dfrac{1}{2\delta^3}(\sinh\delta\eta - \delta\eta)$ | $\dfrac{2\sinh^2\left(\dfrac{\delta\eta}{2}\right)}{\delta^2}$ |
| $(aH)$ | $\dfrac{\delta^3\sinh\delta\eta}{2\sinh^4\left(\dfrac{\delta\eta}{2}\right)}$ | $\dfrac{\delta^2\cosh\delta\eta}{\sinh^2\delta\eta}$ |
| $\dfrac{8\pi\rho a^2}{3}$ | $\dfrac{\delta^6}{\sinh^6\left(\dfrac{\delta\eta}{2}\right)}$ | $\dfrac{\delta^4}{\sinh^4(\delta\eta)}$ |
| $\Omega$ | $\dfrac{2}{1+\cosh\delta\eta}$ | $\dfrac{1}{\cosh\delta\eta}$ |
| $q_0$ | $\dfrac{1}{1+\cosh\delta\eta}$ | $\dfrac{1}{\cosh\delta\eta}$ |

The formulae in this complex notation are listed in Table 45.3. Note that the $\delta$s are placed so that the limit $\delta \to 0$ always makes sense and gives the correct equations for the critical models.

## PROBLEMS

45.1. (20) How do you find the density parameter $\Omega$ for a general Friedmann universe from the scale factor and the density?

45.2. (12) For a dust universe, how do you find $\eta_0$ if you are given the age and the Hubble constant? What about a radiation universe?

45.3. (17) Rip van Winkle goes to sleep and sets his alarm clock to go off when the microwave background radiation temperature hits 2°. Calculate how long he will sleep for three universe models with $H_0$ = 50 km/sec/mpc and:

(i) dust, $\Omega = .05$;

(ii) dust, $\Omega = 2$;

(iii) radiation, $\Omega = 1$.

45.4. (28) The universe has length scales corresponding to galaxies $(2 \times 10^{12}$ sec), the distances between galaxies $(3 \times 10^{14}$ sec), and clusters of galaxies ($10^{15}$ sec). Convert these to arc-distance in the above three models.

45.5. (17) What are the observed angular sizes of the above length scales at $z = .1$, 1, and 10 in an $\Omega = 2$ dust universe?

45.6. (28) Suppose that the universe contains objects having a continuous existence, such as galaxies. The number density $\mu$ as a function of redshift is the number of objects $\Delta n$ found over the entire sky between redshifts $z$ and $z + \Delta z$ according to

$$\mu \, \Delta z = \Delta n.$$

What is the number density in dust and radiation models?

45.7. (33) For an open dust universe, the spacetime

$$\mathcal{G} = -dt^2 + t^2(d\chi^2 + \sinh^2 \chi \, d\Omega^2)$$

is found by taking the limits

$$a \to 0,$$

$$\eta \to \infty,$$

appropriately. Show how to take these limits, and show that this is an empty universe. Since it is empty, it must be Minkowski spacetime. Find a coordinate transformation into familiar coordinates.

45.8. (30) Do the same as in 45.7 for a radiation universe.

45.9. (26) A galaxy on the line of sight to a more distant quasar causes a narrow absorption line in the quasar spectrum. From a small redshift $z$ for the galaxy, compute the rate of change of $z$ with our proper time in terms of $H_0$ and $q_0$.

45.10. (35) Use the dispersion relation for nonrelativistic particles, and the fact that the wavelength of a wave packet should expand along with the universe, to show that a free particle moving with respect to the universe slows down according to

$$v \propto \frac{1}{R}.$$

45.11. (20) The random motions of galaxies today have a speed of about $10^{-3}$. When would these speeds have been comparable to the speed of light? Comment.

45.12. (35) Repeat Problem 45.11 without restricting attention to nonrelativistic particles. Check that the case of light is correctly included.

45.13. (30) What would you say if a believable measurement gave $q_0$ to be $1.001 \pm .002$?

## 46. The Hubble Diagram

An expanding curved spacetime does not change only the frequency of light; it also changes the intensity, by bending the rays toward or away from focus. Since the early days of general relativity, astronomers have hoped to use this effect to find out what kind of a universe we are in and at what stage in its evolution we are living. That is, they want to measure $q_0$. In practice one can observe only the apparent brightness of distant objects and their redshift. The relation between these quantities is called the Hubble diagram. Each cosmological model predicts a specific shape for the Hubble diagram, and here we calculate that shape.

*Observations*

The available data will be the redshift, $z$, and the spectral energy distribution of the light from a number of objects at different redshifts. This spectral energy distribution, $F_{\nu_0}$ is the energy-flux density measured at the receiver, and has conventional units of watts/meter²/hertz. The total energy flux is the integral of this flux density over frequency. The power would then be the further integral of the flux over a given collecting area. The energy would finally be the integral of the power over time.

*Source properties*

The light source itself is characterized by a luminosity $L_\nu$, which describes the spectral energy distribution of the emitted light. The true sticking point in using the luminosity/redshift relation to study the universe is in finding something out about the intrinsic luminosity $L_\nu$. This requires some knowledge of the objects and a model for their behavior. For example, to use light from an entire galaxy, one needs to be able to predict $L_\nu$ from the optical appearance, age, and spectrum of the galaxy. Absolute accuracy is not needed, as long as we can identify a class of objects that all have the same $L_\nu$. The intrinsic luminosity $L_\nu$ involves difficult and not fully known astrophysics. We will discuss it no further, and proceed to discuss the geometric aspects of the problem as if a set of objects of fixed $L_\nu$ at a variety of different redshifts were available.

The main theoretical result needed for this problem is that, as light propagates through an expanding universe, the number of photons stays constant. This result can be derived from a generalization of Maxwell's equations to curved spacetime, studied in the same high-frequency limit used in Section 27.

*A photon bunch*

Let us concentrate on a group of photons sent out in a narrow spectral band, from a frequency $\nu$ to $\nu + \Delta\nu$, over which the luminosity of the source $L_\nu$ is constant. Let $N_\nu$ be the number of photons sent out per second per hertz. Then we have

$$L_\nu = h\nu N_\nu, \qquad (46.1)$$

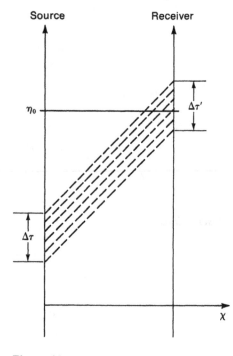

**Figure 46.1.**

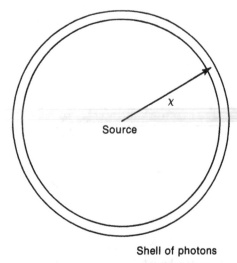

**Figure 46.2.** A section through our space-time at time $\eta_0$ indicated in Figure 46.1 finds our photons in a shell of coordinate radius $\chi$.

using Planck's law for the energy of a photon

$$E = h\nu. \tag{46.2}$$

Nearly all cosmological discussions assume that this law is true at all times in the evolution of the universe. A time-dependence for Planck's constant $h$ would change the delicate interrelationships within the spectrum of a distant object. Since the interrelationships are the same for both nearby and distant sources, very little change in $h$ can be tolerated over the age of the universe. Also, let $n_{\nu_0}$ be the observed rate of arrival of photons per unit area. We have

$$F_{\nu_0} = h\nu_0 n_{\nu_0}. \tag{46.3}$$

The received frequency $\nu_0$ is lower than the emitted frequency $\nu$ because of the cosmological redshift:

$$\frac{\nu}{\nu_0} = (1 + z). \tag{46.4}$$

The photons sent out in a spectral band of width $\Delta\nu$ are received in a smaller bandwidth $\Delta\nu_0$ given by

$$\frac{\Delta\nu}{\Delta\nu_0} = (1 + z), \tag{46.5}$$

since the redshift law is linear.

Now look at the photons that are sent out thus during a time interval $\Delta t$. They will be received during a different time interval, $\Delta t_0$, and this time interval will also be redshifted:

$$\frac{\Delta t_0}{\Delta t} = (1 + z). \tag{46.6}$$

We are now talking about a definite number of photons, sent out in a time interval $\Delta t$ and in a specific spectral bandwidth $\Delta\nu$, as shown in Figure 46.1.

In Figure 46.2 we show these same photons in a solid-ball model of our spacetime. They have traveled an arc-distance $\chi$ from the source, which we have placed at the North pole, $\chi = 0$. The arc-distance is given by

$$\chi = \eta_0 - \eta, \tag{46.7}$$

and in this the arc-time $\eta$ is found from

$$(1 + z) = \frac{R(\eta_0)}{R(\eta)}. \tag{46.8}$$

The appropriate function $R$ must be used here. These photons all cross a 2-sphere, well-shown in Figure 46.2, which has a surface area

$$A = 4\pi R_0^2 S^2(\chi). \tag{46.9}$$

The number of photons sent out must equal the number received at the 2-sphere. Thus we must have

$$L_\nu \frac{\Delta\nu}{\nu} \Delta t = F_{\nu_0} \frac{\Delta\nu_0}{\nu_0} \Delta t_0 A, \tag{46.10}$$

and this leads us to our fundamental equation

$$F_{\nu_0} = \frac{L_\nu}{4\pi R_0^2 (1 + z) S^2(\chi)}. \tag{46.11}$$

This expression relates the observed flux to the intrinsic luminosity for a source and observer at rest in any Robertson-Walker spacetime.

Equation 46.11 is not in the most convenient form for actual use. The scale factor $R_0$ is completely unknown, and little may be known about the absolute luminosity as well. One usually plots the logarithm of $F_{\nu_0}$ to turn this multiplicative uncertainty into a simple translation. Further, for small $z$ we have                                   *New variables*

$$S(\chi) \approx \chi \approx \frac{z}{(R_0 H_0)}, \tag{46.12}$$

and for nearby objects of constant luminosity, the flux observed will vary as $1/z^2$. This is the familiar inverse-square law. For large $z$ we are probing the early universe. The arc-distance $\chi$ goes to a limit $\eta_0$, and the flux varies like $(1 + z)^{-1}$ because of the redshift. To isolate the effects of curved spacetime from these known effects, consider the quantity

$$\varphi \equiv \frac{1}{2} \log_{10} \left[ \frac{z^2 F_{\nu_0}}{(1 + z)} \right]. \tag{46.13}$$

For small $z$ it behaves like $z^2 F_{\nu_0}$; so it is constant in the limit $z \to 0$. For large $z$ it goes like $z F_{\nu_0}$ and again is constant. In Equation 46.13, $\varphi$ is

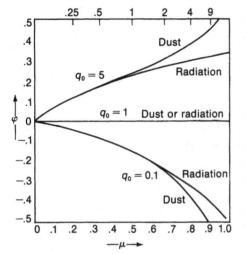

**Figure 46.3.**

[Since $\varphi$ and the $(\mu, \varphi)$-diagram are a recent invention of mine, you will not find them in common use yet.]

defined by purely observed quantities. The astronomer can compute $\varphi$ without using any cosmological theory. In terms of a specific cosmological model, we can predict the form of the $(\chi, z)$ relationship. From Equation 46.11,

$$\varphi = \log_{10} \frac{z}{(1+z)S(\chi)} + \frac{1}{2}\log_{10}\frac{L_\nu}{4\pi R_0^2}, \qquad (46.14)$$

and the unknowns $L_\nu$ and $R_0$ appear now only in the constant term. Astronomers love these logarithmic quantities. Their magnitude, for example, is an increment of 0.4 in $\log_{10}$. Here $\chi$ is a function of $z$ through Equations 46.7 and 46.8. Its dependence on the cosmological model enters here in the parameter $\eta_0$.

Since $\varphi$ goes to a finite limit for large $z$, the curves for large $z$ contain little information. A better redshift variable is the equally natural variable

$$\mu \equiv \frac{z}{(1+z)} = \frac{\lambda_0 - \lambda}{\lambda_0}, \qquad (46.15)$$

which ranges from zero to one over the observable universe. I will take the liberty of calling a plot of $\varphi$ against $\mu$ a Hubble diagram even though Hubble drew no such diagram.

In Figure 46.3 we give some theoretical Hubble-diagram curves. Note how the open and closed models fit smoothly together along the curve for the critical model. The shape of these curves does not depend at all on $H_0$. Different values of $q_0$ show up in the different over-all slopes of the curves. The vertical position is arbitrary unless $L_\nu$ and $R_0$ are known.

*Prospects*   To use such observations to find $q_0$ will be difficult. A very good theoretical understanding of the intrinsic luminosity $L_\nu$ and of its possible dependence on $z$ will be needed. In addition, the light from distant objects is bent by all the local gravitational fields in the universe, as well as by the smooth cosmological field. Such small-scale bending can alter the Hubble diagram systematically as well as increase its scatter. A tentative Hubble diagram for quasars in the form given here can be found in the paper by Baldwin *et al.*, which lists some earlier references.

PROBLEMS

46.1.  (21) Show that

$$\left.\frac{d\phi}{d\mu}\right|_{\mu=0} = \frac{\log e}{2}(q_0 - 1).$$

46.2. (24) What is the Hubble diagram for the empty universe of Problems 45.8 and 45.9?

46.3. (28) Show that the surface brightness, received flux divided by solid angle, does not depend on $z$.

46.4. (29) Suppose one were to have a class of objects of known proper size, and observations of their apparent diameter and redshift. Show how much data may also be plotted directly on a Hubble diagram.

46.5. (28) Show that number-count data as in Problem 45.7 cannot be plotted on a Hubble diagram. It is a different measurement.

46.6. (39) Invent a more useful form of the Hubble diagram for the case

$$\Omega \ll 1.$$

46.7. (31) Assume that all the objects in Problem 45.6 are sources of radio waves of equal strength. What number density in terms of apparent radio flux would be observed in various models?

# *47. Newtonian Cosmology*

Some of the properties of general relativistic cosmology can be found from quasi-Newtonian arguments. Such arguments are used by many introductory cosmology texts to avoid the extensive investment needed to honestly describe a general-relativity spacetime. Unfortunately, these quasi-Newtonian arguments give the wrong answers in many places; so the student cannot safely use them.

It is instructive to look into the reasons why these arguments sometimes work, not only to appreciate better our more careful treatment, but also to shed some light on the validity of our approximation in which the matter in the universe was treated as a smooth fluid. The actual mass distribution in the universe is very far from uniform. The mass density inside a star can be $10^{36}$ or more times the density between the stars. An important question is whether or not the galaxies expand with the universe. We assume that they do not, but this is a difficult question, and few good calculations on it have been done. The argument of this section will be that the local dynamics follows Newtonian laws, and is not affected by the expansion of the universe.

***Dust universe and dust cloud***

Let us compare the dynamics of a Friedmann universe with the dynamics of an isolated cloud of matter. In both, let us take the matter to be dust, that is, matter with negligible pressure. If the density distribution in the cloud is uniform, and if for any point in the cloud the velocity at any time is proportional to the radius of that point, then the density remains uniform as the cloud expands and contracts under its own self-gravity.

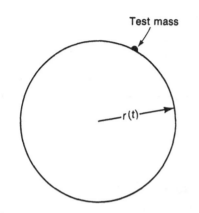

Test mass

$r(t)$

**Figure 47.1.**

***Proof***

The acceleration at any radius is given by $m/r^2$ where $m$ is the mass inside that radius. For a uniform density, we have

$$m = \frac{4}{3}\pi\rho r^3,\qquad(47.1)$$

and hence an acceleration

$$\frac{d^2r}{dt^2} = -\left(\frac{4\pi\rho}{3}\right)r\qquad(47.2)$$

which increases linearly with radius. This ensures that the ratios of the distances from the center of any two particles remains constant. The sphere thus expands and contracts uniformly.

Under these assumptions, the cloud is completely described by its total mass $M$ and its radius $r(t)$.

***Collapse of dust cloud***

To find the dynamics of this dust cloud, consider a test particle right on the surface of the cloud, as drawn in Figure 47.1. This test particle is like a little piece of the surface and will follow the cloud as it contracts. Thus we have

$$\frac{d^2r}{dt^2} = -\frac{M}{r^2},\qquad(47.3)$$

which is Newton's law of gravitation with $G = 1$. A constant of the motion, energy conservation, can be found if we multiply Equation 47.3 by $r'$:

$$\frac{1}{2}\frac{d}{dt}(r'^2) = \frac{d}{dt}\left(\frac{M}{r}\right),\qquad(47.4)$$

which can be integrated to give

$$r(r')^2 - 2Er = 2M,\qquad(47.5)$$

where $E$ is the constant of integration. It is the total energy of the cloud, kinetic plus negative gravitational energy.

Next we compare this to a sphere of matter inside an expanding Friedmann dust universe. To do this we follow the matter inside a small sphere of fixed arc-radius $\chi_0 \ll 1$. The proper radius will be

$$r = R(t)\chi_0, \tag{47.6}$$

and the dynamics of this proper radius will be given by multiplying the Friedmann equation

*Dust universe*

$$R(R')^2 + \kappa R = a \tag{47.7}$$

by $\chi_0{}^3$,

$$r(r')^2 + (\kappa\chi_0{}^3)r = (a\chi_0)^3. \tag{47.8}$$

This is the same equation as that for the Newtonian sphere, 47.5, if we take for the mass of the sphere

$$M = \frac{a\chi_0{}^3}{2}. \tag{47.9}$$

From the definition of $a$, this mass is

$$M = \frac{4\pi}{3}\rho(R\chi_0)^3 = \frac{4\pi}{3}\rho r^3, \tag{47.10}$$

and this is indeed the mass inside a small sphere in a Friedmann universe. The constant $E$ is the total energy of the cloud, and depends on its size and the initial conditions. If $E < 0$, then the sphere is bound and can only expand to some maximum size and then collapse. This corresponds to the $\kappa = 1$, closed-universe behavior.

[This is true only for a small sphere. For a large sphere we have difficulty defining the total mass, since the proper radius and the proper circumference are not related by the Euclidean expression.]

Thus we see that the dynamics of a small cloud of matter is unaffected by the expanding universe. This is an extension of the Newtonian result that a spherically symmetric mass distribution has no effect on the matter inside it. One can then consider carving out a spherical cavity inside a Friedmann universe. Reform the matter obtained into a galaxy of $10^{11}$ stars. The dynamics of this galaxy should be governed by Newtonian dynamics, and it will not expand as the universe expands. Although this is a pretty loose argument, one really cannot do much better. Incidentally, in a radiation-dominated universe, one cannot hope to carve out such a hole, since there would always be the energy density of radiation inside the galaxy. The loss of this radiation energy would cause the galaxies to expand slightly with the expansion of the universe.

*Ignore the universe*

The more one thinks about this result, the less obvious and the more surprising it becomes. One would have expected the gravitational potential energy of the entire universe to come in somehow. Often general relativity acts as if the gravitational field had no energy. This is necessary if it is to produce a finite description of an infinite universe.

PROBLEMS

47.1. (19) Complete the argument in the text to show that a uniform density distribution is preserved.

47.2. (36) Suppose we lived in a radiation-filled, $\Omega = 2$ universe. As it expands, the mass of this radiation inside the Earth's orbit changes. How much would the Earth's orbital period change in a million years? (This problem requires more classical-mechanics skills than many of you have. You need to know about adiabatic invariants.)

## 48. Do-It-Yourself Cosmology

The preceeding cosmological models have used unrealistic equations of state so that we could solve the resulting equations. One should not expect to be able to solve a system of ordinary differential equations in terms of ordinary functions. Usually they will define new functions; so there will be no alternative to numerical calculation. Using a programmable hand calculator, you can easily construct cosmological models for any equation of state. Here I will show you how to do this. This is your chance to make for yourself a better model of the universe than any I have provided. Also, putting a calculation into the organized form necessary for computation often gives you a better understanding of the problem than you can gain from just analytic solutions for special cases.

[The book by Acton has a lot of good material on numerical methods if you really get interested in this.]

*Friedmann equations*

The system of equations for the general Friedmann universe is

$$2\frac{R''}{R} + \left(\frac{R'}{R}\right)^2 + \frac{\kappa}{R^2} + 8\pi p = 0, \qquad (48.1)$$

$$\left(\frac{R'}{R}\right)^2 + \frac{\kappa}{R^2} = \frac{8\pi\rho}{3}. \qquad (48.2)$$

In addition, a third relation is needed for the pressure $p$. It may be an analytic function, a table, or a computer algorithm. Equations 48.1 and

48.2 always contain an energy-conservation law, which we can use, as before, to reduce the order of one equation. The conservation law is *General conservation law*

$$\frac{d}{dt}(\rho R^3) + p\frac{d}{dt}(R^3) = 0. \qquad (48.3)$$

This is a familiar law, saying that

$$\frac{d}{dt}(\text{energy}) = -\text{pressure} \times \frac{d}{dt}(\text{volume}). \qquad (48.4)$$

**Proof** | Multiply Equation 48.2 by $R^3$ and differentiate to find

$$(R')^3 + 2RR'R'' + \kappa R' = \frac{d}{dt}\left(\frac{8\pi}{3}\rho R^3\right). \qquad (48.5)$$

Multiply Equation 48.3 by $R^2 R'$ to find

$$(R')^3 + 2RR'R'' + \kappa R' = -8\pi p R^2 R'. \qquad (48.6)$$

From these the conservation law follows easily.

Our system of equations is then the first-order ordinary differential equations:

$$\frac{dR}{dt} = \left(\frac{8\pi\rho R^2}{3} - \kappa\right)^{1/2}, \qquad (48.7) \quad \textit{First-order system of equations}$$

$$\frac{d\rho}{dt} = (-)\frac{3(\rho + p)}{R}\frac{dR}{dt}; \qquad (48.8)$$

and the equation for arc-time,

$$\frac{d\eta}{dt} = \frac{1}{R}. \qquad (48.9)$$

In the early universe a lot happens during a very short interval of proper time. If we use arc-time as a variable instead, then we can slow down our computation in this region. Compare Figures 42.1 and 42.2. Our equations in terms of arc-time are

$$\frac{dR}{d\eta} = R\left(\frac{8\pi\rho R^2}{3} - \kappa\right)^{1/2}, \qquad (48.10) \quad \textit{Arc-time equations}$$

$$\frac{d\rho}{d\eta} = \frac{-3(\rho + p)}{R}\frac{dR}{d\eta}, \qquad (48.11)$$

$$\frac{dt}{d\eta} = R. \qquad (48.12)$$

From these functions, we can answer the common cosmological questions.

If we try to integrate these equations forward from $t = 0$, we will run into a lot of trouble. Not only is $\rho$ infinite and $R$ zero, but the question of starting values is tricky. Instead, let us compute backward from the present. What will we use for initial values? The proper-time scale is shift-invariant; so we can call the present time zero if we wish. Different models result from different values or $\rho_0$, the current energy density. It will be one useful free parameter in our models. Another useful free parameter is the Hubble constant, $H_0$. From Equation 48.7, we have

*Starting values*

$$H_0^2 = \frac{8\pi\rho_0}{3} - \frac{\kappa}{R_0^2}; \qquad (48.13)$$

so the Hubble constant and the density determine a starting value for $R$. The initial conditions could also be given simply in terms of the density parameter

$$\Omega_0 = \frac{8\pi\rho_0}{3H_0^2}. \qquad (48.14)$$

*Numerical algorithm*   Once we have equations and initial values, we need an algorithm for solving the differential equations. A reasonable algorithm to use on a small calculator is a second-order Runge-Kutta scheme. For a differential equation

$$\frac{dy}{dt} = f(y), \qquad (48.15)$$

This scheme goes from values at time $t$ to those at a time $t + \Delta t$ according to

$$y(t + \Delta t) = y(t) + \Delta t f\left\{y(t) + \frac{\Delta t}{2} f[y(t)]\right\}. \qquad (48.16)$$

This scheme computes a crude new value of $y$ half a step ahead from

$$y(t) + \frac{\Delta t}{2} f[y(t)], \qquad (48.17)$$

which is just the first term of a Taylor's series. Then it uses the derivative there to compute $y$ a full step ahead. It is equivalent to taking second derivatives, but with more computing and less programming. It is easy to extend this to any system of equations.

*Example*

Let us see how a mixture of dust and radiation behaves. We will compare two different models, one with pure dust, having today

$$H_0 = 50 \text{ km/sec/mpc},$$

$$\Omega = 0.043,$$

(48.18)

and one having today as much matter present as radiation,

$$\Omega = 0.086.$$

(48.19)

The energy density of the dust universe will be

$$\rho = \frac{3\Omega}{8\pi} H_0^2$$

$$= 1.35 \times 10^{-38} \text{ sec}^{-2}.$$

(48.20)

The initial value for $R$ comes from Equation 48.13,

$$(R_0 H_0)^2 = \frac{\kappa}{(\Omega - 1)};$$

(48.21)

for pure dust,

$$R_0 = 6.31 \times 10^{17} \text{ sec}.$$

(48.22)

The pure-dust model provides a check on our computation, since we already know that solution. We need some way to compute the pressure. Assume that the dust and the radiation do not interact. Let $\epsilon$ be the fraction of the energy in radiation at $t_0$. Then the radiation energy increases like $R_0/R$ relative to the dust energy. The pressure will be $\frac{1}{3}$ times this radiation energy, that is,

$$p = \frac{\epsilon R_0 \rho}{3(R + \epsilon R_0)},$$

(48.23)

where $\rho$ is the total energy density.

In Figures 48.1 to 48.4, we show the pasts of these two models. The relations $\chi(z)$, $t(z)$, $P(z)$, and $H(z)$ plotted are the most generally useful. They determine all others. Again we have used $\mu = z/(1 + z)$ as a variable to better represent the early universe.

In the real universe, the pressure does not depend only on the density if the expansion proceeds fast enough to keep the matter out of thermodynamic equilibrium; so a more detailed theory of the matter must be used. This will involve further ordinary differential equations for the reaction rates and decay rates of nuclei and particles. One can no longer run the computation backward in time. Some clever tricks are needed to calculate away from the initial singularity, and to find the initial conditions.

*Complications*

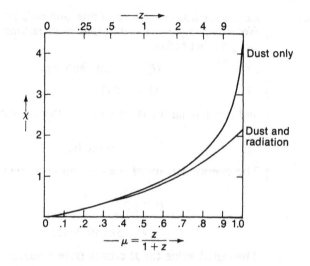

**Figure 48.1.** The arc-distance as a function of redshift for two universe models: dust, and dust and radiation that do not interact.

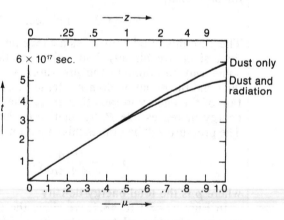

**Figure 48.2.** The proper time between redshift $z$ and the present for our two models.

PROBLEMS

48.1. (12) Why do the two curves in Figure 48.1 coincide for small $z$? Do they coincide exactly?

48.2. (10) An object of known luminosity is observed at $z = 9$ in both of the models shown in Figure 48.1. How different is the apparent brightness?

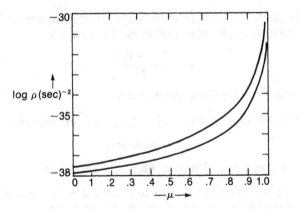

**Figure 48.3.** The density as a function of redshift for our two models.

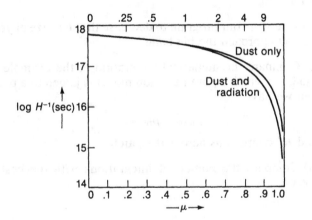

**Figure 48.4.** The expansion time-scale (inverse Hubble constant) as a function of redshift.

48.3. (12) Why do the two curves in Figure 48.2 coincide for small $z$? Do they coincide exactly?

48.4. (18) How do you draw Hubble diagrams for models specified by the curves given here?

48.5. (15) Show that the density parameter $\Omega$ at any time can be computed from

$$\Omega = \left[1 - \frac{3\kappa}{8\pi\rho R^2}\right]^{-1}.$$

**48.6.** (28) Show that in the early stages of the expansion, when the matter content acts like pure radiation, the function

$$\hat{\eta} \equiv \left[ \frac{|1 - \Omega|}{\Omega} \right]^{1/2}$$

estimates the arc-time since the big bang.

**48.7.** (12) Show that in the early stages of the universe, we have

$$z\hat{\eta} = \text{constant},$$

where $\hat{\eta}$ is defined as in Problem 48.6.

**48.8.** (17) Relate $(z\hat{\eta})$ to the size of a critical radiation universe that describes the early evolution of any model containing radiation.

**48.9.** (29) Find $(z\hat{\eta})$ exactly for the models having noninteracting dust and radiation.

**48.10.** (24) Use $\hat{\eta}$ in your program to decrease the step size of your calculation as you approach the big bang.

**48.11.** (32) Compare the numerical integration of the example with a patched model in which a pure radiation model is joined to a pure dust model when we have

$$\rho_{\text{RAD}} = \rho_{\text{DUST}}.$$

Keep $R$ and $R'$ continuous across the patch.

**48.12.** (33) Compare the numerical integrations with the best-fitting empty universes.

## 49. The Big Bang

*"Sturgeon's Law: Ninety per cent of science fiction is crud, but then ninety per cent of everything is crud."*

THEODORE STURGEON,
as quoted by Frederick Pohl

What can we say about the early stages of the universe? What if we make the (outrageous?) assumption that our current ideas of the contents of the universe are complete and accurate? Although we would be crazy to take the answers too seriously, it is certainly an interesting question to ask. The resulting scenario is called the "standard big-bang model" of the universe.

The present universe appears to contain matter in galaxies estimated to be about

$$\rho_{m0} \approx 2 \times 10^{-31} \text{ gm/cm}^3. \tag{49.1}$$

*Known matter*

This mass is in slowly moving particles, and is nearly all rest-mass energy. With a Hubble constant of 50 km/sec/mpc, this gives us a density parameter

$$\Omega_0 = 0.043. \tag{49.2}$$

[This is not known to better than a factor of two. A density parameter as small as 0.01 might just be consistent with observed matter.]

This leads to an open universe in its late stage of evolution

$$\eta_0 \approx 5. \tag{49.3}$$

*Open model*

There is now also a small amount of energy in the form of microwave radiation. Its present energy density is smaller by a factor of about 500. I will now try to describe for you the early stages of such a universe under the assumption that the Friedmann models are an accurate description everywhere.

At early times the radiation comes to dominate because it satisfies

*Early universe*

$$\rho_\gamma R^4 = \text{constant}, \tag{49.4}$$

as opposed to the matter-conservation law,

$$\rho_m R^3 = \text{constant}. \tag{49.5}$$

For times earlier than $z = 500$, we have a radiation-dominated universe. At even earlier times, the particles themselves become relativistic. Since there are about $10^8$ microwave photons per nucleon today, and since this number is roughly constant, the particles will not be important.

At around $z = 1,000$, the universe changed from a uniform plasma of ionized hydrogen to its present content of mostly neutral hydrogen plus helium plus contamination with heavier atoms. This recombination of the ions takes place at a temperature of about 3,000 K. The microwave radiation comes from this plasma. The neutral hydrogen is transparent, and the radiation comes freely from $z = 1,000$, and today appears at a temperature of 3K.

*Primeval fireball*

The physics of this early universe is dominated by two effects. One is the high temperature, which is roughly

$$T \approx T_0 z, \tag{49.6}$$

where $T_0$ is the present temperature of the microwave background. The other effect is the very short amount of proper time available. In a critical radiation universe, the Hubble time vanishes according to

$$\frac{1}{H} = a\eta^2. \tag{49.7}$$

[The physics of the early universe is very difficult. I regret having to merely quote results at you, but it is not possible to deduce these facts from any simple theory. This unpleasant pontifical mode will persist only for this section.]

***Equilibrium phase***

The earliest stage that we can reliably discuss had a temperature of about $10^{12}$K, and consisted of a plasma of elementary particles: neutrons, protons, electrons, positrons, muons, neutrinos, photons, and so on. Before then the particle interactions involve physics too far beyond the range of our laboratory experience to predict. Even though the early universe was changing on a time-scale of seconds, the nuclear reactions at those high temperatures were fast enough to keep the particles in thermodynamic equilibrium. All particles were present in numbers proportional to the Boltzman factor

$$e^{-E/kT}. \tag{49.8}$$

For early times the energy of the particles depended very little on their masses, since they were all moving at almost the speed of light. One had a rather democratic plasma, with all types of particles in it. As the universe expanded and cooled, the heavy particles slowed down and eventually reached a minimum energy content determined by their rest mass.

[The mass of a proton has an energy corresponding to a temperature $kT$ of $10^{13}$K.]

Below $10^{13}$ K there were only enough of such heavy particles to satisfy the conservation laws, such as the law of conservation of baryon number. Below $10^{13}$ K protons and neutrons continued to exist only because there happened to be more matter than antimatter. All our present-day matter results from that imbalance. It was either put in at the beginning, or arose from a matter-antimatter asymmetry in the unknown physics of the early universe. Between $10^{13}$ K and $10^{10}$ K, the energy of neutrons and protons is nearly all rest-mass energy, the neutron-proton ratio is determined by their mass difference, and we have

[The neutron-proton mass difference is 1.29 Mev.]

$$\frac{n_n}{n_p} = \frac{e^{-M_n/kT}}{e^{-M_p/kT}} = e^{-(M_n - M_p)/kT}. \tag{49.9}$$

At a temperature of about $10^{10}$ K, the interactions that keep the neutrons and protons in equilibrium with each other are no longer able to keep up with the conversion of neutrons to protons needed to satisfy Equation 49.9. At that temperature, the ratio is one neutron for every five protons. For lower temperatures, the number of neutrons will decrease, primarily because the neutrons decay with a lifetime of 1,000 seconds.

From $10^{10}$ K to $10^9$ K there was competition between nuclear reactions that combine a neutron and a proton to form deuterium, the disintegration of that deuterium by all the photons, and the formation of helium from the deuterium. Above $10^9$ K the photodisintegration wins, but at $10^9$ K the balance shifted abruptly, and all the remaining neutrons were rapidly converted into helium. That helium should still be around today, and it is important to know how much was made. Most of the above physics is complicated nuclear physics, to which you and I are helpless spectators. Cosmology enters only in determining how long it takes the universe to cool from $10^{10}$ K to $10^9$ K, since this determines how many of the neutrons decay. The calculation of this proper time is a nice exercise in cosmology. At those temperatures not only was the universe radiation-dominated, but it was also young ($\eta \ll 1$), so that it behaved approximately like a critical radiation universe. We are faced here with a universe that makes a transition between the two types of models so far considered. To describe the critical radiation phase of the model, we need two parameters. The scale factor at a redshift $z \gg 1$ is given by

*Helium production*

[You should be relieved to come at last to a calculation within your skills.]

$$R = \frac{R_0}{z}. \tag{49.10}$$

The other parameter will be the energy at that redshift. This energy will be due solely to the compressed radiation. We have, for the present matter density,

$$\frac{8\pi\rho_{m0}R_0^2}{3} = (R_0 H_0)^2 \Omega_0 \approx \Omega_0, \tag{49.11}$$

since for the low-density universe we have

$$R_0 H_0 \approx 1. \tag{49.12}$$

It is in its late phase. At present we have a fractional amount of radiation $\epsilon_0$ such that

$$\frac{8\pi\rho_{\gamma 0}R_0^2}{3} = \epsilon_0\Omega_0, \tag{49.13}$$

and we are taking

$$\epsilon_0 \approx \frac{1}{500}. \tag{49.14}$$

At any time for this radiation we have

$$\rho_\gamma R^4 = \text{constant}, \tag{49.15}$$

so that we have at a redshift $z$

$$\frac{8\pi\rho_\gamma R^2}{3} = \frac{8\pi\rho_\gamma R^4}{3R_0^2}z^2 = \epsilon_0\Omega_0 z^2. \tag{49.16}$$

From the density and radius, we can determine the parameters $a$ and $\eta$ of the critical phase of the universe.

*Details*   We calculate $a$ and $\eta$ from the equations in Table 45.2. We have

$$R = a\eta \tag{49.17}$$

and

$$\frac{8\pi\rho a^2}{3} = \frac{1}{\eta^4}. \tag{49.18}$$

From Equations 49.10 and 49.16, we have

$$a\eta = \frac{R_0}{z} \tag{49.19}$$

and

$$\frac{8\pi\rho R^2}{3} = \frac{1}{\eta^2} = \epsilon_0\Omega_0 z^2. \tag{49.20}$$

The second gives us

$$\eta = \frac{1}{z\sqrt{\epsilon_0\Omega_0}}, \tag{49.21}$$

and the first gives us

$$a = R_0\sqrt{\epsilon_0\Omega_0}. \tag{49.22}$$

Note that $a$ is independent of $z$, as it must be. The result depends only on $\epsilon_0\Omega_0$, the radiation density, and so requires only a knowledge of the Hubble constant, but not of the present matter density. This is only true in a low-density universe

$$\Omega_0 \ll 1, \tag{49.23}$$

where we can take

$$R_0 H_0 \approx 1. \tag{49.24}$$

The proper time from $10^{10}$ K will be the time

$$\Delta t = \frac{R_0}{2\sqrt{\epsilon_0 \Omega_0}} \left( \frac{1}{z_9^2} - \frac{1}{z_{10}^2} \right),$$

(49.25)

[Here $z_9$ is the redshift $z$ at a temperature of $10^9$K.]

and this is approximately

$$\Delta t \approx \frac{R_0}{2\sqrt{\epsilon_0 \Omega_0}} \left( \frac{3}{10^9} \right)^2.$$

(49.26)

In our standard model this leads to a time of 300 seconds.

This is the time that is available for the neutrons to decay. It must be a cosmic accident that this time is roughly comparable with the neutron lifetime of 1,000 seconds. After 300 seconds, the fractions of neutrons remaining will be

$$e^{-300/1,000} = 0.75.$$

(49.27)

Of the original 20 per cent by mass of neutrons, one quarter of them decay, and the other three quarters merge with an equal number of protons to form helium. Our big bang thus turns the hydrogen into 30 per cent by mass of helium. Far more careful calculations have been done, and their result is that about 23 per cent is turned into helium.

Remarkably, when one surveys the universe, one always finds helium mixed with the hydrogen in about the above ratio. Sometimes there is more, because hydrogen is burned to helium in stars, but in general the helium content of the universe more or less agrees with the above calculation. This is one of the triumphs of big-bang cosmology. There were not very many assumptions in the above calculation beyond the one that even at $z = 10^{10}$ the universe was homogeneous and isotropic enough to be described by a Friedmann model.

[I am indebted to the excellent review article by Harrison for help with the above scenario. Look there for a much more extensive discussion, or in the book by Silk.]

## PROBLEMS

49.1. (19) One might expect that the condensations of matter that will ultimately form galaxies would have some effect on the microwave background radiation at $z = 1,000$. Estimate the angular size of this for various models.

49.2. (28) Compare the amount of helium formed in an $\Omega = 1$ dust model with the amount formed in an $\Omega = 1$ radiation model.

49.3. (35) Discuss the helium formation in a universe whose equation of state is

$$p = \rho$$

at early times.

## 50. Observations of the Real Universe

*"The contemplation of things as they are,
without error or confusion, without substitution
or imposture, is itself a nobler thing than a
whole harvest of invention."*

                                    FRANCIS BACON

*"The principle of strategy is, having one thing,
to know ten thousand things."*

                                    MUSASHI

We can now face the question of how much we know about the universe.
Do these Friedmann models work? What can they tell us? You will
probably be disappointed to learn how little we really know. We can
find models consistent with the observations. The infinite universe is
no longer a problem, nor is Olber's paradox. The microwave background radiation fits in nicely, as does the cosmic helium abundance.
Some things do not fit naturally. The most embarrassing problem is that
galaxies do not seem to form in these models. To get them in, we have
to insert them at the beginning. Neither can we decide today between
open models, which expand forever, and closed models, which end in
gravitational recollapse. This section is a brief review of the observational evidence relevant to these models.

*Symmetry*    The cornerstone of our models is symmetry, which we need to examine carefully. Did we assume symmetry only to be able to solve the
resulting equations? How symmetric is the universe now? We have a
chance to observe the isotropy around us, but the homogeneity can
only be inferred indirectly. We also want to know how symmetric the
universe was at earlier times. Small departures from symmetry are
called perturbations, and their study is a very active area of research.
*Perturbations*    Growing perturbations are important because they will form galaxies.
A good theory should explain why galaxies have the masses they are
observed to have. Decaying perturbations are also important. Their
behavior lets one infer, from a measurement of present symmetry,
something about the past symmetry. The study of these perturbations
is difficult and not at all finished. The physics involves the hydrody-
namics of a relativistic, radiation-dominated, self-gravitating plasma in
a curved, expanding spacetime. So far we have learned that, in the
radiation-dominated era, perturbations do not grow very fast. This

hinders galaxy formation, but lets us feel that the early universe could not have been too asymmetric. Initial perturbations that survive through the radiation era grow afterward by a factor of $10^2$ or $10^3$. If we put in perturbations of around $10^{-3}$ initially, these coast through the radiation era, and then grow after $z = 1,000$ to form galaxies at around $z = 10$. But where did they come from initially?

The prime observational evidence for the symmetry of the universe is the isotropy of the microwave background radiation. Studies made from high-altitude aircraft show that it has the same temperature in all directions to one part in a thousand. This is a remarkable observation. Since the radiation was formed in the hot plasma at $z = 1,000$, and has then come freely to us (we think), we can say that the universe was symmetric to one part in a thousand at $z = 1,000$. These experiments also measured the velocity of our galaxy with respect to a frame in which the microwave background radiation is isotropic. The velocity is about $10^{-3}$, which is just about the observed random velocities of other galaxies. This is consistent with our galaxy being typical, as demanded by homogeneity. Had our velocity turned out to be $10^{-6}$, then we would have been in trouble.

Upon further thought, the isotropy of the microwave background is even more remarkable. One consequence of the fact that the universe has a finite age in terms of arc-time is that different parts of the universe can interact only after a definite amount of time has passed. Two points that are an arc-distance $\chi$ apart cannot interact until an arc-time equal to $\chi$ has passed. When we look at the microwave background radiation in two different directions, we are looking at two different points in the plasma at $z = 1,000$. Have these points had enough time to interact? The answer is no, not if they are very far apart in direction.

*Isotropy of the microwave background radiation*

[To be careful, this observation requires only that there be a surface of spherical symmetry. We still must assume homogeneity to extend the argument. See the paper of Ellis *et al.* for an amusing counterexample.]

*Horizons*

**Example**

It is a fun geometry problem to calculate how far apart these points have to be on the sky. Look at the models used in the do-it-yourself calculations. Table 50.1 gives

TABLE **50.1.** HORIZON DATA FOR THE MODEL UNIVERSES.

| Type | Arc-distance to $z = 1,000$ | Arc-time at $z = 1,000$ | Angular size of horizon, e |
|------|------|------|------|
| Dust | 4.2 radians | 0.3 radians | 0.009 radians |
| Dust and Radiation | 2.2 | 0.004 | 0.001 |

the arc-time $\chi$ back to $z = 1,000$ and also the arc-time $\eta_{BB}$ from there to the beginning. The geometric situation is the same as that in Section 43 (pages 274ff) except that, instead of having a specified proper distance between the points, we have an arc-distance. This removes the factor of $R(\eta)$. Also, this model is an open universe; so $\sin \chi$ becomes $\sinh \chi$. The angle $\epsilon$ seen on the sky today between two points at $z = 1,000$ that have just come into physical contact is

$$\epsilon = \frac{\eta_{BB}}{\sinh \chi}, \qquad (50.1)$$

and this is also listed in Table 50.1.

This separation between unrelated points depends somewhat on your model of the universe. Models with less radiation have a more leisurely early expansion, but even in the pure-dust model points which are further apart than $\frac{1}{2}°$ are still out of communication. The limit beyond which an event has no possible physical contact is called the horizon. The horizon for any observer is at $z = \infty$.

*Example* | In a closed universe at arc-time 90°, one can never see more than half the universe no matter how powerful the telescope.

It is remarkable to find that all parts of the hot plasma, as we can tell from the microwave background radiation, have the same redshifted temperature to one part in a thousand, when they had yet had no chance to interact. The big bang must have started at the same time everywhere. And how did the different parts agree on a temperature? Nothing in known physics could have acted to smooth out the temperature fluctuations. If these Friedmann models are correct, this symmetry must have been present right at the beginning.

*Entropy*    There is also an indirect and subtle argument which says that the early universe had to be nearly as symmetric as a Friedmann model. If fluctuations and structure in the early universe were to be washed out by some dissipative process, then that dissipation would increase the entropy of the universe. We can estimate that entropy, since most of it lies in the microwave background radiation (and also in the associated neutrino background, which must be there but which is not detectable). The entropy found in the universe is not large enough to accommodate a very chaotic early universe.

[I doubt that my readers are familiar enough with entropy to appreciate much of this argument. I cannot really follow it myself in any detail.]

Let us grant, with some caution, that a Friedmann universe is a good model for the universe. The step from symmetry, which implies Robertson-Walker spacetimes, to Friedmann spacetimes assumes that general relativity is the correct theory of gravitation. Once cannot expect cosmology itself to tell us the answer to this. If it is a Friedmann universe, which one is it? There are lots of them, corresponding to a choice of equation of state, such as for dust or radiation, three possibilities for the spatial sections, the over-all size parameter a, and our present location within the model. Let us finish this section by discussing how well these possibilities can be narrowed down.

*Two-parameter family*

A better choice for the two parameters for our Friedmann model will be $H_0$ and $\Omega$. Using $\Omega$ for an age parameter automatically covers the three spatial possibilities. If we let $\Omega$ be a free parameter, we allow for some as-yet undiscovered matter in the universe. It will be difficult or impossible to choose between different equations of state for such missing matter. We know so little that any further undetermined parameters will kill us. One usually takes the missing matter to be dust, since the extra pressure of radiation would force the universe to expand so rapidly that there would be barely enough proper time to include the known ages of globular clusters. It might also be the conjectured Lorentz-invariant matter described by the cosmological constant. Such matter has a universal tension rather than a pressure. This slows up the expansion, and can patch up a model that otherwise is too young. Let us stick with dust. Then each point in $(H_0, \Omega)$ space specifies a unique cosmological model.

[To my mind, using such matter without any hint that it exists is on a par with cheating at solitaire.]

[I am following here the review article of Gott *et al.*, and like them I will discuss the observations in the $(H_0, \Omega)$ plane. Although I follow their arguments, I am being more generous with the limits. The author of a book must prepare for a longer period of potential embarrassment, and be somewhat more cautious in his pronouncements.]

*Hubble constant*

$H_0$ can be measured directly, but not very well. It requires a direct estimation of the distances to objects. Unless the objects are far away, the redshifts caused by the expansion of the universe will be confused with the Doppler shifts caused by their random motions. But for distant objects we have many other problems. For us to see anything at a great distance, it must be bright and therefore scarce locally. Furthermore, it is hard to measure the distance to distant objects. Finally, objects at large distances are younger. To use them we need to build a model to predict the evolution of their properties. Decades of work have given us limits,

$$30 \leq H_0 \leq 120 \text{ Km/sec/mps}, \qquad (50.2)$$

and a feeling that 50 is not such a bad value. I have used these limits to shade off parts of the $(H_0, \Omega)$ plane drawn in Figure 50.1.

*Age*

Another datum is the age of the universe. Conservative limits for the ages of stars, star clusters, and the elements themselves are

$$8 \leq t_0 \leq 18 \times 10^9 \text{ yrs.} \qquad (50.3)$$

These age limits appear to rule out nearly the same regions as the Hubble-constant limits. Any such internal consistency is comforting.

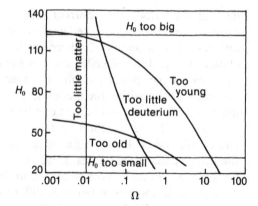

**Figure 50.1.** A conservative resumé of cosmological knowledge in the late 1970s.

*Mass density*     One can find a lower limit to $\Omega$ just from the observed matter content of the universe. The masses are not observed directly, however, and there is always some indirect inference here. Usually masses are inferred from brightness, on the assumption that galaxies are filled with familiar objects and all have similar mass-to-light ratios. One can also estimate the masses of self-gravitating clusters of galaxies. In order to have enough mass there to hold together the cluster, given the random velocities it is observed to have, one needs about ten times more mass than one guesses from the first argument. Most workers would agree on a lower limit of about

$$\Omega \geq 0.01, \tag{50.4}$$

with $\Omega \approx .05$ being a comfortable value for the observed matter. This closes off one side of the allowed region.

*Hubble diagram*     The Hubble diagram has not produced any really good results. The brightness of galaxies changes with time, both because the stars age and grow dim, and because from far away one can see only giant galaxies, which occur in clusters and can grow brighter by sweeping up smaller galaxies. This latter effect is nearly impossible to model. Still it is hard to push the interpretation of the present data higher than a limit of

$$\Omega < 5. \tag{50.5}$$

Crucial information may be available in the cosmic deuterium abun-
*Deuterium*     dance. As helium is produced, some deuterium is also made. The

amount depends on just how fast the universe passed through the nucleosynthesis era. The deuterium limits come from detailed numerical studies. You can see that this can give important information. However, the deuterium production is so sensitive to conditions during the first few minutes of the universe that the figures may be unreliable. Perhaps some small isolated inhomogeneities have produced the deuterium that we see. This would weaken the limits. We can really say very little about the first few minutes of the universe.

Figure 50.1 is a conservative view of our knowledge of the universe. If the deuterium argument is correct, then we have $\Omega < 1$ ($q_0 < \frac{1}{2}$ for dust); so the universe is open, will expand endlessly, cool off, and end quietly. But something may be wrong with the deuterium message, and $\Omega$ may be larger than 1; if so the universe is closed, and we will all go out in a flash in the big crunch.   *Summary*

### PROBLEMS

50.1. (15) How were the lines of constant age in Figure 50.1 plotted?

50.2. (35) If the neutrinos became free particles at an earlier redshift than the microwave photons, what is their present temperature?

50.3. (15) Check the numbers in line 1 of Table 50.1 by using the analytic expressions.

50.4. (25) Check the numbers in line 2 of Table 50.1 if the last point calculated at

$$z \doteq 35.7$$

was an arc-distance

$$\chi = 2.04$$

away and the predicted age of the universe (Problem 48.6) was

$$\hat{\eta} \doteq 0.125.$$

Assume a critical radiation universe for earlier times.

50.5. (31) Suppose the microwave background was formed at $z = 20$ in an $\Omega = 2$ universe. What angle on the sky do two points now make if at $z = 20$ they were just within each other's horizon? Note that you cannot assume small angles here.

## 51. Singularities

*"Ideas came into my mind quite unrelated to
graphic art, notions which so fascinated me that
I longed to communicate them to other people.
This could not have been achieved through words,
for these thoughts were not literary ones, but
mental images of a kind that can be made
comprehensible to others by representing them as
visual images."*

M. C. ESCHER

At this point everyone asks about what happened before the big bang.
The physics that we know today gives no answer to this. Nor is there
any answer to what happens in a closed universe after the end. Let
me finish by discussing some speculations about this problem. Such
problems, where physics is simply unable to proceed, are fascinating.
There must be new laws of physics hiding somewhere nearby.

*Early universe*    The early moments of all Friedmann universes look like radiation
universes unless there is some bizarre form of matter with an ultrastiff equation of state. In the early moments of a radiation universe,
we have

$$R(\eta) \approx a\eta, \tag{51.1}$$

and a metric

$$\mathscr{G} \approx a^2\eta^2[-d\eta^2 + d\chi^2 + S^2(\chi)\, d\Omega^2]. \tag{51.2}$$

[The set $\eta = 0$ is a three-dimensional
subspace, and these are usually called
hypersurfaces.]

In Figure 51.1 we sketch the behavior of this metric, giving the metric
figures at several places. The $\eta$ coordinate was constructed to give a
simple light-cone structure. The metric figures expand as we approach
$\eta = 0$. The distance between any two points goes to zero as we approach
that hypersurface.

$R = 0$    Such behavior is not unusual. Even ordinary polar coordinates
behave this way. The metric figures for the Euclidean metric

$$\mathscr{E} = dr^2 + r^2\, d\theta^2 \tag{51.3}$$

also expand without limit as $r$ goes to zero, as shown in Figure 51.2.
This representation of the metric $\mathscr{E}$ behaves badly as $r$ goes to zero,

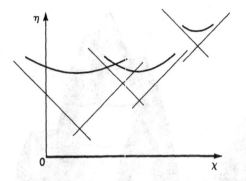

**Figure 51.1.** The behavior of the metric near $R = 0$.

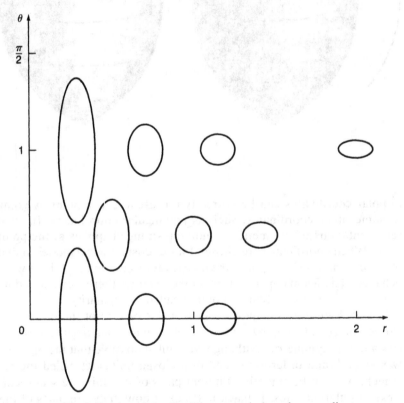

**Figure 51.2.** The behavior of the Euclidean metric in polar coordinates.

because the apparent line $r = 0$ is actually a single point. That could be a problem with our cosmological models as well. Perhaps the closed universe should be drawn as in Figure 51.3. I can say no more than "perhaps" because it makes no sense (with current physical laws) to say that the entire universe started from a single point. The point $R = 0$

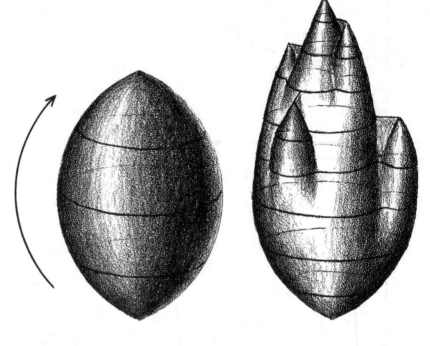

**Figure 51.3.** A representation of a closed universe in which the big bang and the big crunch are represented as single points.

**Figure 51.4.** An attempt to represent the big crunch as the accumulation of a lot of little crunches.

*Initial singularity*

in polar coordinates can be correctly represented as a point by going to some other coordinates, such as rectangular coordinates. In such representations of the metric $\mathscr{E}$, nothing strange happens at the point $r = 0$. When coordinates are chosen for our cosmological model so that $R = 0$ becomes a single point, the metric representation is still not well-behaved. The Riemann curvature tensor diverges there, corresponding to infinite tidal forces. Such a point is called a singularity.

*The big crunch*

In fact, Figure 51.3 cannot be an accurate picture of the future of a closed universe. Clumping caused by self-gravity is making the universe less and less symmetric. Although we think it possible that the big bang was smooth and orderly, the end of a closed universe, called the big crunch, will not be smooth. Different parts of the universe will recollapse at different rates. If black holes exist now, they are parts of the universe that have already been crunched. In Figure 51.4 I try to represent better the chaotic nature of the big crunch. The lack of past-future symmetry is quite interesting. The direction of time seems to be from smooth singularities to rough ones.

*Horizon dilemma*

The dilemma of the finite horizon was mentioned in the last section. Figure 51.5 shows it well. There is no point in the universe that can send a signal to both event *A* and event B. Event *B* is over the horizon from *A*, and vice versa. Despite this, the universe at *A* and *B* is so similar that somehow they must have "talked it over."

One attempt at a resolution is to extend the universe, and to allow events before $\eta = 0$. Point *C* in Figure 51.6 can interact with both *A* and *B*. The hope is that the universe can collapse and then bounce back again. This is referred to as an *oscillating universe.* The oscillating universe extends the picture of Figure 51.3 to the picture of Figure 51.7. The trouble here is that the big crunch is not properly represented by Figure 51.3. If Figure 51.4 is used to represent the big crunch, then a nice, orderly oscillating universe is not possible. The best that we can do is the nightmare of Figure 51.8. In this picture, our whole universe would be the inside of a minor part of a larger universe. Furthermore, our universe would itself spawn smaller universes. We would end up with a state best described as turbulent spacetime. Now, Figures 51.4 and 51.8 are probably not good representations of the big crunch either. They show the universe evolving into an infinite number of disconnected pieces. It is more reasonable for all the little crunches to collect together and have one big crunch as a finale. Figure 51.9 is an attempt to represent this idea.

The preceding fanciful scenarios are nothing but wild-eyed speculation. We know of no laws of physics that will allow us to pass through the $R = 0$ hypersurface. Such passage is not allowed by general rela-

[You might call this the conspiracy theory of cosmogenesis.]

### Before the beginning

[Some speculations on this line have been carried out in quite some detail by Hoyle and Narlikar. A popular description of them can be found in Hoyle's *Astronomy and Cosmology.*]

[See the book on fractals by Mandelbroit for more on such bizarre sets.]

[The complexity suggested in these drawings is beyond what our mathematics can now describe. It is an active area of research. The drawings were done by Chris Shaw.]

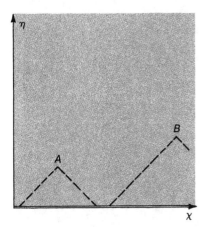

**Figure 51.5.** The lack of causal interaction in the early universe.

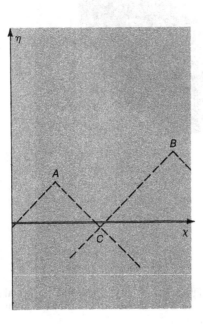

**Figure 51.6.** Events before the time when $R = 0$ could interact with the disconnected points shown in Figure 51.5.

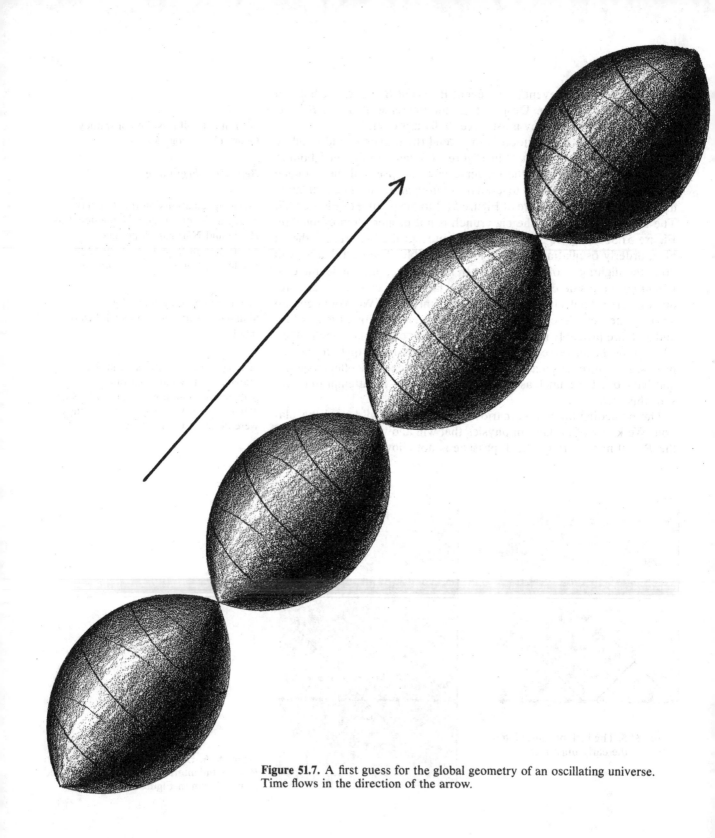

**Figure 51.7.** A first guess for the global geometry of an oscillating universe. Time flows in the direction of the arrow.

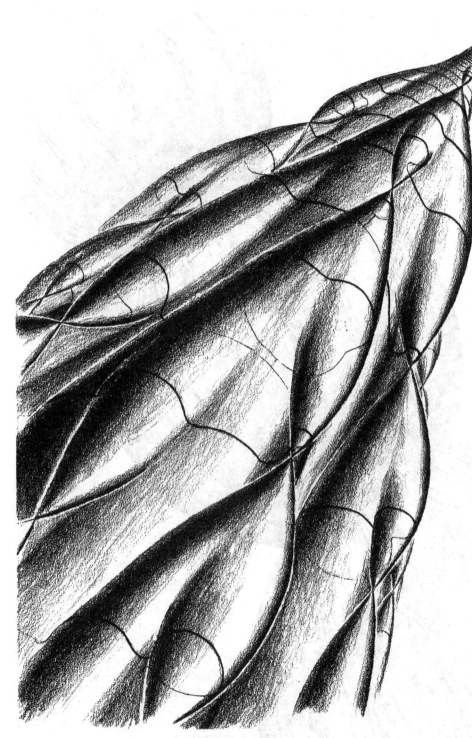

**Figure 51.8.** A schematic representation of a recurrent universe suggested by Figure 51.4.

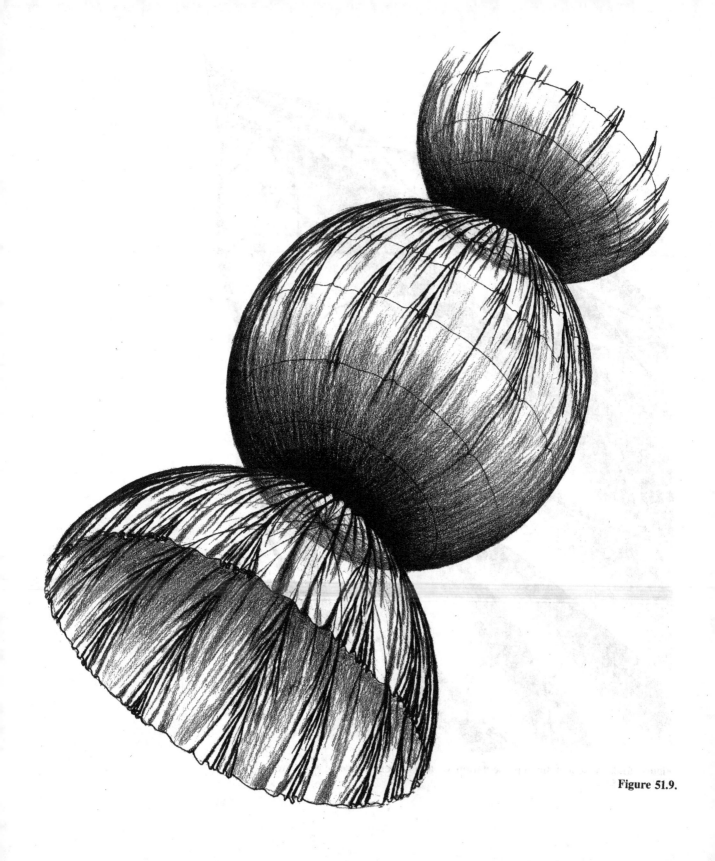

Figure 51.9.

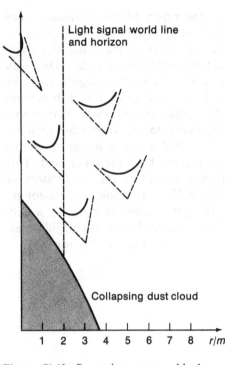

Light signal world line
and horizon

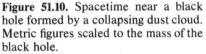

Collapsing dust cloud

1 2 3 4 5 6 7 8 *r/m*

**Figure 51.10.** Spacetime near a black hole formed by a collapsing dust cloud. Metric figures scaled to the mass of the black hole.

[With the tools at your disposal, you are prepared to study this in much greater detail if you wish. The expressions for the metric from which the above picture was drawn can be found in Misner, Thorne, and Wheeler, Section 31.4. They are called Eddington-Finkelstein coordinates. MTW also discusses several other representations. What you are not prepared for is a discussion of why general relativity leads to a spacetime such as that.]

tivity. But, then, no one expects general relativity to be rigorously true. No doubt it is only an approximate theory, and will break down under the extreme conditions near the $R = 0$ hypersurface.

A preview of the big crunch is available. General relativity can describe the collapse of a cloud of matter whose pressure cannot support it against its self-gravity. The final state of such a collapse is called a black hole. In this case, too, the objects collapse to zero radius. Figure 51.10 is a sketch of the spacetime describing such a situation. Notice in particular the behavior of the light cones in that spacetime. An observer falling into the black hole who passes the curve labeled "horizon" cannot escape being carried right into the singularity at the origin. To avoid this would require speeds faster than light. The language of metric figures and spacetime diagrams gives us a concise picture of the collapse. This is really the only honest way to describe a black hole.

The crucial feature of a black hole is that its gravity is so strong that nothing, not even light, can escape. The collapse to a singularity might seem to be an accidental result, caused by our assumption of perfect spherical symmetry. This is not so. There are some surprising theorems

*Gravitational collapse*

*Black holes*

in general relativity which say that under quite liberal conditions, even without spherical symmetry, some kind of singularity must form.

The singularities in general relativity, such as the one at the origin of a Friedmann universe, are not well-understood today. One way to extend our physical laws through the big-bang singularity is by interpreting $\mathscr{G}$ not as a metric, but only as a generator of dispersion relations. Then there is no need to shrink the $t = 0$ surface down to a point. It remains a surface of singularity for the dispersion relation. Viewed this way, the masses of the particles appear to vanish on this surface, and they can be cheaply created by a radiation field. This approach, developed by Fred Hoyle, cannot be the whole story of singularities, however. The black-hole dilemma is that one cannot get out of the horizon once inside. The guess today is that only a fully quantum treatment can resolve the questions about these singularities, and no such quantum treatment is yet within our capabilities.

# Bibliography

Acton, F. S. *Numerical Methods That Work*. Harper and Row, 1970.

Allen, C. W. *Astrophysical Quantities*. University of London Press, 1973.

Arnold, V. I. *Ordinary Differential Equations*. MIT Press, 1973.
_____. *Mathematical Methods of Classical Mechanics*. Springer, 1978.

Arnold, V. I. and A. Avez. *Ergodic Problems of Classical Mechanics*. W. A. Benjamin, 1968.

Baldwin, J. A., W. L. Burke, C. Martin Gaskell, and E. J. Wampler. *Nature* **273**, 431–435 (1978).

Bondi, H. *Cosmology*. Cambridge University Press, 1961.

Davies, P. C. W. *Space and Time in the Modern Universe*. Cambridge University Press, 1977.

Dodson, C. T. J., and T. Poston. *Tensor Geometry*. Pitman, 1977.

Ellis, G. F. R., R. Maartens, S. D. Nel. *Monthly Notices R.A.S.* **184**, 439–466 (1978).

Feynman, R. P., R. B. Leighton, and M. Sands. *The Feynman Lectures on Physics*. Addison-Wesley, 1963.

Gingerich, O., ed. *Cosmology + 1*. W. H. Freeman, 1977.

Gott, J. R., J. E. Gunn, D. N. Schramm, and B. M. Tinsley. *Astrophys. J.* **194**, 543–553 (1974).

Harrison, E. R. *Ann. Rev. Astron. Astrophys.* **11**, 155–186 (1973).

Hoyle, F. *Astronomy and Cosmology*. W. H. Freeman, 1975.

Hoyle, Fred, and Geoffrey Hoyle. *Into Deepest Space*. Harper and Row, 1974.

Lanczos, C. *The Einstein Decade.* Academic Press, 1974.

LeGuin, U. K. "Direction of the Road," reprinted in *Orbit 12*, D. Knight, ed. Berkeley Medallion Books, 1973.

Lighthill, J. *Waves in Fluids.* Cambridge University Press, 1978.

Loomis, L. H., and S. Sternberg. *Advanced Calculus.* Addison-Wesley, 1968.

Lorrain, Paul, and Dale Carson. *Electromagnetic Fields and Waves.* W. H. Freeman, 2nd ed., 1970.

Mandelbroit, B. *Fractals,* W. H. Freeman, 1977.

Misner, C. W., K. S. Thorne, and J. A. Wheeler. *Gravitation.* W. H. Freeman, 1973.

Ohanian, H. C. *Gravitation and Spacetime.* W. W. Norton, 1976.

Peebles, P. J. E. *Physical Cosmology.* Princeton University Press, 1971.

Pierce, J. R. *Almost All About Waves.* MIT Press, 1974.

Polya, G. *Induction and Analogy in Mathematics.* Princeton University Press, 1954.

———. *Patterns of Plausible Inference.* Princeton University Press, 1954.

Porteous, I. R. *Topological Geometry.* Van Nostrand-Reinhold, 1969.

Poston, T., and I. Stewart. *Catastrophe Theory and Its Applications.* Pitman, 1978.

Reichenbach, H. *The Philosophy of Space and Time.* Dover, 1958.

Rindler, W. *Essential Relativity.* Springer, 2nd ed., 1977.

Robertson, H. P., and T. Noonan. *Relativistic Cosmology.* W. B. Saunders, 1968.

Sciama, D. W. *Modern Cosmology.* Cambridge University Press, 1971.

Silk, J. *The Big Bang,* W. H. Freeman, 1980.

Synge, J. L. and A. Schild. *Tensor Calculus.* University of Toronto Press, 1956.

Synge, J. L. *Proc. Roy. Irish Acad.* **63**, 1–34 (1963).

Rosen, Joe. *Symmetry Discovered.* Cambridge University Press, 1975.

Taylor, E. F., and J. A. Wheeler, *Spacetime Physics.* W. H. Freeman, 1966.

Weinberg, G. M. *An Introduction to General Systems Thinking.* Wiley, 1975.

Whitham, G. B. *Linear and Nonlinear Waves.* Wiley, 1974.

# Index

*"Half-way through the labour of an index
to this book I recalled the practice of my
ten years' study of history; and realized
I had never used the index of a book fit to read.
Who would insult his* Decline and Fall, *by
consulting it just upon a specific point?"*

T. E. LAWRENCE

aberration, 67
absolute-time clock, 12
acceleration, 60
  effect on clock rates, 24
  of a wave packet, 217
action, 224
  minimal principle for, 225
age
  of closed dust universe, 269
  of parts of the universe, 238–239
arc-distance, 260
arc-time, 259–260

basis 1-forms and tangent vectors, 92
basis vectors from coordinates, 124
big bang, 302ff
big crunch, 316
big splash, 110
black hole, 321
black-hole spacetime, 195

canonical reference frame, 20, 29
  operational definition, 45
  for water waves, 174
capillary waves, 171
Cartesian product, 116
catastrophe theory, 101
celestial sphere, aberration, 69
characteristics, 148
chart, 151
  defective, 152
  for $S^2$, 248–249
  for $S^3$, 251ff
clock, 9
  accelerated, 24
  observations of real clocks, 23
  paradox, 54
  rate of moving clock, 37
  special relativity, 29
  water wave, 172
collapse, gravitational, 321
conductivity tensor, 118
configuration space, 96

conservation of 4-momentum, 71
constraints, smooth, 97
coordinates, light signal, 46
cosmological constant, 264
cotangent space, 89
covariance, 14, 17, 21
    of covectors under linear
        transformations, 80
covectors, 77
    free, 78
    linear structure for, 79
critical Friedmann universe, 282
curve, parametrized, 84
curvature, 228ff
    of black-hole spacetime, 202–204
    measured by Riemann tensor, 232
    of pseudosphere, 235
    of 3-sphere, 257

deceleration parameter, $q_0$, 273
density parameter, 272
deuterium
    formation, 305
    limits for cosmological models, 312
dihedral product, 61
directional derivative, 90, 94
disc, rotating, 65
dispersion relation, 105, 162
    for common wave equations, 106
    for particles in curved space-
        time, 220
    for water waves with surface
        tension, 171–173, 181
dispersive waves, contrast with
    no dispersion, 198
Doppler shift, 61
    alternative derivation, 133
dot product
    Euclidean, 15
    4-vector, 57
duality, 81
dummy index, 137
dust
    closed-universe model, 264ff
    equation of state, 262

Earth, metric near, 199
elastic waves, dispersion relation, 106

electromagnetism, 133
    as a vector field in infinite
        dimensions, 146
energy-momentum space, 72
entropy of the universe, 310
envelope, 134
    construction for hyperbola, 134
    construction for parabola, 170
Eötvös experiment, 218
Escher, "Circle Limits," 280
Euclidean geometry, 15
    in general linear frame, 16
Euclidean metric tensor, 125
    analytic, 117
    graphical, 115
Euclidean universe, 282
event, 4

faster-than-light motion, 54
flips, 157
flow, 146
4-momentum, 70
    conservation, from wave diagram,
        188–189
4-velocity, 57
free index, 137
free particle, 8
    redefined in general relativity, 193
Fresnel equations, 182
frequency, 104
Friedmann equations, 263
    conservation law for, 297
    numerical integration of, 296ff
Friedmann universe, 261ff
    dust, table, 282
    numerical integrations, 300–301
    radiation, table, 286
    universal, table, 287

galaxy formation, 308
Galilean symmetry, 20
general relativity, simple geometric
    statement of, 194–195
generalized force, 97
geodesic, 220ff
geodesic square, 229
geometric units, table, 196
global structure of a space, 249
globular clusters, 239

gradient, 86, 92
gravitational redshift. *See* redshift
group velocity, 102, 108, 163

Hamiltonian mechanics, as a vector
    field, 144
harmonic oscillator, 144
helium, production in big bang,
    239, 305–307
homogeneity, 242
Hooke's law, 118
horizons, 309
Hubble constant, $H_0$, 271
Hubble diagram, 289ff
    observational, 312
Hubble flow, 239, 270–271
Huygens' construction, 210ff

inertial reference frame, 10, 20
integral curve, 142
invariance, 21
inverse tensor, 140
isometry, 242
isotropy, 242

Kepler's Law, 198
Klein-Gordon equation, 103
    dispersion relation for, 106
Kronecker delta, 139

Lagrange multipliers, 82–83, 98
Laue equations, 185
length, proper, 38
light
    new, vs. old, 27
    bending by massive object, 234
light cone, 27
light second, 24
light signals
    in general relativity, 196
    group velocity, 197–198
    postulates, 26
linear transformations, 13
linear vector space, 5

Lorentz invariance, 44
Lorentz transformation, 48
    and 4-vector dot product, 57

manifold, 150
maps, 7
mass, geometric unit, 195, 196
metric figure, 16, 127
    for 3-sphere, 257
Michelson-Morley experiment, 39
microwave background radiation,
    240, 261
    observed isotropy, 309
    in primeval fireball, 303
Minkowski spacetime, 129, 134
    and wave equation, 197
mirage, 213–214
multilinearity, 117
muon observations, 24

Newtonian cosmology, 293ff
non-Euclidean geometry, 52

Olber's paradox, 238
$\Omega$, density parameter, 272
1-form, 86
ordinary differential equation, 141
orthonormal basis, 55
orthonormal coordinates, 16
oscillating universe, 317

paradoxes, 52
partial evaluation, shorthand for, 118
pendulum, 98
pentagon, construction of, 235
phase velocity, 102
photons, 72
Planck universe, 284
Poincaré half-plane, 53, 242–243
preferred reference frame, 30
principle of equivalence, 218
principle of virtual work, 96
projective transformation, 19
proper time, 60
protractor, for special relativity, 43

pseudosphere, 244, 276ff
pseudospherical trigonometry, 281
pullback, 111
push-forward, 111

$q_0$, 273
quasars, 240
quasiparticle, 102

radiation, equation of state, 262, 263
rapidity, 43
    as coordinate on the
        pseudosphere, 277
redshift, cosmological, 260–261
    of galaxy random velocities, 288
redshift, gravitational, 207–208
    from energy conservation, 208–209
    from Huygens' construction, 218
reference frames, specialized, 29
reflection from moving obstacle, 185
rest, 29
rest mass, 71
Riemann curvature tensor, 232
rigid rods, 39, 53
ripples, 171
Robertson-Walker spacetimes, 241ff
rotations, space of, 156
Runge-Kutta integration, 298

semiclassical mechanics, 102
semiclassical models, 161
shear, 13
simultaneity, 33
    clock transport, 34
    light signal, 35
    equal-angles construction, 48
singularity, 314ff
Snell's law, 182, 187
spacetime, 4
spacetime diagram, 4
special relativity
    consistency, 52ff
    "proofs," 54
statics, 95
stereographic projection, 219
    for pseudospheres, 278–280

stiffness tensor, 118
summation convention, 138
symmetry, 20
    of black-hole spacetime, 201
    critical discussion, 308
    defined via transformations, 147
    of Robertson-Walker spacetimes,
        241–246
    time-shift, 205

tachyon, 189–191
tangent space, 89
tangent vector, 83
    to a straight world line, 56
    to a curved world line in a vector
        space, 59
    in a manifold, 153
tensor, 114ff
    symmetric, 118
tensor product, 122
3-sphere, 244
    metric, 255ff
    solid-ball model, 253
tides, 233
time in expanding universe, 246
time dilation, 37
torus as a manifold, 158
transformation, 13
    infinitesimal, 147
twin paradox, 54
    for water-wave clocks, 181

ultrastiff matter, 264

Van der Pol oscillator, 144
vector
    basis, 55
    covariance under linear
        transformations, 56
    free, 55
vector field, 141
vector space, 5
velocity, relative, 41
velocity addition law
    analytic, 59
    graphical, 42
velocity dispersion, redshift of, 276

water waves
   dispersion relation for deep
      water, 106
   dispersion relation for finite
      depth, 181
   group velocity for deep water, 163
   reflection from a moving obstacle,
      185–187
   relativity symmetry of, 54, 180
wave diagram, 164--166
wave equation
   dispersion relation for, 106

wave equation (*continued*)
   relation to Minkowski
      spacetime, 132
   wave diagram for, 168
wave number, 104
wave packet, 102
   equations of motion, 162
waveguide, cutoff, 192
world line, 4

x-ray diffraction, 185, 188–189